W9-CNV-197

Science and Technology of Silicones and Silicone-Modified Materials

ACS SYMPOSIUM SERIES **964**

Science and Technology of Silicones and Silicone-Modified Materials

Stephen J. Clarson, Editor
University of Cincinnati

John J. Fitzgerald, Editor
Exatec

Michael J. Owen, Editor
Michigan Molecular Institute

Steven D. Smith, Editor
Procter and Gamble

Mark E. Van Dyke, Editor
Wake Forest University

**Sponsored by the
ACS Division of Polymer Chemistry, Inc.**

American Chemical Society, Washington, DC

Library of Congress Cataloging-in-Publication Data

Science and technology of silicones and silicone-modified materials / Stephen J. Clarson... [et al.], editors ; sponsored by the ACS Division of Polymer Chemistry, Inc.

p. cm.—(ACS symposium series ; 964)

Includes bibliographical references and index.

ISBN : 978–0–8412–3943–2 (alk. paper)

1. Silicones—Congresses.

I. Clarson, Stephen J.

QD383.S54S35   2007
668.4′.227—dc22

2006052661

The paper used in this publication meets the minimum requirements of American National Standard for Information Sciences—Permanence of Paper for Printed Library Materials, ANSI Z39.48–1984.

PRINTED IN THE UNITED STATES OF AMERICA

Foreword

The ACS Symposium Series was first published in 1974 to provide a mechanism for publishing symposia quickly in book form. The purpose of the series is to publish timely, comprehensive books developed from ACS sponsored symposia based on current scientific research. Occasionally, books are developed from symposia sponsored by other organizations when the topic is of keen interest to the chemistry audience.

Before agreeing to publish a book, the proposed table of contents is reviewed for appropriate and comprehensive coverage and for interest to the audience. Some papers may be excluded to better focus the book; others may be added to provide comprehensiveness. When appropriate, overview or introductory chapters are added. Drafts of chapters are peer-reviewed prior to final acceptance or rejection, and manuscripts are prepared in camera-ready format.

As a rule, only original research papers and original review papers are included in the volumes. Verbatim reproductions of previously published papers are not accepted.

ACS Books Department

Dedication

We humbly dedicate this book to our good friends and colleagues

Dan Morse, Mark Hildebrand, Carole Perry, Siddharth Patwardhan, Manfred Sumper, and Nils Kroger

for moving the siloxane bond forward into a new era.

Contents

Copolymers

Networks

Surfaces and Interfaces

Silsesquioxanes

Silica and Related Systems

xi

Indexes

Preface

The third *Silicones and Silicone-Modified Materials* Symposium was held March 29–April 1, 2004 at the American Chemical Society (ACS) National Meeting in Anaheim, California. The four full-day symposium consisted of 66 oral and an evening poster session of 26 presentations. The summaries of the contributions to the symposium can be found in *Polymer Preprints*, Volume 45, Number 1, 2004.

Our chosen title this time is a respectful bow in the direction of Walter Noll and his classic book in our field. The cover is based on an original computer simulation carried out at the University of Cincinnati by David Rigby and Stephen Clarson. It depicts a low energy conformation of the D_{11} dimethylsiloxane ring and was photographed from the Silicon Graphics monitor by our good friend and colleague Jay Yocis. This volume was acquired and kindly moved towards publication by Stacy Vanderwall, Dara Moore, and Bob Hauserman in acquisitions. Margaret Brown in editing/production at the ACS Books Department is thanked again for carefully overseeing our transition to production. Michelle Bishop is thanked for producing the final cover for the book

The books from our earlier *Silicones and Silicone-Modified Materials* symposia in Dallas, Texas (1998) and San Diego, California (2001) are: *Silicones and Silicone-Modified Materials;* edited by S. J. Clarson, J. J. Fitzgerald, M. J. Owen, and S. D. Smith; ACS Symposium Series 729; American Chemical Society: Washington, D.C.; ISBN 0-8412-3613-5 and *Synthesis and Properties of Silicones and Silicone-Modified Materials;* edited by S. J. Clarson, J. J. Fitzgerald, M. J. Owen, S. D. Smith, and M. E. Van Dyke; ACS Symposium Series 838; American Chemical Society: Washington, D.C.; ISBN 0-8412-3804-9. Fortunately, the American Chemical Society and Oxford University Press have kept these books in press for all of us to enjoy.

From the ACS Division of Polymer Chemistry, Inc. (POLY), we thank President Ken Carter, Program Chair Doug Kiserov, and Treasurer Bob Moore for invaluable help and assistance. We also thank Procter and Gamble and the ACS for kindly providing financial support and thus making it possible for many of our presenters to be able to attend the Anaheim symposium.

The fourth *Silicones and Silicone-Modified Materials* Symposium was held at the ACS National Meeting in San Francisco, California, in September 2006.

Stephen J. Clarson

Professor of Chemical and Materials Engineering
  and Director, NSF I/U CRC Membrane Applied Science
  and Technology Center
601B Engineering Research Center
University of Cincinnati
Cincinnati, OH 45221–0012
513-556-5430 (Telephone)
Stephen.Clarson@UC.Edu (Email)

Science and Technology of Silicones and Silicone-Modified Materials

Introduction

Chapter 1

Say Yes to Silicones! [Si to Si!]

Michael J. Owen

Michigan Molecular Institute, 1910 West Saint Andrews Road, Midland, MI 48640–2696

Polydimethylsiloxane (PDMS) is a truly unique material: it is the dominant product of the large, growing, over 60-year-old silicone market, now approaching $10 billion in world-wide sales. It is the only commercially important "semi-inorganic" polymer available, i.e. one whose backbone is wholly inorganic but with organic, pendent groups or side-chains. By way of introduction to *"Silicones and Silicone-Modified Materials III"*, the fundamental characteristics of silicones that have given rise to this large market are considered.

Although not a precise term, most "silicone" chemists would agree with Noll (*1*) that what is meant is a polymeric material based on a silicon-oxygen backbone with hydrocarbon radicals combined directly with silicon. The term derives from Kipping's early anticipated analogy of silicones with ketones (*2*). The propensity for forming siloxane chains rather than a carbonyl-like double bond has a history that dates back more than 130 years (*3*) and has spawned an industry emerging during the Second World War that now has global sales of the order of ten billion US dollars. Ladenburg's 1872 observation, that $Et_2Si(OEt)_2$ evolves alcohol in acidified water to give a very viscous oil that decomposes only at extremely high temperature and does not freeze at $-15^{O}C$, is remarkable. It was made well before the modern era realization of the nature of polymers yet manages to describe a still viable synthesis route and point out some of the most interesting and useful characteristics of this class of materials.

The successful growth of this industry and its continuing importance are the reasons for our focus on silicones in this series of symposia. Other organosilicon based polymers, such as polycarbosilanes, polysilanes and polysilazanes have been discovered and developed in the last sixty years (4) but the dominance of polysiloxanes, and most particularly polydimethylsiloxane [PDMS], is self-evident. These newer polymers may well be materials of the future but so too are the silicones. We are of the opinion that the growth and diversity of silicone opportunities is far from over; a view strongly reinforced by the contributions to this third ACS Polymer Division symposium on *Silicones and Silicone-Modified Materials*. Even if we broaden the field of comparison beyond silicon to what are known as "semi-inorganic" polymers (5) (inorganic backbones with organic, pendent substituents) there is no other such polymer of comparable scientific and technological importance.

Fundamental Silicone Characteristics

The fundamental polymer characteristics that account for the dominant position of PDMS silicones among semi-inorganic polymers are (6):

* Low intermolecular forces between methyl groups
* Compact size of the methyl group
* High siloxane backbone flexibility
* High siloxane bond energy
* Partial ionic nature of the siloxane bond

The first three properties explain much of the physical behavior of these polymers and the last two account for the chemical consequences of their exposure to various environments. When considering the intermolecular interactions between polymer chains, the spectrum is from strongly polar, hydrophilic materials such as polyacrylamide to low-surface-energy, aliphatic fluoropolymers. Silicones occupy a region low down on this scale between hydrocarbons such as polypropylene [PP] and fluorocarbons such as polytetrafluoroethylene [PTFE], and overlapping both. They are thus potential replacement in certain conditions for such materials and, by the same token, vulnerable to competition from them. At the lowest end the opportunity/threat comes from low-glass-transition fluoropolymers, such as the fluoroethers, rather than conventional rigid fluoropolymers such as PTFE

PDMS chain flexibility benefits from both the compact size of the methyl group (the smallest possible alkane substituent) and the intrinsic flexibility of the siloxane backbone. The alternating architecture of small, unsubstituted oxygen linkages with larger substituted silicon atoms must also play a part. The glass transition is a significant reflection of this flexibility. For PDMS the temperature of this transition is usually given in the 146-150 K range for cross linked linear PDMS chains but this is not the lowest methylsiloxane glass transition reported. Kurian and co-workers (7) have described a polymer of D_5H, a cyclic siloxane pentamer with a methyl group and hydrogen atom on each silicon known as PD_5, that has a glass transition at 122 K. Pt catalysis in the presence of water hydrolyses some of the SiH groups to SiOH which then condense to a polymeric structure of pentamer rings linked by siloxane linkages. For comparison, the lowest reported fluoropolymer glass transition is *ca* 140 K for copoly(oxytetrafluoroethylene-oxydifluoromethylene) (8), but note that this latter polymer has no pendent groups, only fluorine atoms, along its backbone. Table 1 summarizes these and other relevant glass transition data.

Table 1. Selected Values of the Glass Transition Temperature

	T_g (K)
Poly(pentamethylcyclopentasiloxane) (PD_5)	122
Co-poly(CF_2CF_2O-CF_2O)	140
Polydimethylsiloxane	150
Polyethylene	148
Polydimethylsilmethylene	173
Polymethylnonafluorohexylsiloxane	198
Polyisobutylene	200
Polymethyltrifluoropropylsiloxane	203
Polyoxyhexafluoropropylene	207
Polytetrafluoroethylene	293

The combination of low intermolecular forces and high chain flexibility provides the fundamental explanation of the hydrophobic nature of the PDMS surface in air and its facile reorganization at polar, condensed interfaces such as water. A convincing demonstration of this behavior was provided at the symposium by Chen and co-workers (9) who used sum frequency generation (SFG) vibrational spectroscopy to show that PDMS surfaces in air are dominated by the side-chain methyl groups that reversibly reorganize in the presence and absence of water.

Hoffman's comment on the siloxane bond energy is very illuminating (10); he notes that "the bond energies of C-E and Si-E (E = an element) are similar, within 50 kJ/mol, for a large variety of elements. The exception is Si-O (and Si-

F), which is nearly 100 kJ/mol stronger than C-O. Coupled to the unhappiness of Si with unsaturation, silicon's love of oxygen leads to an overwhelming difference between a world of carbon-oxygen multiple bonds and the universe of silicates. Contrast CO_2 and solid polymeric SiO_2". In the context of this symposium, contrast also acetone and PDMS. The high siloxane bond energy clearly has much to do with the high thermal stability of organosiloxanes first noticed by Ladenburg. The partial ionic nature of the siloxane backbone aids in this too. Methyl groups on silicon are stabilized by electron withdrawal to the positively biased silicon. Since ionic bonds are not spatially directed as covalent bonds are, the partial ionic character may also contribute to chain flexibility. However, there is a downside to the partial polar nature. The Achilles heel of siloxanes in my estimation is hydrolytic instability at extremes of pH from nucleophilic or electrolytic attack. Such heterolytic cleavage is likely with any polymer with alternating backbone atoms. Perhaps this is why polyphosphazenes are less competitive with silicones than might be expected. Their hydrolytic stability range seems more restricted than silicones. Indeed this may be an important reason why no other inorganic backbone of any sort has yet emerged to challenge the dominant position of silicones. The only heterogeneous chain polymers that can stand up to concentrated mineral acids and strong alkalis seem to be the perfluoroethers (*8*)

Macromolecular Architecture

Much of the preceding discussion has involved the molecular architecture of the silicone polymer chain. The next level of organization has been termed macromolecular architecture (*11*) and recognizes four major classes:

- Linear [e.g. flexible coils, rigid rods, cyclics]
- Cross-linked [e.g. elastomers, resins, interpenetrating networks]
- Branched [e.g. random branches, comb-branched, stars, cages?]
- Dendritic [e.g. hyperbranched, dendrimers]

Silicone examples of all four classes are represented in this and the preceding two *Silicones and Silicone-Modified Materials* symposia. However, it is instructive to see which architectures are not well represented and consider whether such gaps indicate opportunities or limitations. The order presented above is broadly a historical order so the paucity of dendritic material contributions is simply a function of the relatively recent recognition of this class. There is much current interest in dendritic polymers and no reason to believe that silicone dendrimers will not play a significant role in this future growth. Some of the papers given at the symposium concerned organosilicon hyperbranched materials including a novel catalysis of MeHSi(OMe)$_2$ by B(C$_6$F$_5$)$_3$ catalysis (*12*), but no actual dendrimer papers were presented. Interestingly, there was a poly(amidoamine-organosilicon) (PAMAMOS) coplymeric dendrimer paper presented at the Anaheim meeting (*13*) but not in the "Silicones and Silicone-Modified Materials" symposium. Expectedly, this symposium did not contain all the papers on silicones and siloxanes at the meeting, although it did account for *ca* 90% of such papers.

The apparent lack of interest in branched silicones is rather more surprising. The best known example is the rake or comb block polymers of polydimethylsiloxane and polyethers that are used to stabilize polyurethane foam and in other surfactant applications. The symposium contained several such copolymer materials, for example, a side-chain liquid crystal copolymer (*14*). Almost all reported silicone surfactants are nonionic; Coo-Ranger et al. offered us an unusual example of an anionic silicone surfactant (*15*). There seems to be little exploitation of homopolymeric organosiloxane randomly-branched or star structures. It was pleasing to note that the symposium contained one contribution on the topic of star polymers (*16*). If we consider silsesquioxane cages to be a sub-set of the branched category then we have a clear scientifically active example with considerable technological potential, nearly a quarter of the contributions to this symposium could be classified in this POSS (polyhedral oligomeric silsesquioxane) category. Interestingly, there was also a contribution on ladder-like silsesquioxanes (*17*). Mabry and co-workers (*18*) reported fluorosilicone-POSS materials with surface energies as low as any fluorosilicones yet discovered.

Expectedly, there is much that is new even in the more familiar categories such as linear and cross-linked siloxanes. For example, Mark's review of polysiloxane elastomers (*19*) noted a number of important developments, some of which were the subject of contributions to this symposium including thermoplastic elastomers (*20*) and fluorosiloxane chains (*21*). One sub-set that seems poised for new contributions are silicone-modified materials based on interpenetrating networks; three examples of this genre were contributed (*22-24*). One also wonders with the wealth of knowledge regarding linear and cyclic silicones why there has not been more interest in polyrotaxane structures with cyclics threaded along linear chains.

The Future

Although a broadening of the field of organosilicon-based polymers is apparent in research and development activities, polydimethylsiloxane continues to dominate the silicone polymer arena. This growth of opportunities for PDMS will continue as a direct consequence of its remarkable polymer architecture. It remains our opinion, amply supported by the considerable diversity of contributions to this symposium, that PDMS and other silicones have a bright future in the twenty first century. Ongoing technical ingenuity is clearly evident that will provide a basis for new applications and support existing ones.

The siloxane backbone is the most flexible chain available, but the aliphatic fluorocarbons provide a more surface-active pendent group than methyl. There has been renewed technical and commercial interest in fluorosilicones in recent years that ought to continue in the future although it must be recognized that there were relatively few such contributions to this present symposium. A more flexible inorganic polymer backbone than the siloxane backbone is not anticipated. The nearest rival seems to be the polyphosphazenes, the dimethyl version of which has a 60 K higher glass transition temperature than PDMS. Perfluoroether polymers come closest to fluorosilicones in terms of high chain flexibility and low intermolecular forces and we can anticipate more overlap of interest between fluorosilicones and fluoroethers than with other fluoropolymers.

There is abundant evident growth in copolymers of silicones with other materials, a trend reflective of the polymer field generally. Because of their low environmental impact we have long predicted the rise of hybrids or copolymers of silicones and various natural products. This has not happened yet to the extent that might have been expected but one clearly evident aspect of this is the current interest in bio-derived silica and other silicon containing biomaterials. Contributions such as controlled protein deposition (*25*) and DNA attachment (*26*) to silicones are also part of this theme from an interface perspective.

Our focus is on silicones but the broader field of silicon-containing materials, particularly silicon inorganic materials, is still in its infancy. The current considerable interest in silsesquioxanes is consistent with a trend towards more "inorganic" silicones.

We have already noted that certain silicone polymer architectures, notably branched, hyperbranched, dendrimer and star architectures seem underutilized. However, interest is growing and we can reasonably predict that this will also be an area of increasing exploration and utilization in the near future.

Acknowledgements

The title of this paper was suggested by Tom Lane of Dow Corning Corporation more than a decade ago; I have used it gratefully since in numerous presentations. I would also like to collectively thank all my companions in the course of a thirty five year R&D career at Dow Corning Corporation.

8

References

1. Noll, W.; *Chemistry and Technology of Silicones*, Academic Press Inc., New York, 1968, p. 2.
2. Kipping, F. S.; *Proc. Chem. Soc.* **1904**, *20*, 15.
3. Ladenburg, A.; *Ann. Chem.* **1872**, *164*, 300.
4. Jones, R. G.; Ando, W.; Chojnowski, J.; *Silicon-Containing Polymers*, Kluwer Academic Publishers, Dordrecht, The Netherlands, 2000.
5. Elias, H. G.; *Macromolecules*, Vol. 2, Plenum Press, New York, 1977.
6. Owen, M. J.; in *Silicon-Containing Polymers*, Jones, R. G.; Ando, W.; Chojnowski, J. Eds., Chapter 8, Kluwer Academic Publishers, Dordrecht, The Nerherlands, 2000.
7. Kurian, P.; Kennedy, J. P.; Kisliuk, A.; Sokolov, A.; *J. Polym. Sci. Part A, Polym. Chem. Ed.* **2002**, *40*, 1285.
8. Scheirs, J.; in *Modern Fluoropolymers*, Schiers, J.; Ed., Chapter 24, John Wiley and Sons Inc., New York, 1997, p. 435.
9. Chen, C,; Wang J,; Chen, Z. *Polym. Preprints* **2004**, *45(1)*, 639.
10. Hoffman, R.; *Chem. Eng. News* **2003**, *81(36)*, 56.
11. Dvornic, P. R.; Tomalia, D. A.; *Curr. Op. Colloid Interface Sci.* **1996**, *1*, 221.
12. Rubinsztajn, S. *Polym. Preprints* **2004**, *45(1)*, 635.
13. Kohli, N.; Validya, S.; Ofoli, R.; Worden, M.; Lee, I. *Polym. Preprints* **2004**, *45(1)*, 124.
14. Zhao, Y.; Dong, S.; Schuele, D. E.; Nazarenko, S.; Rowan, S.; Jamieson, A. M. *Polym. Preprints* **2004**, *45(1)*, 572.
15. Coo-Ranger, J. J.; Zelisko, P. M.; Brook, M. A. *Polym. Preprints* **2004**, *45(1)*, 674.
16. Cai, G.; Weber, W. P. *Polym. Preprints* **2004**, *45(1)*, 710.
17. Gunji, T.; Arimitsu, K.; Abe, Y. *Polym. Preprints* **2004**, *45(1)*, 624.
18. Mabry, J. M.; Vij, A.; Viers, B. D.; Blanski, R. L.; Gonzales, R. I.; Schlaefer, C. E. *Polym. Preprints* **2004**, *45(1)*, 648.
19. Mark, J. E.; *Polym. Preprints* **2004**, *45(1)*, 574.
20. Schafer, O.; Weis, J.; Delica, S.; Csellich, F.; Kneissel, A. *Polym. Preprints* **2004**, *45(1)*, 714.
21. Grunlan, M. A.; Lee, N. S.; Weber, W. P. *Polym. Preprints* **2004**, *45(1)*, 581.
22. Fichet, O.; Laskar, J.; Vidal, F.; Teyssie, D. *Polym. Preprints* **2004**, *45(1)*, 577.
23. Darras, V.; Boileau, S.; Fichet, O.; Teyssie, D. *Polym. Preprints* **2004**, *45(1)*, 680.
24. Zhao, L.; Clapsaddle, B. J.; Shea, K. J.; Satcher, J. H. *Polym. Preprints* **2004**, *45(1)*, 672.
25. Ragheb, A.; Chen, H.; Marshall, M.; Hrynyk, M.; Sheardown, H.; Brook, M. A. *Polym. Preprints* **2004**, *45(1)*, 602.
26. Vaidya, A.; Norton, M. *Polym. Preprints* **2004**, *45(1)*, 606.

Synthesis

Chapter 2

Tertiary Silyloxonium Ions in the Ring-Opening Polymerization (ROP) of Cyclosiloxanes: Cationic ROP of Octamethyltetrasila-1,4-dioxane

Marek Cypryk, Julian Chojnowski, and Jan Kurjata

Center of Molecular and Macromolecular Studies, Polish Academy of Sciences, Sienkiewicza 112, 90–363 Łódź, Poland

Cationic ring-opening polymerization of octamethyltetrasila-1,4-dioxane ($^2D_2$) initiated by strong protic acids leads to polymer which is the structural silicon analogue of polyoxyethylene. The study of the polymerization of $^2D_2$ using the new initiating system, $R_3SiH+Ph_3C^+ B(C_6F_5)_4^-$, indicates that tertiary silyloxonium ions participate in propagation. The polymerization shows kinetic features consistent with those observed for the process initiated by protic acids. The broad molecular weight distribution indicates that extensive transfer occurs in the polymerization system. Quantum mechanical calculations suggest the involvement of transient silylium ions in the chain transfer reactions.

Octamethyl-1,4-dioxatetrasilacyclohexane (^{2}D$_2$) is a permethylated silicon analogue of 1,4-dioxane which, in contrast to this cyclic ether, has a thermodynamic ability to polymerize according to the ring-opening mechanism, (Figure 1). The cationic ROP of ^{2}D$_2$ initiated by a strong protic acid, such as CF$_3$SO$_3$H leads to an equilibrium mixture of linear polymer with cyclic oligomers, containing ca. 60% of the linear fraction when the reaction is carried out in a 50% v/v solution (1,2).

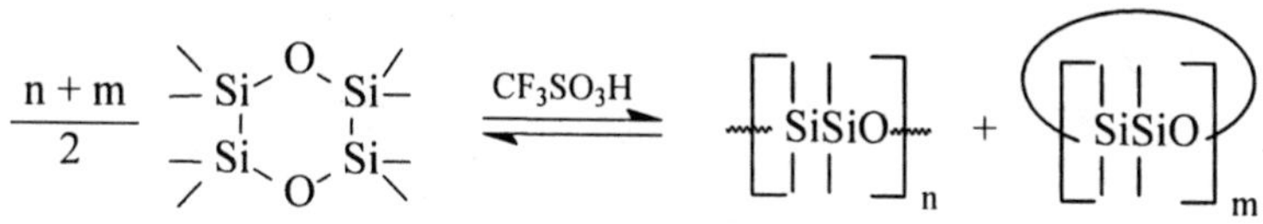

$$^2D_2$$

Figure 1. Ring-opening polymerization of ^{2}D$_2$ (symbol ^{2}D stands for the Me$_2$SiMe$_2$SiO unit).

The resulting polymer, having an equal number of SiOSi and SiSi linkages alternatively arranged along the chain, is the permethylated silicon analogue of polyoxyethylene. The polymer has interesting morphological, thermal and chemical properties, which offer some perspectives of practical application (3). Cationic ROP of this monomer initiated by strong protic acids was intensively studied and has been shown to occur exclusively via the cleavage and reformation of the siloxane bond, analogously to the ROP of cyclosiloxanes (1,2).

Kinetics of polymerization of ^{2}D$_2$ initiated by CF$_3$SO$_3$H shows close similarities to the polymerization of D$_4$ (1,2). Both processes involve simultaneous formation of cyclic oligomers and polymer and lead to equilibrium with similar proportions of cyclic to linear fractions (1,2,4). They also show similar thermodynamic parameters and similar effect of water addition on the initial rate of polymerization. The specific feature of the polymerization of ^{2}D$_2$ is that cyclic oligomers, dodecamethyl-1,4,7-trioxahexasilacyclononane, (^{2}D$_3$) and hexadecamethyl-1,4,7,10-tetraoxaoctasilacyclododecane, (^{2}D$_4$) are formed simultaneously with the polymer but they equilibrate with monomer much faster than the open chain polymer fraction (1). This behavior is best rationalized assuming formation of the tertiary oxonium ion intermediate, which can isomerize by a ring expansion-ring contraction mechanism (2). Polymerization of ^{2}D$_2$ initiated by CF$_3$SO$_3$H is considerably faster than that of D$_4$. The higher reactivity of ^{2}D$_2$ compared to D$_4$ may be explained by a higher basicity of this monomer, which should result in higher concentration of the active propagation species, presumably tertiary silyloxonium ions.

The structure of the active propagation center and the detailed mechanism of the cationic ROP of cyclosiloxanes is still controversial. Trisilyloxonium ions generated from hexamethylcyclotrisiloxane, D_3, and octamethylcyclotetrasiloxane, D_4, were observed in ^{29}Si NMR at -70 °C by Olah et al. (5). The direct precursors of these ions ($R_3SiH+Ph_3C^+$ $B(C_6F_5)_4^-$) initiate cationic ROP of cyclosiloxanes, which was taken as the proof that tertiary silyloxonium ions are the true propagation centers (6). However, the main objection against this mechanistic concept is a relatively low rate of polymerization in the systems of very low nucleophilicity allowing formation of persistent tertiary oxonium ions (7). It should be mentioned that the polymerization was studied at room temperature whereas the oxonium ions were observed only at very low temperatures (5,6). Because of peculiar kinetics of the formation of cyclic oligomers in polymerization of $^2D_2$, this process seems to be an interesting model reaction for solving mechanistic problems of the cationic ROP of cyclosiloxanes.

Experimental

Materials

Dichloromethane (Fluka) was shaken with conc. H_2SO_4, then washed with water and Na_2CO_3, dried with $CaCl_2$ and distilled from CaH_2. It was stored in the dark in a glass ampoule with Rotaflo stopcock over CaH_2. n-Hexane was distilled from CaH_2. Dichloromethane–2D (Dr Glaser AG) was stored over calcium hydride and distilled from it in a vacuum line before use. Triphenylmethylium chloride (Aldrich, 98%) was crystallized from dry hexane. Potassium tetrakis(pentafluorophenyl)borate, $K^+B(C_6F_5)_4^-$, was kindly offered by Rhodia Silicones. It was dried by keeping at 60 °C under high vacuum for 24 hrs. Octamethyl-1,4-dioxatetrasilacyclohexane ($^2D_2$) was synthesised by hydrolytic condensation of 1,2-dichlorotetramethyldisilane (8). It was purified by recrystallization from ethanol and sublimation. Triphenylmethylium (trityl) tetrakis(pentafluorophenyl)borate (TPFPB), $Ph_3C^+B(C_6F_5)_4^-$, was prepared from triphenylmethylium chloride and $K^+B(C_6F_5)_4^-$ as described previously (9). Triethylsilane (Fluka, pure grade, >97%) was dried over CaH_2, distilled under vacuum to an ampoule with Rotaflo stopcock and stored under argon.

Instrumentation

Size exclusion chromatography (SEC) was used to determine molecular weights and molecular weight distributions, M_w/M_n, of polymer samples with respect to polystyrene standards (Polysciences Corporation). The system configuration was LDC Analytical refractoMonitor IV instrument equipped with a constaMetric 3200 RI detector and two columns: SDV 8×300, 5 µm, 10^4 Å, and SDV 8×300, 5 µm, 100 Å (PSS). Solvent: toluene.

^{29}Si NMR spectra of the polymers were obtained on a Bruker DRX 500 spectrometer in 5 mm o.d. tubes, using the INEPT technique. Preparation of samples has been described previously (10).

The infrared spectra were determined using a Philips Analytical PU9800 FTIR spectrometer at 2 cm^{-1} spectral resolution. The spectrum of each compound was determined alone in solution in carbon tetrachloride and in CCl_4 containing 0.01 M phenol. A path length of 1 cm. was used.

Gas chromatography analyses were performed on a Hewlett-Packard 6890 apparatus with thermal conductivity detector and 30 m capillary column HP-1 HP 190592-023, using He as a carrier gas. The column temperature program was: 60-240 °C at 10 °C/min; detector temperature: 250 °C; injector temperature: 250 °C.

Polymerization of Octamethyl-1,4-dioxatetrasilacyclohexane ($^2D_2$) Initiated with $Et_3SiH+Ph_3C^+B(C_6F_5)_4^-$. A Typical Procedure.

To a solution containing 20 mmol of $^2D_2$, 0.01 mmol of trityl TPFPB (10^{-3} mol·L^{-1}) and 10 mL of CH_2Cl_2 under dry argon 0.02 mmol of Et_3SiH was added with a syringe. The reaction mixture was kept at constant temperature under magnetic stirring. The reaction progress was followed by taking aliquots of the mixture at regular intervals and monitoring the conversion by GC. After the reaction equilibrium was reached, the sample was taken for SEC and NMR analyses.

Computational Methods

Geometries of model compounds were calculated using density functional theory (DFT) at the B3LYP/6-31G* level. Stationary points were confirmed by vibrational analysis. Final electronic energies were calculated using the B3LYP/6-311+G(2d,p) method. This level of theory is denoted as B3LYP/6-311+G(2d,p)//B3LYP/6-31G* and is referred to as DFT further in the text. Vibrational components of the thermal energy were scaled by 0.98. Enthalpies and Gibbs free energies of model reactions were corrected to 298 K. Calculations were performed for the gas phase conditions, using the Gaussian 03 program (11).

Results and Discussion

Studies of the Basicity of Octamethyl-1,4-dioxatetrasilacyclohexane

The high basicity of $^2D_2$ relative to other siloxanes was supported by the IR studies of hydrogen bond formation of model cyclosiloxanes with phenol in CCl_4. The absorption maximum of the phenol O-H stretching vibration band is shifted upon formation of the hydrogen bond between phenol and $^2D_2$ by $\Delta v = 211$ cm^{-1}, while for D_4 the corresponding shift is only 159 cm^{-1} (Figure 2). The complex formation constant in the case of $^2D_2$ is by one order of magnitude greater than that for D_3 (Table I).

Table I. The shift of the O-H stretching frequency in phenol in CCl_4 upon the hydrogen bond formation with model siloxane bases and the corresponding phenol-siloxane complex formation constants

Siloxane base		v (cm^{-1}) PhOH complexed	Δv (cm^{-1}) free- complexed	K_{298} (l mol^{-1})	Relative basicity[a] (kcal mol^{-1})
$(Me_3Si)_2O$	M_2				1.3
$\lfloor Me_2SiO \rfloor_3$	D_3	3438	174	1.84	0
$\lfloor Me_2SiO \rfloor_4$	D_4	3453	159		3.5
$-[Me_2SiO]_n-$	PDMS	3480	132		-
$\lfloor Me_2SiMe_2SiO \rfloor_2$	$^2D_2$	3400	211	15.1	8.7

[a]) Relative basicity calculated by DFT method as the negative Gibbs free energy of siloxane protonation in the gas phase.

Quantum mechanical DFT calculations of proton affinity, (negative enthalpy, PA) and basicity (negative Gibbs energy, ΔG_{prot}) for the protonation reaction of both hydrogen- (12) and methyl-substituted analogues of $^2D_2$, D_4 and D_3 confirmed that tetrasila-1,4-dioxanes, $(R_2SiR_2SiO)_2$ (R = H, Me), are considerably more basic than the other model siloxanes (Table I). $^2D_2$ has also a higher silicophilicity compared to the other siloxanes (by 3.3 and 2.8 kcal·mol^{-1} higher than that of D_3 and D_4, respectively). The silicophilicity, calculated as the

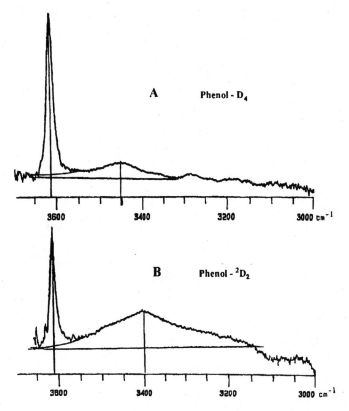

Figure 2. IR spectra of a phenol-siloxane mixture ([PhOH]=2·10⁻³ mol·dm⁻³, [D₄]=[²D₂]=0.2 mol·dm⁻³ in CCl₄, temp. 25 °C)

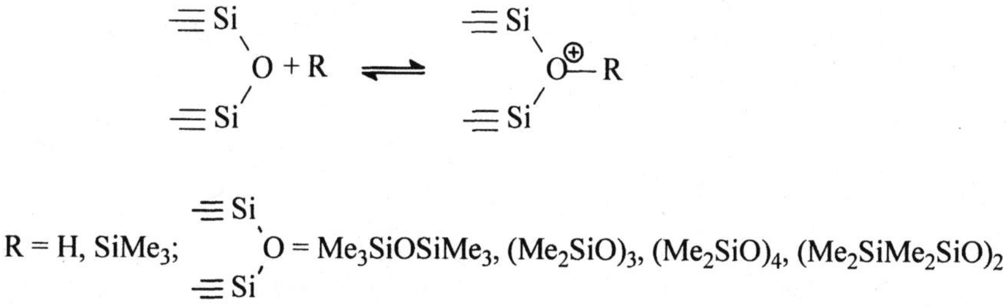

$$R = H, SiMe_3; \quad \begin{array}{c} \equiv Si \\ \diagdown \\ O \\ \diagup \\ \equiv Si \end{array} = Me_3SiOSiMe_3, (Me_2SiO)_3, (Me_2SiO)_4, (Me_2SiMe_2SiO)_2$$

Figure 3. Model reactions of protonation and silylation of siloxanes.

negative free energy of addition of the trimethylsilylium cation to model siloxanes (Figure 3) may be regarded as the approximate measure of the reactivity of siloxanes toward the silyloxonium ion formation. The reason for the higher basicity and silicophilicity of $^2D_2$ compared to D_3 and D_4 is a different orbital interaction pattern involving the lone electron pair of oxygen, due to the difference in structures of their ring skeletons. The contribution from the hyperconjugation effect $n_O \rightarrow \sigma^*_{SiO}$ in polysiloxanes significantly reduces the electron density on oxygen. This effect is less important in $^2D_2$ because the $n_O \rightarrow \sigma^*_{SiSi}$ interaction is insignificant (*13*).

Synthesis of Tertiary Silyloxonium Ions Derived From $^2D_2$

To synthesize in situ the initiator, we repeated the procedure of Lambert (*14*) and Olah (*6*) involving the reaction of hydrosilane with the carbenium ion in the presence of the very weakly nucleophilic tetrakis(pentafluoro-phenyl)borate counter-ion. The hydride transfer leads to transient formation of the tricoordinate silylium cation, which reacts instantaneously with $^2D_2$, giving the cyclic trisilyloxonium ion, according to Figure 4.

$$\text{Ph}_3\text{CCl} + \text{KB}(\text{C}_6\text{F}_5)_4 \xrightarrow[-\text{KCl}]{} \text{Ph}_3\text{C}^+ \text{B}(\text{C}_6\text{F}_5)_4^- \xrightarrow[-\text{Ph}_3\text{CH}]{\text{HSiEt}_3} [\text{Et}_3\text{Si}^+ \text{B}(\text{C}_6\text{F}_5)_4^-]$$

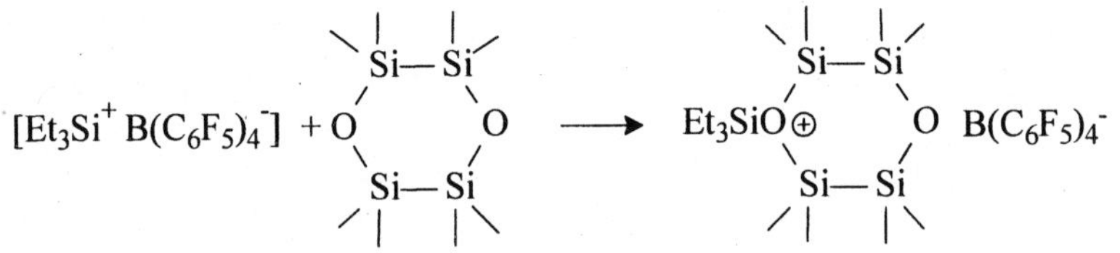

Figure 4. Generation of the primary trisilyloxonium ion (initiation).

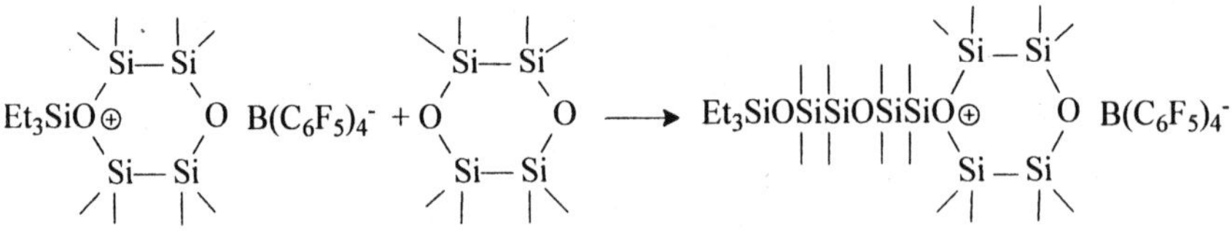

Figure 5. Formation of the secondary trisilyloxonium ion (propagation).

The resulting trisilyloxonium ion and the ion being the product of addition of monomer (Figure 5) were identified by ^{29}Si NMR at -45 °C and by quantum mechanical NMR calculations (*10*). Signals of silicon atoms at the oxonium center appear in the range of 50-60 ppm, in accord with previous observations for trisilyloxonium ions derived from D_4 and D_3 (*5*). The relative intensities of signals changed with time, as the signals corresponding to the original oxonium ion (Figure 4) disappeared, while those attributed to the oxonium ion at the polymer chain end rose (Figure 5) (*10*). Thus, the original trisilyloxonium ion formed by reaction 4 was transformed by reaction 5 into the cyclic trisilyloxonium ion at the polymer chain end. This ion is therefore likely to be the active propagation center, or one of the possible propagation centers, in the cationic ROP of $^2D_2$.

It may be concluded that $^2D_2$, due to its higher basicity and nucleophilicity, forms much more stable oxonium ions than permethylcyclosiloxanes. Thus, it is a good model for the study of the role of cyclic trisilyloxonium ions in the cationic polymerization of cyclic siloxanes.

Polymerization of $^2D_2$ initiated by $Et_3SiH + Ph_3C^+ B(C_6F_5)_4^-$

Polymerization of $^2D_2$ initiated by $Et_3Si^+ B(C_6F_5)_4^-$ generated in situ using the above procedure was studied by ^{29}Si NMR and by gas chromatography. The process is fast and proceeds in a clean way, giving linear polymer, polyoxybis(dimethylsilylene), $-(Me_2SiMe_2SiO)_n-$, and the related cyclic oligomers, $(Me_2SiMe_2SiO)_n$ (Figure 6, 7, and 8). Polymerization leads to an equilibrium mixture which consists of linear polymer and small cyclics appearing in proportions similar to those observed for the polymerization initiated by CF_3SO_3H. Secondary reactions, such as the Si-Si and Si-C cleavage were not observed. In the presence of oxygen, slow oxidation of products to SiOSiOSi was observed, analogously to the system initiated by protic acids.

Kinetics of the polymerization followed by ^{29}Si NMR at -45 °C proved that silyloxonium ions maintain throughout the entire process. Polymerization of $^2D_2$ followed by GC at room temperature was very fast (half-time of monomer conversion was ca. 250 s, $[^2D_2]_0 = 1.5$ mol·L^{-1}, $[Et_3Si^{-2}D_2^+ B(C_6F_5)_4^-]_0 = 10^{-3}$ mol·L^{-1}). The kinetics of monomer conversion in polymerization at 0 °C is shown in Figure 6. A considerable deviation from the first internal order is observed as the monomer consumption strongly slows down at ca. 60% of monomer conversion. A similar course of the reaction was observed in the polymerization of $^2D_2$ initiated by protic acids. In the first stage of the reaction the monomer is mainly transformed into higher cyclic oligomers, particularly $^2D_3$ and $^2D_4$. The fast reaction period extends until the equilibrium between monomer and the cyclic oligomers is established. In the second stage, a slower transformation of cyclics into polymer occurs.

The equilibration of cyclics proceeds via the ring expansion-ring contraction mechanism of the tertiary silyloxonium ion at the end of polymer chain (Figure 9).

18

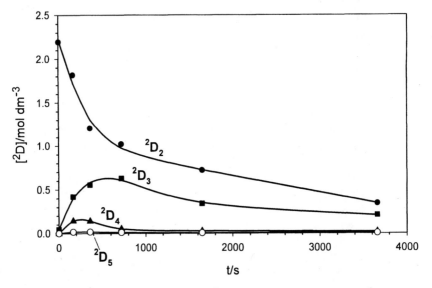

Figure 6. Kinetics of polymerization of $^2D_2$ at 0 °C followed by GC ($[^2D_2]_0$ = 1.1 mol·L^{-1}, $[Et_3Si$-$^2D_2^+$ $B(C_6F_5)_4^-]_0$ = 2.2·10^{-3} mol·L^{-1})

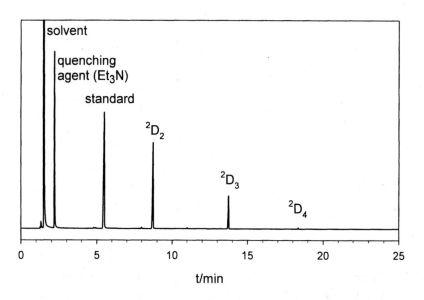

Figure 7. Gas chromatogram of the equilibrium polymerization mixture (parameters as in Figure 6).

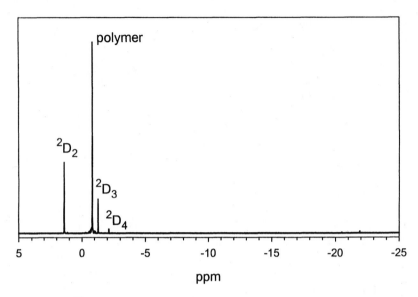

Figure 8. ^{29}Si NMR spectrum of the equilibrium polymerization mixture (parameters as in Figure 6).

The yield of the polymer obtained at 0 °C was ca. 70%, M_n = 5900, M_w/M_n = 2.3. Theoretical M_n was ca. 18000 (taking into account that the excess of Et_3SiH in the polymerization mixture may act as the transfer agent). The GPC trace is shown in Figure 10. Too low MW and broad molecular weight distribution point to extensive chain transfer and are consistent with the data reported by Olah for the polymerization of D_3 and D_4 (6).

Theoretical calculations of relative stability of silyloxonium and silylenium ions

Tertiary silyloxonium cation may be considered as the complex of the silylium ion with a siloxane. The "free" silylium ion would exist in much lower concentration than its complex but, being much more reactive, could significantly contribute to propagation. Recent ab initio calculations show that silylium ions such as Me_3Si^+ and Et_3Si^+ are strongly complexed even in solvents of such low nucleophilicity as CH_2Cl_2. In the gas phase, in the presence of bases even as weak as siloxanes, they are practically completely converted into

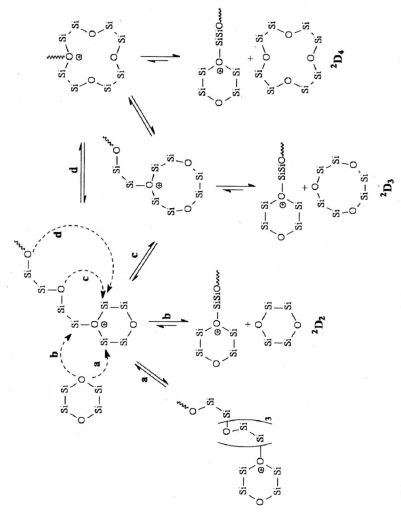

Figure 9. Mechanism of inter-conversion of cyclic oligomers in the polymerization of $^2D_2$.

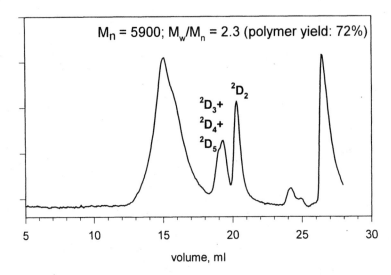

$M_n = 5900$; $M_w/M_n = 2.3$ (polymer yield: 72%)

$^2D_2$
$^2D_3+$
$^2D_4+$
$^2D_5$

volume, ml

Figure 10. SEC trace of polymer obtained by polymerization of $^2D_2$ at 0 °K in toluene. MW calibration according to PS standards.

silyloxonium ions (*12*). However, quantum mechanical calculations for these conditions neglect interactions of a cation with solvent and with counter-ion. These interactions stabilize both silylium and silyloxonium ions but the stabilization energies for these two species may be different, largely affecting the equilibrium in solution.

It is very difficult to estimate the solvation energies in the condensed phase. The accuracy of available theoretical methods is limited. However, comparison of the energies of silyloxonium ion formation for the bare and CH_2Cl_2-complexed silylium cations shows that the equilibrium constants are indeed different (Figure 11). Additional stabilization of cations may be provided by weak interactions with the counterion $B(C_6F_5)_4^-$.

$$Me_3Si^+ + O(SiMe_3)_2 \rightleftharpoons (Me_3Si)_3O^+$$
$$\Delta G^{298} = -12.0 \text{ kcal/mol } (K_1 = 6.1 \times 10^8)$$

$$Me_3Si^+ \text{-}CH_2Cl_2 + (Me_3Si)_2O \rightleftharpoons (Me_3Si)_3O^+ + CH_2Cl_2$$
$$\Delta G^{298} = -7.4 \text{ kcal/mol } (K_2 = 2.6 \times 10^5)$$

$$Me_3Si^+ \text{-}CH_2Cl_2 + (Me_3Si)_2O \rightleftharpoons (Me_3Si)_3O^+ \text{-}CH_2Cl_2$$
$$\Delta G^{298} = -2.6 \text{ kcal/mol } (K_3 = 80)$$

Figure 11. Gibbs free energies and equilibrium constants for the formation of the free and CH_2Cl_2-complexed silyloxonium ion in the gas phase.

Thus, the solvated Si^+ cation may play a significant role in the polymerization initiated by $R_3Si^+ B(C_6F_5)_4^-$, participating in propagation and in side reactions (chain transfer), analogously to the carbenium ions in the polymerization of cyclic ethers and acetals (*15*). These reactions lead to broadening of polydispersity of the polymer and may affect the polymer structure (branching).

First, it may form silyloxonium ions with oxygen atoms inside the polymer chain in the intra- or inter-molecular process, leading to back-biting or chain scission, respectively. Gibbs free energy of the direct exchange of silylium group between the monomer and the polymer chain calculated for the model reaction (Figure 12) is 11.6 kcal·mol^{-1}, which corresponds to the equilibrium constant $K_{298} = 3.2 \times 10^{-9}$. Thus, the transfer of the silylium group from monomer to polymer is very unfavorable, because of a much lower basicity of oxygen atoms in the silaether chain, compared to those in the monomer.

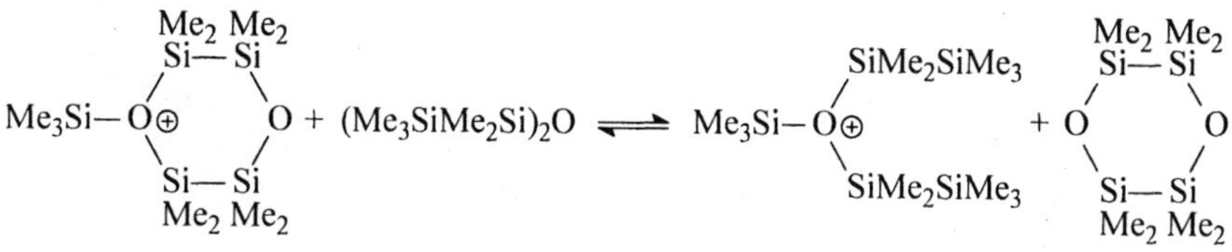

$$
\text{Me}_3\text{Si}-\overset{\overset{\displaystyle \text{Me}_2\ \text{Me}_2}{\underset{\displaystyle \underset{\text{Me}_2\ \text{Me}_2}{\text{Si}-\text{Si}}}{\text{Si}-\text{Si}}}{\text{O}\oplus} \quad \text{O} + (\text{Me}_3\text{SiMe}_2\text{Si})_2\text{O} \rightleftharpoons \text{Me}_3\text{Si}-\overset{\text{SiMe}_2\text{SiMe}_3}{\underset{\text{SiMe}_2\text{SiMe}_3}{\text{O}\oplus}} + \text{O}\overset{\overset{\displaystyle \text{Me}_2\ \text{Me}_2}{\text{Si}-\text{Si}}}{\underset{\underset{\text{Me}_2\ \text{Me}_2}{\text{Si}-\text{Si}}}{}}\text{O}
$$

Figure 12. Silylium ion exchange between the monomer ($^2D_2$) and open polymer chain fragment.

Second, it may split off the methyl group from silicon in the polymer chain generating the silylium cation via the inter- or intra-molecular Me transfer. According to calculations, there is almost no difference in the stability of the terminal silylium ions in silaether and siloxane chain. In contrast, the stabilization of the cation located inside the chain is significant (ca. 8 kcal/mol, Figure 13).

The energy barriers for the intramolecular 1,2- and 1,4-migrations of the Me group are very similar (Figure 14). They reflect mostly the angular strain in these transition states. Substitution proceeding through the unstrained transition state

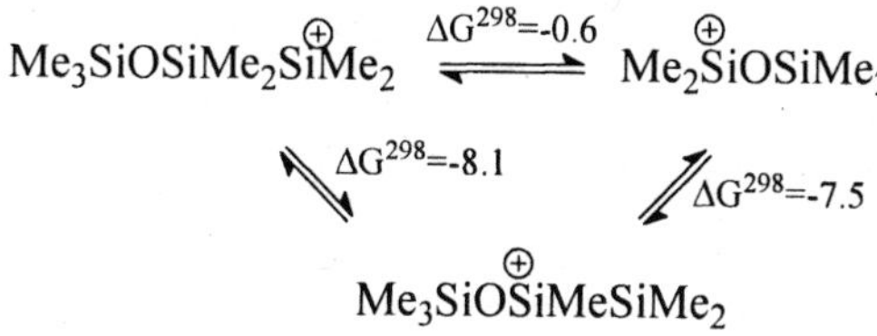

$$
\text{Me}_3\text{SiOSiMe}_2\text{Si}\overset{\oplus}{\text{Me}}_2 \xrightleftharpoons[\quad]{\Delta G^{298}=-0.6} \text{Me}_2\overset{\oplus}{\text{Si}}\text{OSiMe}_2\text{SiMe}_3
$$

$$
\Delta G^{298}=-8.1 \qquad \Delta G^{298}=-7.5
$$

$$
\text{Me}_3\text{SiO}\overset{\oplus}{\text{Si}}\text{MeSiMe}_2
$$

Figure 13. Stability of silylium ions resulting from the methyl group migration

(i.e., intermolecularly) in the gas phase is essentially barrierless, like in the case of hydride transfer, Si-H⁺-Si. The energy barrier for the hydride transfer in solution due to partial desolvation of reagents was estimated to be ca. 6 kcal·mol⁻¹ (16). The barrier for the methyl group transfer should be similar or even lower, taking into account that the hydrogen-substituted silyl cation is more strongly solvated than the methyl-substituted one, due to a more localized charge. The methyl group transfer, postulated also by Olah (5), would lead to branching of the polymer.

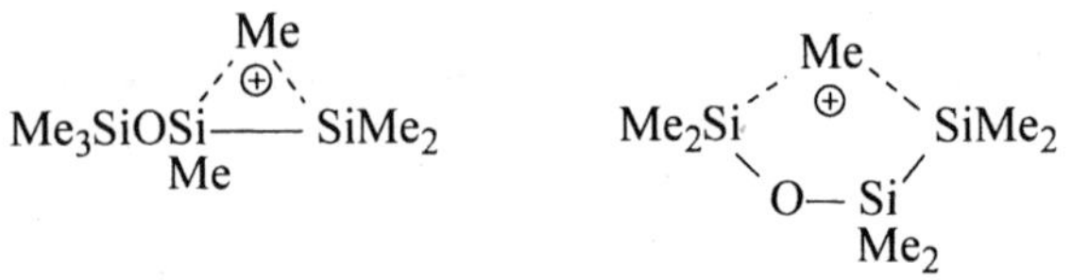

1,2-migration: $\Delta G^{\ddagger} = 11.5$ 1,4-migration: $\Delta G^{\ddagger} = 10.1$ kcal mol⁻¹

Figure 14. Structures of transition states and free energies of activation for the 1,2- and 1,4-migration of the methyl group in Me₃SiOSiMe₂SiMe₂⁺.

Finally, a cationic center at the end of the chain, either silyloxonium or silylium, may be the subject of the back-biting reaction, resulting in the observed interconversion of cyclic oligomers, as shown in Figure 9. Similar mechanism of ring expansion was postulated to be responsible for the formation of D₆ in the cationic polymerization of D₃ (17). DFT calculations of the ring expansion of model cyclic hydrogen-substituted siloxonium ion suggest that the energy barrier of the nucleophilic substitution reaction proceeding via the transition state with pentacoordinate silicon is rather high, $\Delta G^{\ddagger} = 25.0$ kcal·mol⁻¹ at 25 °C (Figure 15). Thus, the conversion of cyclic oligomers is likely to proceed through the very fast reversible dissociation-association of siloxonium ion involving transient formation of silylium cation.

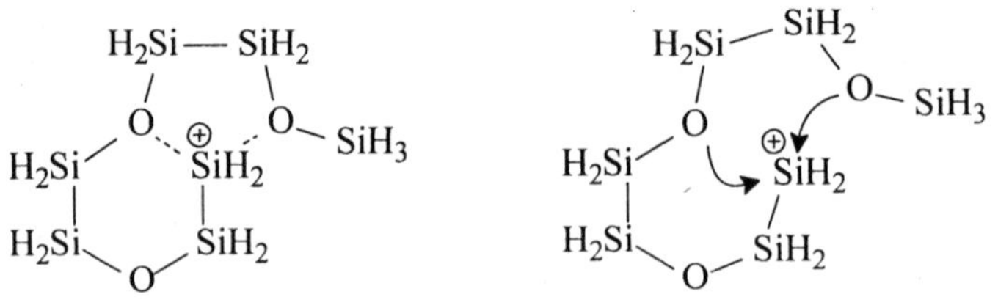

transition state (Si^{V})   intermediate (Si^{III})

Figure 15. Transition state (Si^{V}) and intermediate (Si^{III}) structures in the ring expansion of the model silyloxonium ion.

Conclusions

Octamethyltetrasila-1,4-dioxane, $^2D_2$, due to its higher basicity and nucleophilicity, forms more stable oxonium ions than permethylcyclosiloxanes, such as D_4. This monomer is therefore a good model for study of the role of cyclic trisilyloxonium ions in the cationic polymerization of cyclosiloxanes. In the $^2D_2$ + Et_3SiH + Ph_3C^+ $B(C_6F_5)_4^-$ system at low temperatures, not only the generation of Et_3Si^{+}-$^2D_2$ is observed but also its transformation to the corresponding tertiary silyloxonium ion at the chain end, i.e., $Et_3Si~O(SiMe_2)_2O(SiMe_2)_2^{+}$-$^2D_2$. Thus, the polymerization may be followed by ^{29}Si NMR at low temperatures. Cationic polymerization of $^2D_2$ at room temperature initiated by Et_3SiH + Ph_3C^+ $B(C_6F_5)_4^-$ proceeds fast and chemoselectively with the exclusive siloxane bond cleavage and reformation. The same kinetics of the cyclics formation in the polymerization of $^2D_2$ initiated by Et_3SiH + Ph_3C^+ $B(C_6F_5)_4$ and by protic acids reflects the same mechanism of polymerization. The mechanism of polymerization of cyclic siloxanes is consistent with the observed features of polymerization of $^2D_2$. The kinetic differences between the polymerizations of D_3, D_4 and $^2D_2$ can be explained by differences in basicity and ring strain of these monomers. Theoretical calculations point to the possible involvement of silylium ions in side reactions which affect the polydispersity and microstructure of the polymer.

References

1. Kurjata, J.; Chojnowski, J. *Makromol. Chem.* **1993**, *194*, 3271-3286.
2. Chojnowski, J.; Kurjata, J. *Macromolecules* **1994**, *27*, 2302-2309.
3. Chojnowski, J.; Cypryk, M.; Kurjata, J. *Prog. Polym. Sci.* **2003**, *28(5)*, 691-728.
4. Wright, P.V.; Beevers, M.S. in: *Cyclic Polymers*; Semlyen, J.A., Ed.; Elsevier Applied Science Publ.: London, 1986, p 85.
5. Olah, G.A.; Li, X.-Y.; Wang, Q.; Rasul, G.; Prakash, G.K.S. *J. Am. Chem. Soc.* **1995**, *117*, 8962-8966.
6. Wang, Q.; Zhang, H.; Prakash, G.K.S.; Hogen-Esch, T.E.; Olah, G.A. *Macromolecules* **1996**, *29*, 6691-6694.
7. Toskas, G.; Moreau, M.; Masure, M.; Sigwalt, P. *Macromolecules* **2001**, *34*, 4730-4736.
8. Kumada, M.; Yamaguchi, M.; Yamamoto, Y.; Makajima, I.; Shina, K. *J. Org. Chem.* **1956**, *21*, 1264-1268.
9. Chien, J.C.W.; Tsai, W.-M.; Rausch, M. D. *J. Am. Chem. Soc.* **1991**, *113*, 8570-8571.

10. Cypryk, M.; Kurjata, J.; Chojnowski, J. *J. Organomet. Chem.* **2003**, *686*, 373-378.
11. Frisch, M.J.; Trucks, G.W.; Schlegel, H.B. et al. *Gaussian 03, Revision C.02*, Gaussian, Inc., Wallingford CT, 2004.
12. Cypryk, M.; Chojnowski, J.; Kurjata, J. In *Organosilicon Chemistry VI: From Molecules to Materials*; Auner, N.; Weis, J., Eds., VCH-Wiley, Weinheim, 2005; pp 82-92.
13. Cypryk, M. *Macromol. Theory Simul.* **2001**, 10, 158-164.
14. Lambert, J.B.; Zhang, S.; Stern, C.L.; Huffman, J.C. *Science* **1993**, *260*, 1917-1920.
15. Penczek S.; Kubisa, P. in *Comprehensive Polymer Chemistry*; Pergamon Press: Oxford, 1989; pp 751-786.
16. Apeloig, Y.; Merin-Aharoni, O.; Danovich, D.; Ioffe, A.; Shaik S. *Isr. J. Chem.* **1993**, *33*, 387-402.
17. Nicol, P.; Masure, M.; Sigwalt, P. Macromol. Chem. Phys. **1994**, *195*, 2327-2334.

Chapter 3

Synthesis of Novel Silicon-Containing Monomers for Photoinitiated Cationic Polymerization

James V. Crivello and Myoungsouk Jang

Department of Chemistry and Chemical Biology, Rensselaer Polytechnic Institute, Troy, NY 12180

The regioselective monohydrosilation of α,ω-hydrogen-functional oligodimethylsiloxanes has been used to prepare a series of well-characterized α-hydrogen-ω-epoxy functional oligodimethylsiloxanes. Using these latter compounds as substrates, a wide variety of interesting cationically polymerizable monomers can be prepared a) by further hydrosilation b) oxidative coupling and b) by reaction with vinyl functional photosensitizers. These monomers possess outstanding reactivity. The reactivities of the new monomers and oligomers were examined using Fourier transform real-time infrared spectroscopy and optical pyrometry. Those monomers containing epoxycyclohexyl groups displayed excellent reactivity in cationic ring-opening polymerization in the presence of lipophilic onium salts photoinitiators.

Background

Cationic ring-opening crosslinking photopolymerizations of epoxide and oxetane monomers have found wide commercial usage in many applications (*1*). Among these are rapidly curing decorative and protective coatings, printing inks and adhesives for metals, paper and glass, photoresists and stereolithography.

There are several major motivating factors driving the adoption of this technology. First, the ability to conduct these crosslinking polymerizations very rapidly, with low energy and without the use of an inert atmosphere provides important economic incentives. Second, since solvents are not employed, there are no emissions and consequently, the environmental consequences of these polymerizations are minimal. Furthermore, these systems typically display low orders of toxicity which is important for worker exposure considerations. Third, one-component systems can be formulated using epoxide and oxetane monomers with very long shelflives. Lastly, the thermal, mechanical, chemical resistance, low shrinkage, adhesion, and high durability characteristics of the network polymers that are formed are excellent. As a result of the above listed considerations, the industrial impact of photoinitiated cationic polymerizations in particular, is expected to increase markedly in the future as this novel technology undergoes further industrial development and implementation *(2)*.

Depicted in Scheme 1 is a generalized mechanism for the photoinitiated cationic ring-opening polymerization of epoxide monomers using a diaryliodonium salt photoinitiator *(3)*. Irradiation of the photoinitiator generates a number of reactive species (eq. 1) that subsequently react with the solvent or monomer to give the protonic acid, HMtXn (eq. 2). Ring-opening polymerization takes place by protonation of the monomer (eq. 3) followed by repetitive addition of additional molecules of the monomer to the oxonium ion polymer chain end (eq. 4).

Reactivity in cationic photopolymerizations involves a complex interplay between both the photoinitiator and the monomer. In this article, we will mainly focus on the contributions of monomer structure to reactivity.

Scheme 1

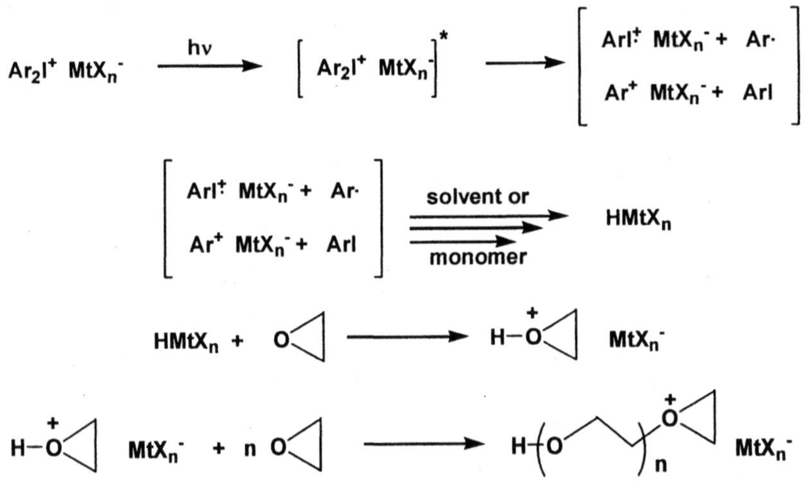

$$Ar_2I^+ \ MtX_n^- \xrightarrow{\ h\nu\ } \left[Ar_2I^+ \ MtX_n^- \right]^* \longrightarrow \left[\begin{array}{l} ArI^{+\cdot} \ MtX_n^- + \ Ar\cdot \\ Ar^+ \ MtX_n^- + \ ArI \end{array} \right] \quad (1)$$

$$\left[\begin{array}{l} ArI^{+\cdot} \ MtX_n^- + \ Ar\cdot \\ Ar^+ \ MtX_n^- + \ ArI \end{array} \right] \underset{\text{monomer}}{\overset{\text{solvent or}}{\rightleftarrows}} HMtX_n \quad (2)$$

$$HMtX_n + \ O{\triangleleft} \longrightarrow H-\overset{+}{O}{\triangleleft} \ MtX_n^- \quad (3)$$

$$H-\overset{+}{O}{\triangleleft} \ MtX_n^- + \ n \ O{\triangleleft} \longrightarrow H{-}\left(O\diagup\diagdown\right)_n\overset{+}{O}{\triangleleft} \ MtX_n^- \quad (4)$$

Epoxy functional silicones

Silicones are well known for their excellent thermal and photostability as well as their unique electrical resistance and low surface energy characteristics. The ability to deposit and rapidly photocrosslink silicone resins is interesting and highly attractive from a commercial standpoint. Considerable progress has already been achieved by functionalizing silicone polymers and oligomers with epoxy (*4*) and oxetane (*5*) groups. Particularly noteworthy is the use of the hydrosilation reaction to attach vinyl functional epoxy groups to the backbone of linear and branched silicone polymers as depicted in equation 5. When combined with a soluble lipophilic diaryliodonium salt photoinitiator, these silicones are very rapidly UV cured to give release liners used for adhesive labels as well as many other applications (*6*). The use of 4-vinyl cyclohexene-1,2-oxide in this synthesis provides resins with extremely high rates of photopolymerization. This chemistry has virtually revolutionallized the previously solvent-based release liner industry due to its inherent excellent economics and environmental safety.

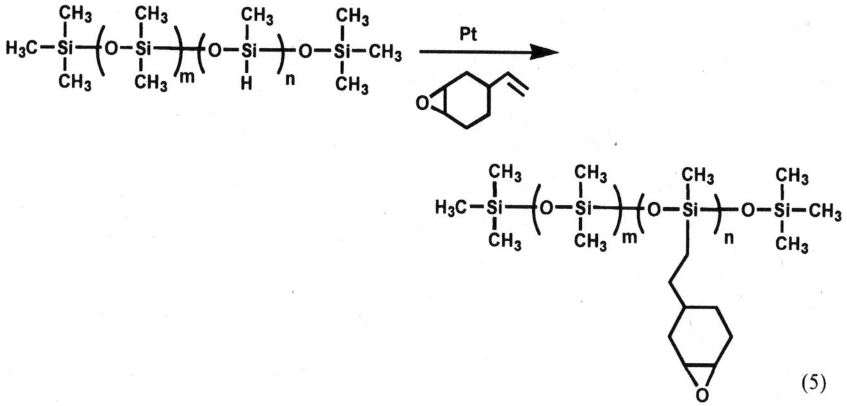

(5)

Prevously, we have also described the use of a similar approach to prepare di- and multifunctional monomers (*7,8,9*). For example, the synthesis of a highly reactive difunctional monomer **I** is depicted in equation 6.

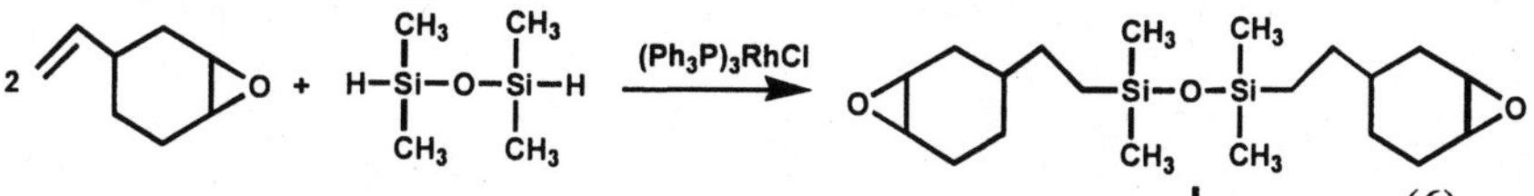

(6)

A careful study (*10*) of the reaction shown in equation 6 showed that it proceeds in two discrete steps as detailed in Scheme 2.

Scheme 2

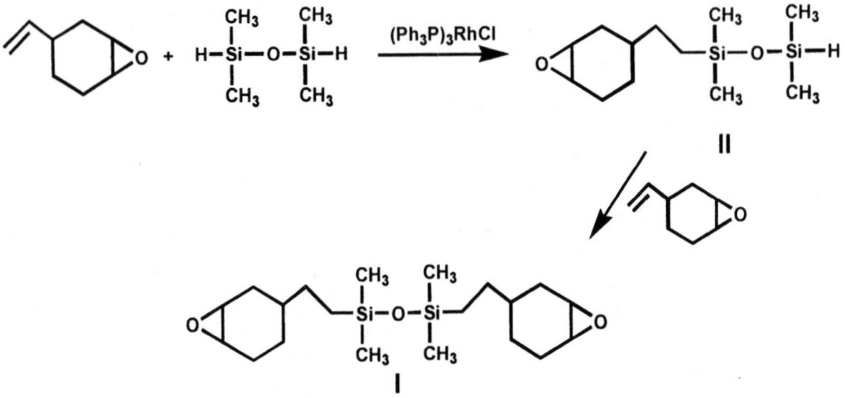

Further work demonstrated that **II** bearing an epoxy and a Si-H group can be isolated in high yield. Extension of this chemistry using α,ω-Si-H difunctional siloxanes showed that monofunctional epoxy analogs of **II** with up to four silicon atoms in the siloxane spacer could be readily prepared.

α-Hydrogen-ω-epoxy silane **I** and its analogs with various types of epoxy and oxetane groups and with different siloxane chain lengths are highly versatile synthons that can be used for the preparation of a wide variety of interesting photopolymerizable monomers and oligomers. For example, as depicted in equation 7, it has recently been found (*11*) that **II** undergoes facile dehydrodimerization to give **III** by simply bubbling air into the reaction mixture in the presence of the rhodium-containing Wilkinson's catalyst. **III** can also be synthesized by the direct double hydrosilation of 4-vinylcyclohexene with 1,1,3,3,5,5,7,7-octamethyltetrasiloxane oxide but the poor availability and high cost of this α,ω-Si-H difunctional siloxane precursor precludes most potential practical uses.

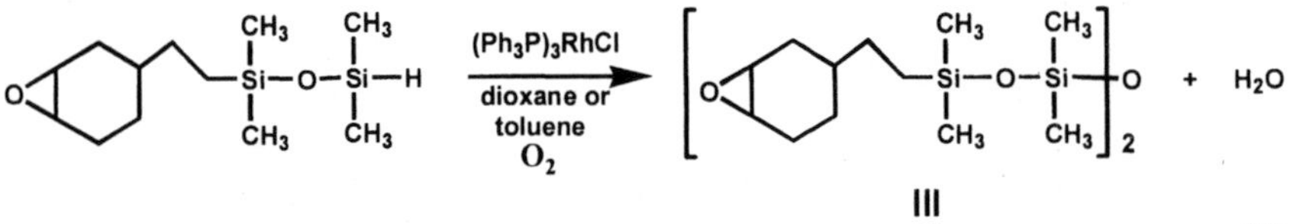

(7)

Using the above synthetic approach, a series of novel symmetrical difunctional monomers were generated and their structures are depicted in Table 1.

Ring-opening cationic photopolymerization of **I** bearing a two-silicon atom spacer gives a highly rigid, brittle glassy polymer. In contrast the photopolymerization of analogs **III** and **IV** with a longer siloxane spacers gives flexible, crosslinked polymers with considerable elongation. Figure 1. shows a

Table 1
Structure and Yields of Dehydrodimerization Monomers

Structure	Yield (%)	Structure	Yield (%)
III	94	**IV**	93
V	86	**VI**	82
VI	76	**VII**	82

study of the cationic photopolymerization of **III** using real time infrared spectroscopy (RTIR) carried out in the presence of (4-n-decyloxyphenyl)phenyliodonium SbF$_6^-$ (IOC-10) (*12*) as a photoinitiator. The incorporation of the long alkoxy chain into this lipophilic photoinitiator renders it soluble in this and other highly nonpolar silicon-containing monomers. Identical results were obtained from **III** prepared by dehydrodimerization and by the direct double hydrosilation method.

Compared to other epoxy monomers, those monomers bearing epoxycyclohexyl groups such as **I**, **III** and **VI** display exceptionally high reactivity in photoinitiated cationic ring-opening polymerizations. This results from the high epoxide ring strain in these monomers. Consequently, the photopolymerization rates of **I**, **III** and **VI** as can be seen in Figure 2 are very similar and greatly exceed those of the other monomers appearing in Table 1. In this study we have employed optical pyrometry (OP), a novel technique developed in this laboratory to monitor the photopolymerizations (*13,14*). This method employes a sensitive optical pyrometer to remotely sense the temperature of a thin film sample of monomer undergoing exothermic polymerization as a function of time.

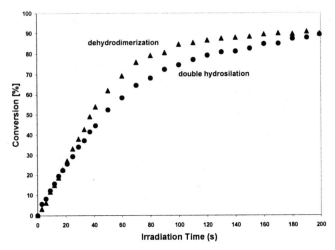

*Figure 1. RTIR comparison of the cationic photopolymerization of **III** produced by double hydrosilation and by dehyrodimerization. (1.0 mol% IOC 10 as photoinitiator, light intensity: 305 mJ/cm^2 min).*

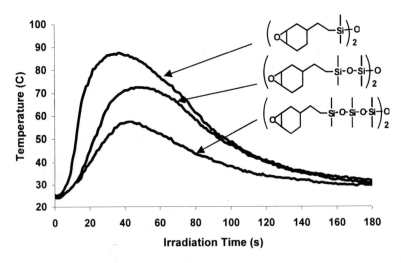

Figure 2. OP study of the photopolymerization of three related difunctional silicone-epoxide monomers using 1 mol% IOC-10 (light intensity 497 mJ/cm^2 min)

Oligomeric Silicone-epoxides

The regioselective dihydrosilation of 1,3,5,7-tetramethylcyclotetrasiloxane as shown in equation 8 leads to the synthesis of epoxycyclohexyl disubstituted compound **VIII** (*15*). Oligomerization of this latter compound (eq. 9) using the rhodium catalyzed dehydrodimerization reaction described earlier yields the highly reactive oligomer **IX** (n = 4-6). Upon cationic photopolymerization this slightly viscous oligomer yields a colorless, transparent glassy crosslinked polymer.

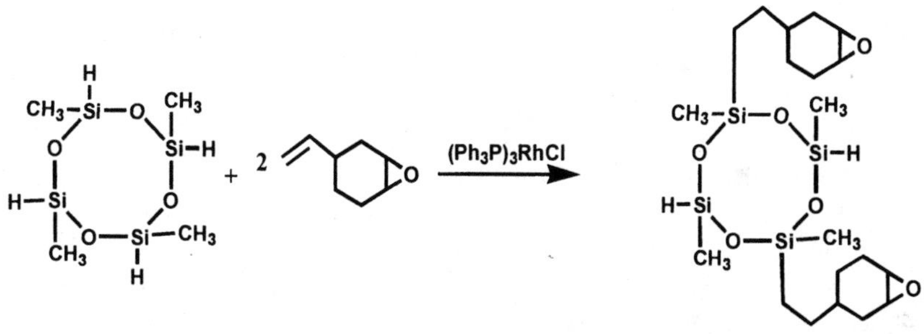

VIII (8)

IX (9)

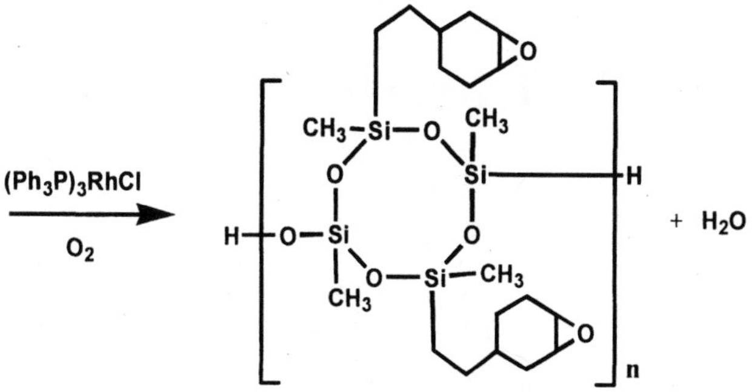

Monomer-bound photosensitizers

For a photopolymerization to proceed efficiently, the wavelengths of the irradiating light must be absorbed by the photoinitiator. Diaryliodonium and triarylsulfonium salts employed as photoinitiators in cationic polymerizations typically absorb most strongly at wavelengths below 300 nm. Longer wavelength irradiation is not effective for such systems. For this reason, many lasers and LEDs are inefficient light sources for conducting cationic photopolymerizations. Fortunately, these photoinitiator systems are readily sensitized to longer wavelength light through the addition of electron-transfer photosensitizers. For example, anthracenes, carbazoles and phenothiazines are excellent electron-transfer photosensitizers for diaryliodonium and triarylsulfonium salts. Unfortunately, oftentimes these compounds have undesirable toxic effects and are readily leached from the crosslinked polymer after polymerization. To avoid these complications, we have fitted these photosensitizers with a polymerizable epoxide or oxetane group so that during polymerization it becomes permanently bound into the polymer matrix (*16*). An example of the synthesis of one such monomer-bound photosensitizer incorporating a carbazole group is depicted in equation 10.

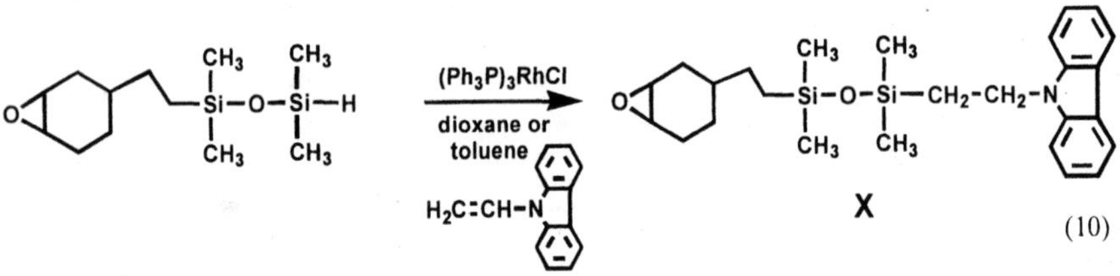

(10)

In a similar fashion, the preparation of **XI** containing an anthracene photosensitizing chromophor was also prepared.

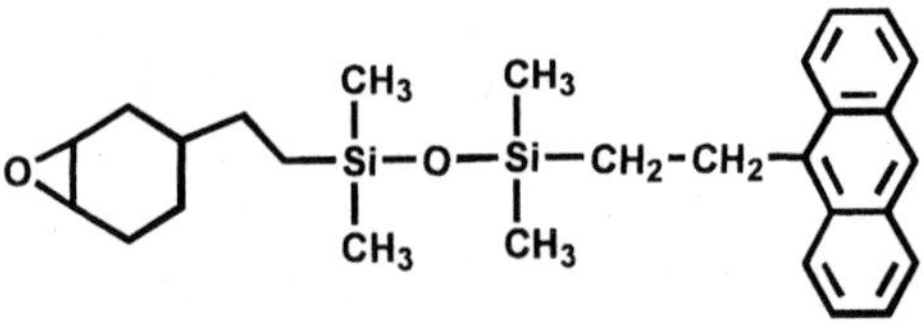

XI

Figure 3 shows an OP comparison of the photoinitiated cationic polymerization of **I** conducted in the presence of the low molar mass photosensitizer, 9-vinylcarbazole, with monomer bound photosensitzer **X** and without a photosensitizer.In the presence of the monomer-bound photosensitizer, **X**, the rate of polymerization of monomer **I** is markedly enhanced and the induction period is eliminated. At the same time, the carbazole photosensitizer moiety is bound into the matrix of the crosslinked polymer that is formed and cannot be removed even after exhaustive solvent extraction.

Conclusions

Novel synthetic methodology has been developed for the synthesis of a new series of silicon-containing monomers and oligomers as well as monomer-bound photosensitizers. These silicone-epoxy compositions display outstanding reactivity in photoinitiated cationic polymerization.

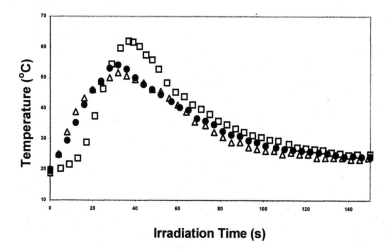

Irradiation Time (s)

*Figure 3. OP study of the photopolymerization of **I** with 1.0 mol% IOC10 carried out in the absence of a photosensitizer (□) and in the presence of 3.0 mol% 9-vinylcarbazole (△) and 3.0 mol% **X** (●) (light intensity 475 mJ cm⁻² min⁻¹).*

References

1. Crivello, J.V. *Developments in Polymer Photochemistry-2* Allen, N.S., Ed, Appl. Sciences Pub., London, 1981; pp. 1-38.
2. Casatelli, L.M. *American Ink Maker*, **2001**, *79*, 18.
3. Crivello, J.V. *Ring-Opening Polymerization*; D.J. Brunelle, Ed., Hanser, Munich, 1993; p.157/196ff.
4. Crivello, J.V.; Lee, J.L. *ACS Symposium Series No. 417*, C.E. Hoyle and J.F. Kinstle, Eds., **1989**, p. 398.
5. Crivello, J.V.; Sasaki, H. *J. Macromol. Sci., Pure Appl. Chem.*, **1992**, *A29*, 915.
6. Eckberg, R.P. *Coatings Technology Handbook, (3ʳᵈ. Ed.)* CRC Press LLC, Boca Raton, FL, 2006, 91/1-92/10.
7. Crivello, J.V.; Lee, J.L. *Polym. Mtls. Sci. and Eng. Prepr.*, **1989**, *60*, 217.
8. Crivello, J.V.; Lee, J.L. *J. Polym. Sci., Polym. Chem. Ed.*, **1990**, *28*, 479.
9. Crivello, J.V.; Lee, J.L. *Proc. of the RADTECH '90 North America Conf.*, Chicago, March 25, **1990**, p. 432.
10. Crivello, J.V.; Bi, D. *J. Polym. Sci.,Part A: Polym. Chem.*, **1993**, *31*, 2563.
11. Jang, M.; Crivello, J.V. *J. Polym. Sci. Part A: Polym. Chem.* **2003**, *41(19)*, 3056.

12. Crivello, J.V.; Lee, J.L. *J. Polym.Sci., Polym Chem. Ed.*, **1989**, *27*, 3951.
13. Crivello, J.V.; Falk, B.; Jang, M.; Zonca, Jr. M.R.; Vallinas, S.M *RadTech Report*, 2004, *May/June*, 36.
14. Falk, B.; Vallinas, S.M; Crivello, J.V. *Polym. Mater. Sci. Eng. Prepr.*, **2003**, *89*, 279.
15. Jang, M.; Crivello, J.V. *Polym. Prepr.* **2004**, *45(1)*, 587-588.
16. Crivello, J.V.; Jang, M. *J. Macromol. Sci.: Pure and Appl. Chem.* **2005**, *A42(1)*, 1-19.

Chapter 4

Fluorinated Copoly(carbosiloxane)s: Synthesis, Copolymerization, and Cross-Linked Networks

Melissa A. Grunlan[1,3], Nam S. Lee[1], John A. Finlay[2], James A. Callow[2], Maureen E. Callow[2], and William P. Weber[1,*]

[1]Loker Hydrocarbon Institute, Department of Chemistry, University of Southern California, Los Angeles, CA 90089–1661
[2]School of Biosciences, University of Birmingham, Birmingham, England
[3]Current address: Department of Biomedical Engineering, Texas A&M University, College Station, TX 77843-3120

A desirable combination of properties of siloxane and fluorinated polymers may be achieved by preparation of fluorinated copoly(carbosiloxane)s. A series of novel non-fluorinated (**I**) and fluorinated α,ω-bis(glycidyloxypropyl)-pentasiloxanes (**II-III**) were prepared and subsequently copolymerized with piperazine to form well-defined non-fluorinated (**IV**) and fluorinated copoly(carbosiloxane)s (**V-VI**) with high molecular weights. **I-III** were also reacted with α,ω-diaminoalkanes (**a-c**) to form cross-linked networks or films (**VIIa-c, VIIIa-c, and IXa-c**). Mechanical, surface, and thermal properties of these were studied. Anti-foul and foul-release properties of the films were evaluated by settlement and removal tests with spores of the green alga, *Ulva*. These films represent a novel class of epoxies with an unusual combination of properties: high flexibility, low T_g, hydrophobic surfaces, and good thermal stability.

37

Properties of Fluorosiloxanes

Fluorosiloxane copolymers and their films are of interest for achieving a blend of properties of siloxane and fluorinated polymers. Siloxanes, such as poly(dimethylsilsiloxane) (PDMS), retain their properties over a broad range of temperatures, possess low temperature flexibility (i.e. low glass transition temperature, T_g), hydrophobicity, high dielectric strength, and weather-resistance as well as thermal, oxidative, biological, and chemical stability (1). Fluorinated polymers, including polytetrafluoroethylene (PTFE), have low surface tensions and are stable to corrosive organic solvents, harsh chemicals, and high temperatures (2). Fluorosiloxanes, such as poly[methyl(3,3,3-trifluoropropyl)siloxane] (PMTFPS), possess low T_g's and chemical inertness (3). Fluorosiloxanes are used as aviation fuel tank sealants and gaskets and as release coatings, antifoams for organic liquids (such as crude oil), lubricants, surfactants, gels, and adhesives. We have previously reported the synthesis of several copoly(carbosiloxane)s with pendant 3',3',3'-trifluoropropyl groups (4). An ethyl group is usually placed between the fluoroalkyl group and the Si center because halogens bonded to silicon in the α or β positions are hydrolytically and thermally unstable (5).

Synthesis of Non-fluorinated and Fluorinated α,ω-Bis(glycidyloxypropyl)pentasiloxanes

Introduction of α,ω-bis-epoxy groups to siloxane and fluorosiloxane oligomers affords the opportunity for subsequent copolymerization and cross-linking reactions. Such reactive end-groups may be introduced via by platinum-catalyzed hydrosilylation reactions. In these, the anti-Markovnikov addition of a Si-H bond across a C-C double bond is usually observed (6). The most frequently used catalyst for hydrosilylation is Karstedt's catalyst (1,3-divinyl-tetramethyldisiloxane platinum complex) (7). We have reported the preparation of novel α,ω-bis(Si-H) pentasiloxane oligomers bearing methyl, 3',3',3'-trifluoropropyl and 1'H,1'H,2'H,2'H-perfluorooctyl groups (8, 9). We have also shown that glycidyloxypropyl end-groups may be introduced to these oligomers via Pt-catalyzed hydrosilylation with allyl glycidyl ether (Figure 1) (10). The well-defined microstructures of I-III are useful in establishing structure-property relationships of copolymers and cross-linked films formed in subsequent reactions.

Non-fluorinated and Fluorinated Copoly(carbosiloxane)s

The backbone of polysiloxanes consists of Si-O linkages, whereas polymers whose backbones contain both Si-C and Si-O bonds are referred to as copoly-(carbosiloxane)s (11). Non-fluorinated and fluorinated copoly(carbosiloxane)s

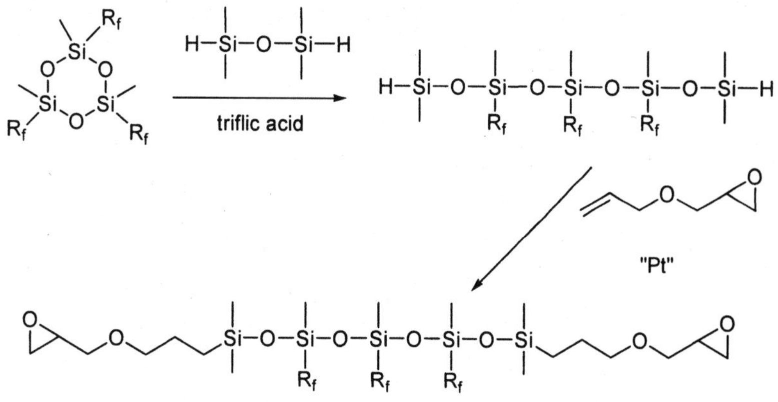

R_f = -CH$_3$ (I), -CH$_2$CH$_2$CF$_3$ (II), or -CH$_2$CH$_2$(CF$_2$)$_5$CF$_3$ (III)

Figure 1. Synthesis of α,ω-bis(glycidyloxypropyl)pentasiloxanes (I-III).

may be produced by copolymerization of **I** or **II-III**, respectively, with functionalized organic segments. The change in backbone composition is anticipated to produce a unique set of properties. For instance, hydroxy-containing organic segments introduced along the siloxane backbone would provide sites for hydrogen bonding for improved adhesion and oxygen barrier properties (*12,13*). Introduction of piperazine groups into the polysiloxane backbone is expected to increase T_g but decrease thermal stability because of its rigid, cyclic structure (*14*). In fact, linear copoly(carbosiloxane)s containing hydroxy-pendant groups and piperazine moieties within the polymer backbone have been prepared via the condensation of α,ω-bis(glycidyloxypropyl)oligodimethylsiloxanes and bisphenol-A diglycidyl ether with piperazine (*15*). We have recently reported the preparation of a series of well-defined non-fluorinated (**IV**) and fluorinated copoly(carbosiloxane)s (**V-VI**) from the reaction of 1,9-bis(glycidyloxypropyl)-pentasiloxane (**I**) and 1,9-bis(glycidyloxypropyl)penta(1'H,1'H,2'H,2'H-perfluoroalkylmethylsiloxane)s (**II-III**), respectively, with piperazine (*10*) (Figure 2).

Condensation reactions between piperazine and **I-III** produced high molecular weight non-fluorinated (**IV**) and fluorinated copoly(carbosiloxane)s (**V-VI**) in quantitative yields. It is critical to have exact stoichiometric conditions to achieve high molecular weights in step-growth copolymerization reactions (*16*). Thus, the high molecular weights obtained for **IV-VI** are noteworthy (Table 1). The expected molecular weight distribution (M_w/M_n) for a high molecular weight linear step-growth polymer is 2 (*17*). The M_w/M_n for the narrower then expected molecular weight distribution is not known. The T_g of PDMS is -123 °C whereas the T_g of PMTFPS is -70 °C (*18, 19*). Introduction

I-III

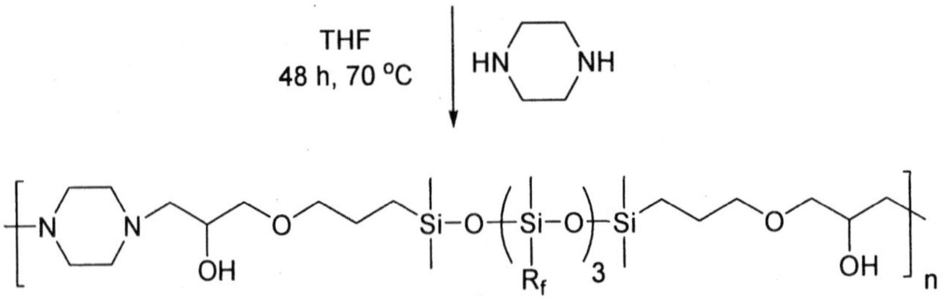

THF
48 h, 70 °C

IV-VI

R_f = -CH$_3$ (I, IV); -CH$_2$CH$_2$CF$_3$ (II, V); -CH$_2$CH$_2$(CF$_2$)$_5$CF$_3$ (III, VI)

Figure 2. Preparation of copoly(carbosiloxane)s (IV-VI).

of rigid piperazine moieties into the siloxane backbone as well as pendant hydroxyl groups should significantly increase T_g of the copolymers. Thus, the T_g's of copolymers **IV**, **V**, and **VI** were -41, -36, and -34 °C, respectively. An increase in T_g of fluorinated copolymers (**V** and **VI**) is anticipated versus the non-fluorinated copolymer (**IV**) as a result of the electronic repulsion between adjacent CF_n groups which leads to side chain rigidity (*5*).

Table 1. Copoly(carbosiloxane)s (IV-VI)

Copolymer	IV	V	VI
M_w	63,000	71,300	48,200
M_n	45,000	56,300	32,700
PDI (M_w/M_n)	1.4	1.3	1.5
T_g (°C)	-41 °C	-36 °C	-34 °C

The thermal stabilities of **IV-VI** were determined by thermal gravimetric analysis (TGA). Fluorosiloxanes typically have lower thermal stabilities than comparable dimethylsiloxanes (*20*). Thus, **IV** was the most thermally stable (Figure 3). **IV** was stable up to nearly 300 °C in both nitrogen and in air. As the level of fluorination increased from 3',3',3'-trifluoropropyl groups (**V**) to 1'H',1'H,2'H,2'H-perfluorooctyl groups (**VI**), thermal stability decreased. In both nitrogen and in air, **V** began to degrade around 250 °C, whereas **VI** was stable only to 200 °C. In nitrogen, catastrophic decomposition of **IV-VI** resulted in virtually no char residue. As expected, the amount of residue remaining after decomposition in air was greater than in nitrogen (*21*). In air, a maximum of 15% residue remained in the case of **IV**.

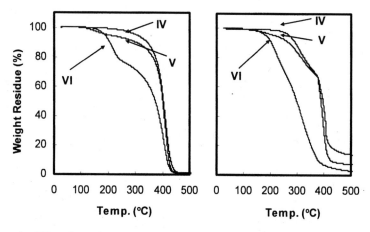

Figure 3. TGA of copoly(carbosiloxane)s (IV-VI) in N_2 (left) and in air (right).

Cross-linking of Non-fluorinated and Fluorinated α,ω-Bis(glycidyloxypropyl)pentasiloxanes with α,ω-Diaminoalkanes

Fluorinated copoly(carbosiloxane)s bearing pendant epoxy end-groups may be cross-linked to form solid networks or films with interesting properties. Epoxy resins are extensively used in adhesives, coatings, composites, and electronics packaging. They demonstrate high resistance to chemicals, solvents, corrosion, and electricity as well as high modulus, tensile strength, and dimensional stability (*22,23*). Upon cure, they form highly cross-linked microstructures with high T_g's. However, contraction during cure generates internal stress which promotes loss of mechanical, thermal, and moisture barrier properties (*24*). Their utility is limited by their brittleness, low thermal stability, flammability, and water absorption. Introduction of silicon or fluorine containing groups into epoxies may decrease brittleness, lower T_g, improve thermal stability, and increase hydrophobicity. Problems associated with blending functionalized rubber or thermoplastic modifiers into epoxy resins (e.g. solubility and stability) may be avoided by covalent attachment of siloxane and fluoroalkyl segments to epoxy moieties. Such materials may combine the useful properties of epoxy resins with those of siloxane and fluoropolymers. Epoxy-type materials containing siloxane or fluoroalkyl moieties have been reported. For example, epoxy-terminated siloxane oligomers have been synthesized and cross-linked (*15,25,26*). Fluorinated epoxy resins have also been synthesized and cured with α,ω-(diamino)oligosiloxanes (*27,28*).

We have reported the cross-linking of α,ω-bis(glycidyloxypropyl)penta-siloxanes (**I-III**) with α,ω-diaminoalkanes (**a-c**) (*29*). In contrast to other studies with siloxane or fluorinated modified epoxies (*15,27,28,30*), these have well-defined microstructures. Molecular distance between cross-links was varied by choice of α,ω-diaminoalkanes (**a-c**). These cure reactions represent the most reactive type of epoxy-amine crosslinking systems: terminal epoxy groups, such as glycidyl ethers, and primary aliphatic amines (*12*). Phenol was used to catalyze the curing of films on glass substrates. Cross-linking of **I**, **II**, or **III** with α,ω-diaminoalkanes (**a**, **b**, or **c**) formed nine films (**VIIa-c**, **VIIIa-c**, and **IXa-c**) (Figure 4).

I-III

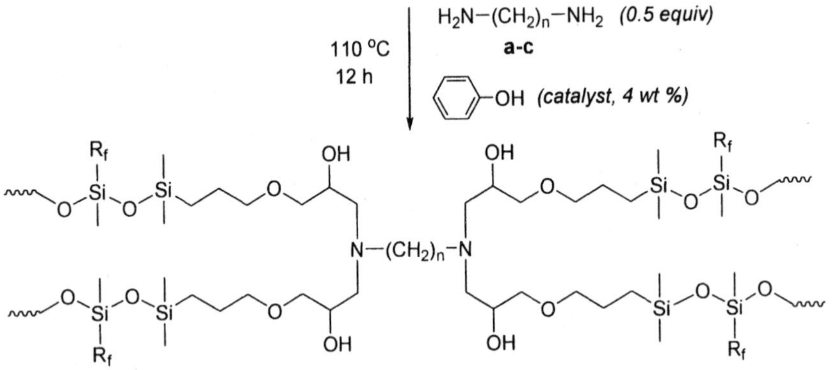

$H_2N-(CH_2)_n-NH_2$ *(0.5 equiv)*

110 °C

12 h

a-c

⬡—OH *(catalyst, 4 wt %)*

Films VIIa-c, VIIIa-c, and IXa-c

R_f = $-CH_3$ (**I, VII**), $-CH_2CH_2CF_3$ (**II, VIII**), or $-CH_2CH_2(CF_2)_5CF_3$ (**III, IX**)

n = 6 (**a**), 8 (**b**), or 12 (**c**)

Figure 4. Preparation of cross-linked films (VIIa-c, VIIIa-c, and IXa-c).

Mechanical, Surface, and Thermal Properties of Films

The mechanical properties of the films (**VIIa-c**, **VIIIa-c**, and **IXa-c**) were determined by dynamic mechanical thermal analysis (DMTA) (Figure 5). Storage modulus (G') increased over the measured temperature range (-130 to 30 °C) in the order: **c** < **b** < **a**. **IXa** displayed the highest overall values of G'. For films prepared with the same α,ω-diaminoalkane (**a-c**), G' typically increased with higher levels of fluorination in the order: **VII** < **VIII** < **IX**.

The loss modulus (G") exhibits a maximum at T_g and $T_β$ (beta transition temperature) (*31*). The values of T_g and $T_β$ for the films are reported in Table 2. For films based on **VII** or **VIII**, T_g increased with cross-linking density in the order: **c** < **b** < **a**. Films based on **IX** deviated from this trend and had T_g's between -45 °C (**IXa, IXb**) and -37 (**IXc**). For films prepared with the same α,ω-diaminoalkane (**a-c**), the T_g was expected to increase with the presence of adjacent CF_n groups due to their mutual electronic repulsion which gives rise to side chain rigidity (*5*). For example, the T_g's of **VIIIc** (-40 °C) and **IXc** (-37 °C) were higher than that of non-fluorinated **VIIc** (-50 °C). The T_g of **VIIIa** (-31 °C)

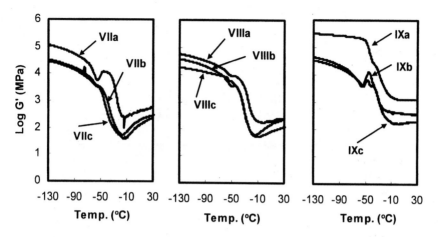

Figure 5. Storage moduli (G') of films (VIIa-c, VIIIa-c, and IXa-c).

was higher than that of non-fluorinated **VIIa** (-36 °C); however, the T_g of **IXa** (-45 °C) was lower. Similarly, the T_g of **VIIIb** (-37 °C) was higher than that of non-fluorinated **VIIb** (-45 °C) but the T_g of **IXb** (-45 °C) was equal to that of **VIIb**.

The presence of a T_β can signify side chain mobility and may be related to toughness (*31*). T_β's of **VIIa-c** (-76 to -70 °C) were lower than for other films, except **IXb** (-77 °C) (Table 2). No T_β was observed for **IXa**.

Surface properties of the films were evaluated with static contact angle measurements of distilled/deionized water droplets (10 μL) on the air (θ_{Air}) and glass (θ_{Glass}) interfaces (Table 2). At the air interface, all films were hydrophobic ($\theta_{Air} > 90$ °) (*32*). Very high values of θ_{Air} were observed for films based on **IX** (124-153 °) which were also quite hazy. Fluoroalkyl groups are known to undergo surface reorganization which could lead to increased hydrophobicity and haziness (*33*). Film **IXc** was extremely hydrophobic ($\theta_{Air} = 153$ °). Its relatively low crosslink density may have enhanced the mobility of fluoroalkyl groups to the surface. θ_{Air} of films based on **VII** (102-112 °) and **VIII** (105-112 °) were quite similar. For all films, θ_{Glass} was lower than the respective θ_{Air}. This effect has been observed for fluoroepoxy films (*20*). θ_{Glass} of films based on **IX** (112-119 °) were quite high compared to those of films based on **VII** (91-95 °) and **VIII** (82-85 °). Thus, films **VIIIa-c** displayed similar θ_{Air} values and lower θ_{Glass} values compared to non-fluorinated films **VIIa-c**. Apparently, the 3',3',3'-trifluoropropyl groups of films based on **VIII** were ineffective in lowering surface energies. θ_{Glass} of **IXc** could not be determined as the droplet would not hold its shape long enough to be measured, indicating that defects formed upon its removal from the glass substrate.

Thermal stabilities of the films (**VIIa-c, VIIIa-c**, and **IXa-c**) under nitrogen and air atmospheres were studied with TGA. In nitrogen, catastrophic decomposition of **VIIa, VIIb**, and **VIIc** began at 200, 275, and 350 °C, respectively (Figure 6).

Table 2. Properties of Films

Film	T_g (°C)	T_β (°C)	θ_{Air} (°)	θ_{Glass} (°)
VIIa	-36	-76	112	92
VIIb	-45	-76	112	95
VIIc	-50	-70	102	91
VIIIa	-31	-61	111	82
VIIIb	-37	-64	105	85
VIIIc	-40	-55	112	82
IXa	-45	----	129	112
IXb	-45	-77	124	119
IXc	-37	-66	153	N/A

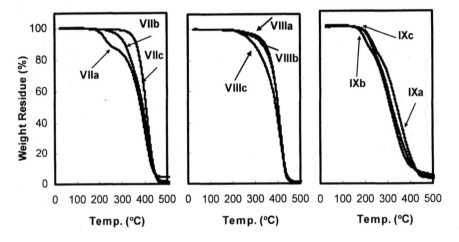

Figure 6. TGA of films in nitrogen.

Films **VIIIa** and **VIIIb** both began to degrade at ~275 °C while **VIIIc** began to decompose at slightly lower temperatures. As expected for highly fluorinated films, those based on **IX** were significantly less thermally stable and began to decompose ~200 °C (*20*). Thermooxidative decomposition of the films in air began at lower temperatures than in nitrogen (*21*). Films **VIIa-c** and **VIIIa-c** began to decompose at ~ 225 °C whereas films **IXa-c** were stable to only ~175 °C. For all films, most of the weight was lost occurred by 450 °C with char yields under ~15 %.

Anti-foul and Foul-release Properties of Films

Polysiloxanes and fluorosiloxanes have demonstrated anti-foul and foul release properties (*9,34-37*). Their minimally adhesive behavior towards marine organisms is a consequence of their hydrophobicity as well as their low polarity, T_g, surface energy, and storage modulus (G') (*9,34,38*). The green macroalga of the genus *Ulva* (syn. *Enteromorpha*) is a cosmopolitan, intertidal macroalga which leads to biofouling of submerged structures, including ships' hulls (*39*). Fouling occurs by settlement of dispersed motile zoospores: quadriflagellate, pear-shaped cells, 5-7 μm in length, and ~5 μm diameter at the widest point (*40*). Spores initially anchor themselves to substratum by excretion of a hydrophilic glycoprotein adhesive which rapidly cures (*41*). The attached spore develops a cell wall and the resultant sporeling eventually matures to form a sessile plant.

Antifoul and foul-release properties of these films were studied by standard settlement and removal tests with *Ulva* spores. Six replicate microscope slides of **VIIa-c**, **VIIIa-c**, and **IXc** and acid-washed uncoated glass slides which served as the controls were tested. Slides were initially leached for 24 h in stirred distilled water. *Ulva* zoospores were first released from fertile *Enteromorpha linza* plants (*41*) and allowed to settle on coated glass slides over 4 h (*42*). Spores settled normally and non-settled spores remained motile, indicating the leachate was nontoxic. Spores which were adhered to the surfaces of **VIIa-c**, **VIIIa-c**, and **IXc** were counted with a Zeiss Kontron 3000 image capture analysis system attached to a Zeiss epifluorescence microscope (*43*) (Figure 7). The surface of **VIIIa** had many rougher patches making counting spores difficult. Also, under fluorescent microscopy, **VIIIa** and some of **VIIIb** contained brightly fluorescing circular bodies which masked the presence of spores. These "bodies" were attributed to phase separation of the α,ω-diaminoalkane (**a-c**). As a result, spore settlement *and* adhesion onto **VIIIa** could not be counted and only the spore settlement onto **VIIIb** could be obtained. Film **IXc** was rough, with a matte appearance. Spore settlement on all films except **VIIIb** was similar or greater to that on uncoated glass. In general, settlement was higher on the non-fluorinated surfaces (**VIIa-c**). Attached spores were then subjected to an automated water jet with a surface pressure of 54 kPa for 5 min (*42,44*). Spore removal by exposure to the water jet was negligible from all non-fluorinated films (**VIIa-c**) and fluorinated film (**IXc**) (Figure 7). The highest spore removal (66%) was from film **VIIIc**.

46

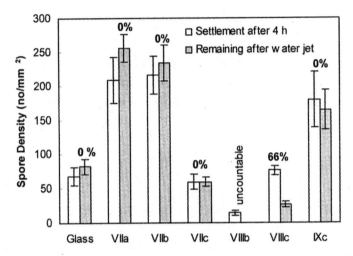

Figure 7. Settlement of Ulva spores on films (after 4 h) and spores remaining after exposure to water jet.

Summary

The preparation of a series of novel non-fluorinated (**I**) and fluorinated α,ω-bis(glycidyloxypropyl)pentasiloxanes (**II-III**) allowed the preparation of copoly(carbosiloxane)s and cross-linked films. **I-III** were copolymerized with piperazine to form well-defined non-fluorinated (**IV**) and fluorinated copoly(carbosiloxane)s (**V-VI**) with high molecular weights. **I-III** were also reacted α,ω-diaminoalkanes (**a-c**) of different molecular weights to form nine films (**VIIa-c, VIIIa-c, and IXa-c**). The mechanical and surface properties of the films were related to their anti-foul and foul-release behavior with *Ulva* spores. In general, films **VIIIb-c**, bearing 3',3',3'-trifluoropropyl pendant groups, demonstrated superior inhibition of *Ulva* spore settlement (anti-fouling) and removal (foul-release) versus non-fluorinated films (**VIIa-c**) or highly fluorinated films (**IXc**). Films **VIIIc** ($T_g = -40$ °C, $\theta_{air} = 112$ °) displayed low G' values, exhibited low spore settlement and the highest percent removal of spores (66 %). Films **VIIc** ($T_g = -50$ °C, $\theta_{air} = 102$ °) and **IXc** ($T_g = -37$ °C, $\theta_{air} = 153$ °) possessed similar low G' values but foul-release properties were worse. The surface roughness of **IXc** may have enhanced the attachment of the spores. These results indicate that a balance of certain properties (T_g, G', and surface energy) is required to minimize adhesion of marine organisms to polymer surfaces.

Acknowledgements

Support from the Office of Naval Research is gratefully acknowledged.

References

1. Odian, G. *Principles of Polymerization*, 3rd Ed.; Wiley: New York, N.Y., 1991; p 138.
2. Odian G. *Principles of Polymerization*, 3rd Ed.; Wiley: New York, N.Y., 1991; p 313.
3. Maxson, M.T.; Norris, A.W.; Owen, M.J. In *Modern Fluoropolymers*; Scheirs, J., Ed.; John Wiley & Sons, Inc.: Chichester, 1997; pp 359-372.
4. Grunlan, M.A.; Mabry, J.M.; Weber, W.P. *Polymer* **2003**, *44*, 981.
5. Pierce, O.R. *Appl. Polym. Symp.* **1970**, *14*, 7.
6. Speier, J.L. *Adv. Organomet. Chem.* **1979**, *1*, 407.
7. Karstedt, B.D. U.S. Patent 3,775,452, 1973.
8. Mabry, J.M.; Paulasaari, J.K.; Weber, W.P. *Polymer* **2000**, *41*, 4423.
9. Grunlan, M.A.; Lee, N.S.; Cai, G.; Gädda, T.; Mabry, J.M.; Mansfeld, F.; Kus, E.; Wendt, D.E.; Kowalke, G.L.; Finlay, J.A.; Callow, J.A.; Callow, M.E.; and Weber, W.P. *Chem. of Mater.* **2004**, *16*, 2433.
10. Grunlan, M.A.; Lee, N.S.; Weber, W.P. *Polymer* **2004**, *45*, 2517.
11. Interrante, L.V.; Shen, Q. In *Silicon-Containing Polymers*; Jones, R.G.; Ando, W.; Chojnowski, J.; Eds.; Kluwer Academic Publishers: Netherlands, 2000; p 309.
12. *Organic Coatings: Science and Technology*; Wicks, Z.W.; Jones, F.N.; Pappas, S.P.; Eds.; John Wiley & Sons, Inc.: New York, N.Y., 1994; Vol. II, pp 163-170.
13. Salame, M.; Temple, E.J. *Adv. Chem.* **1974**, *135*, 61.
14. Odian, G. *Principles of Polymerization*, 3rd Ed.; Wiley: New York, N.Y., 1991; p 32.
15. Riffle, J.S.; Yilgor, I.; Tran, C.; Wilkes, G.L.; McGrath, J.E.; Banthia, A.K. In *Epoxy Resin Chemistry II*; Bauer, R.S.; Ed.; ACS Symposium Series 221; American Chemical Society: Washington, D.C., 1983; pp 21-54.
16. Odian, G. *Principles of Polymerization*, 3rd Ed.; Wiley: New York, N.Y., 1991; p 78.
17. *Organic Coatings: Science and Technology*; Wicks, Z.W.; Jones, F.N.; Pappas, S.P., Eds.; John Wiley & Sons, Inc.: New York, N.Y., 1994; Vol. II; p 24.
18. Andrews, R.J.; Grulke, E.A. In *Polymer Handbook*, 4th Ed; Brandrup, J.; Immergut, E.H.; Grulke, E.A.; Eds.; John Wiley & Sons, Inc.: New York, N.Y., 1999; pp VI-231.
19. Stern, S.A.; Shah, V.M.; Hardy, B.J. *J. Poly. Sci.: Part B: Poly. Phys.* **1987**, *25*, 1263.
20. Knight, G.J.; Wright, W.W. *British Polymer Journal* **1989**, *21*, 199.
21. Dvornic, P.R. In *Silicon-Containing Polymers,* Jones, R.G.; Ando, W.; Chojnowski, J.; Eds.; Kluwer Academic Publishers: Netherlands, 2000; pp 185-212.
22. Ellis, B. *Chemistry and Technology of Epoxy Resins*; Blackie Academic & Professional: London, 1993.

23. May, C.A. *Epoxy Resins: Chemistry and Technology,* 2nd Ed.; Marcel Dekker, Inc.: New York, N.Y., 1988.
24. Ochi, M.; Yamashita, K.; Yoshizumi, M.; Shimbo, M. *J. Appl. Poly. Sci.* **1989**, *38*, 789
25. Crivello, J.V.; Bi, D. *J. Poly. Sci.: Part A: Poly. Chem.* **1993**, *31*, 3109.
26. Crivello, J.V.; Bi, D. *J. Poly. Sci.: Part A: Poly. Chem.* **1993**, *31*, 3121
27. Griffith, J.R. *Chemtech* **1982**, *12*, 290.
28. Twardowski, T.E.; Geil, P.H. *J. Appl. Poly. Sci.* **1990**, *41*, 1047.
29. Grunlan, M.A.; Lee, N.S.; Weber, W.P. *J. Appl. Poly. Sci.* **2004**, *94*, 203.
30. Hou, S.S.; Chung, Y.P.; Chan, C.K.; Kuo, P.L. *Polymer* **2000**, *41*, 3263.
31. Sperling, L.H. *Introduction to Physical Polymer Science,* 3rd Ed.; Wiley: New York, N.Y., 2001; pp 308-335.
32. Sangermano, M.; Bongiovanni, R.; Malucelli, G.; Priola, A.; Policino, A.; Recca, A. *J. Appl. Polym. Sci.* **2003**, *89*, 1254.
33. Uilk, J.; Johnston, E.E.; Bullock, S.; Wynne, K.J. *Macromol. Chem. Phys.* **2002**, *203*, 1506.
34. Singer, I.L.; Kohl, J.G.; Patterson, M. *Biofouling* **2000**, *16*, 301.
35. Baier, R.E.; Meyer, A.E. *Biofouling* **1992**, *6*, 165.
36. Mera, A.E.; Wynne, K.J. U.S. Patent 6,265,515 B1, 2001.
37. Mera, A.E.; Goodwin, M.; Pike, J.K.; Wynne, K.J. *Polymer* **1999**, *40*, 419.
38. Newby, B.Z.; Chaudhury, M.K.; Brown, H.R. *Science* **1995**, *269*, 1407.
39. Callow, M.E. *Bot. Mar.* **1986**, *24*, 351.
40. Fletcher, R.L.; Callow, M.E. *Br. Phycol.* **1992**, *27*, 303.
41. Callow, M.E.; Callow, J.A.; Pickett-Heaps, J.D.; Wetherbee, R. *J. Phycol.* **1997**, *33*, 938.
42. Finlay, J.A.; Callow, M.E.; Schultz, M.P.; Swain, G.W.; Callow, J.A. *Biofouling* **2002**, *18*, 251.
43. Callow, M.E.; Jennings, A.R.; Brennan, A.B.; Seegert, C.E.; Gibson, A.; Baney, R.; Callow, J.A. *Biofouling* **2002**, *18*, 237.
44. Swain, G.W.; Schultz, M.P. *Biofouling* **1996**, *10*, 187.

Chapter 5

Sulfidosilanes: Complex Molecule Systems via Phase Transfer Catalysis

Michael W. Backer[1], John M. Gohndrone[2], Simon D. Cook[3],
Simeon J. Bones[1,†], and Raquel Vaquer-Perez[1,†]

[1]Surface and Interface Solutions Center and [3]Service Enterprise Unit,
Analytical Sciences, Dow Corning Limited, Cardiff Road, Barry,
South Glamorgan, CF63 2YL, United Kingdom
[2]Specialty Intermediates Technology Center, Dow Corning Corporation,
2200 West Salzburg Road, Midland, MI 48686

The complete series of ethoxy and methyl substituted tetra-sulfidosilanes $(EtO)_{3-x}Me_xSi(CH_2)_3-S_4-(CH_2)_3-SiMe_x(OEt)_{3-x}$, with x = 0 - 3, has been prepared via economically favourable and environmentally friendly phase transfer catalysis (PTC) process. UV-absorption spectra and ^1H, ^{13}C, and ^{29}Si Nuclear Magnetic Resonance spectra of these systems, actually mixtures of silanes of various sulfur chain length ranging from disulfane S2 up to dodecylsulfane S12 with an average sulfur value (rank) of about 3.75, are discussed in dependence of the varying substitution pattern. Based on the results, response factors in High Pressure Liquid Chromatography analysis are proposed for each sulfane species per silane reflecting the different weight percentage rates of the sulfur chromophore.

Introduction

During the last three decades sulfur containing organosilicon compounds such as bis(triethoxysilylpropyl)disulfane (TESPD) and bis(triethoxysilyl-propyl)tetrasulfane (TESPT) 1 have become very important coupling agents (*1*) in a variety of industrial applications. In particular, they have become essential components in the production of automotive tires based on filler reinforced rubber vulcanizates. (2) The combination of e.g. precipitated silica and sulfido-silanes is improving the physical properties of the tires such as rolling resistance and wet skidding performance, while maintaining abrasion resistance, due to strong coupling between the inorganic filler and the rubber matrix. (3)

Numerous methods have been described in the art of preparation of sulfur containing organosilicon compounds. While in earlier times reactions of alkali metal sulfides with elemental sulfur and halohydrocarbyl alkoxysilanes had to be carried out mainly in alcohols under strictly anhydrous conditions, nowadays modern phase transfer catalysis (PTC) techniques (4,5) have been developed for the production of sulfur containing organosilicon compounds in the presence of an aqueous phase (6,7,8,9) and steadily improved to an efficient, and environmentally friendly process that can be safely performed on an industrial scale. (10,11,12,13) During the development of the PTC process, pH control of the aqueous phase and the fine-tuning of the correct oligosulfane distribution by adjustment of the accurate amount of elemental sulfur added to an ionic sulfide species have been identified as critical and very sensitive parameters. (14)

Throughout the years, it has been demonstrated that TESPT actually is a mixture of sulfanes with a range from disulfane (S2) to decasulfane (S10) having an average sulfur chain length of around 3.75 - 3.8, and that TESPD normally contains polysulfanes from S2 to S5 with an average sulfur chain length of about 2.1 – 2.15. (15) For quality control purposes and determination of oligosulfane distribution mostly High Pressure Liquid Chromatography (HPLC) techniques are routinely used. (16) However, Nuclear Magnetic Resonance (NMR) and Mass Spectrometry (MS) have also been successfully applied. (17,18)

In this paper, the complete series of ethoxy and methyl substituted tetrasulfidosilanes $(EtO)_{3-x}Me_xSi(CH_2)_3-S_4-(CH_2)_3-SiMe_x(OEt)_{3-x}$, with x = 0 (TESPT 1), 1 (DEMSPT 2), 2 (DMESPT 3), and 3 (TMSPT 4) has been prepared via PTC process. The products have been analysed via HPLC. The UV-absorption spectra of each individual sulfane species per silane series have been recorded and compared. The proton, carbon, and silicon NMR spectra of the silanes have also been taken and discussed in context with the continuously reduced polarity at the silicon center when moving from TESPT 1 to bis(trimethylsilylpropyl)tetrasulfane TMSPT 4. The spectra have been used to conclude on obvious similarities in the sulfidosilane systems in order to calculate reliable response factors for HPLC analysis.

Experimental

Compound preparation

In a typical experiment under PTC conditions, flaked sodium hydrogen-sulfide hydrate has been heated with an equimolar amount of alkali hydroxide solution to elevated temperatures (between 60 and 85 °C) in enough water to keep the mixture soluble. The specifically calculated molar amounts of elemental sulfur have then been added under vigorous stirring, and a short period for equilibration of the generated oligosulfide species has been allowed. After addition of the PTC catalyst, tetrabutylammonium bromide (TBAB), two molar equivalents (to HS⁻) of the according chloropropylsilane have been added over a period that permitted control of the exotherm. The nucleophilic substitution reaction was followed by gas chromatography analysis until chloropropylsilane had disappeared or reached a stable level. The mixture was cooled down and water added to dissolve formed alkali chloride/bromide. The aqueous phase was drained off, and the organic phase was filtered and dried according to various known methods. A typical equipment setup is displayed in Figure 1.

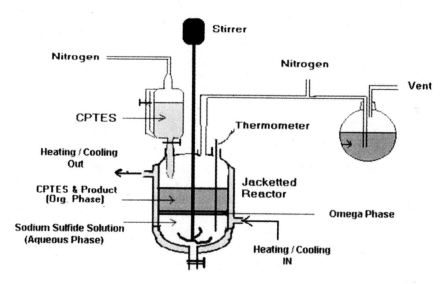

Figure 1. Experimental setup of Phase Transfer Catalysis process.

HPLC analysis

The distribution of the various sulfur containing compounds was analyzed by High Pressure Liquid Chromatography (HPLC). Typical run conditions were as follows: the reaction samples were diluted in cyclohexane and then filtered through a 0.45 μm PTFE membrane (e.g. PURADISC™ 25TF of Whatman®) into a vial. 10 μl sample of filtrate was injected via an autosampler into a HPLC system (e.g. Waters 2690 system with Waters 996 Diode Array detector). The sample was fractionated on a RP18 column (e.g. Phenomenex Gemini) using a water/alcohol gradient mixture as mobile phase. The fractions were investigated via UV-absorption using $\lambda = 254$ nm as the appropriate excitation wavelength. Different UV-sensitivities of every single sulfane species were averaged by division of the respective peak area through specific, empirically evaluated, response factors. The response factors (or inverse values to be multiplied) of TESPT 1 are known from literature (*16*) and listed in Table I. They reflect the hyperchromy with every additional sulfur atom in the chain and elemental sulfur.

Table I. HPLC Response Factors of 1

S_x	S2	S3	S4	S5	S6	S7	S8	S9	S10	S_{el}
RF 1	1.0	3.52	6.39	9.78	13.04	17.39	20.87	26.08	31.30	37.26

NMR Spectroscopy

Liquid state NMR spectra were obtained using a JEOL Lambda 400 MHz spectrometer with a 5 mm autotunable probe operating at the frequencies ω = 400.18 (^{1}H), 100.62 (^{13}C), and 79.50 (^{29}Si) MHz. The deuterated solvent CDCl$_3$ (99.8% deuterated) was dried over molsieve A4 prior to use. The isotropic shifts were calibrated to the signals of CHCl$_3$ at δ = 7.26 (^{1}H) and δ = 77.00 ppm (^{13}C) correlated to Me$_4$Si (TMS, δ = 0.0 ppm). For ^{29}Si chemical shift measurements, one drop of Me$_4$Si can be directly added as an internal standard with the signal to be set at δ = 0.00 ppm. The NMR acquisition parameters for the three nuclei were set in the following manner: (i) ^{1}H 90° pulse length 5.1 μs, sweep width 10 khz, number of scans 32, relaxation delay 3 s; (ii) ^{13}C-{^{1}H} decoupling 90° pulse length 4.50 μs, sweep width 24 khz, number of scans due to the signal-to-noise ratio, relaxation delay 60 s, (iii) ^{29}Si inverse gated {^{1}H} decoupling 90° pulse length 5.45 μs, sweep width 20 khz, number of scans due to the signal-to-noise ratio, relaxation delay 60 s. In order to achieve high spectral resolutions for both the ^{29}Si and ^{13}C spectra, such as 0.07 Hz for optimal separation of peaks, the addition of a relaxation aid as chromium (tris)acetylacetonate [Cr(acac)$_3$] has been relinquished, and a high data point density of 128 k has been applied. The probe was heated to a constant temperature t = 25 °C during acquisition to remove any heating effect the decoupler may have had on the sample.

Results and Discussion

Phase Transfer Catalysis

A simplified scheme of the formation of bis(silylpropyl)polysulfanes is shown in Figure 2 by the reaction of chloropropyltriethoxysilane (CPTES) with a caustic solution of sodium hydrogensulfide and elemental sulfur.

SH$^-$ + OH$^-$ + x S + 2 Cl(CH$_2$)$_3$Si(OEt)$_3$ → (EtO)$_3$Si(CH$_2$)$_3$S$_{1+x}$(CH$_2$)$_3$Si(OEt)$_3$ + 2 Cl$^-$ + H$_2$O

Figure 2. Simplified reaction equation for experiment 1.

As displayed in Figure 3, the first step of the phase transfer catalysis cycle for sulfurization is the ion exchange in the aqueous phase between disodium polysulfide and the quarternary ammonium bromide. The more hydrophobic ammonium polysulfide complex, in concentrated form visible as a thin brown interlayer ("omega phase", see Figure 1), is then transferred into the organic phase consisting of chloropropylsilane and later also of sulfidosilane product. The high nucleophilicity of the sulfide ion leads to the substitution of the chlorine in the bimolecular reaction. The leaving hydrophilic chloride is finally brought back with the ammonium cation into the aqueous layer. Here the catalyst cycle is starting again with the ion exchange step under generation of the catalyst-polysulfide complex and sodium chloride. The exergonic formation of sodium halogenide by-product can be seen as the actual driving force of the whole reaction, and the progress of the reaction can be followed by the increasing amount of sodium chloride precipitating out of the warm saturated aqueous solution.

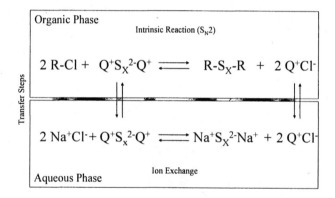

Figure 3. Phase Transfer Catalysis cycle for sulfurization of chloropropylsilanes.

The equilibrium distribution of (poly)sulfide ions (S^{2-}, S_x^{2-}, HS^-) in aqueous solutions under variable pH conditions has been investigated for a long time either by fast protonation to form the analogous, not highly stable sulfanes (*19,20*) for analysis or by optical spectrometric analysis (*21,22*), even at higher temperatures. (*23*) Nevertheless, all results remain effected by method specific uncertainties that do not help to predict accurate sulfidosilane product distributions to support the design of PTC experiments. The estimation of acidity concentration constants (pK_a) in the sequence of $S^{2-} > 14.00$, $S_2^{2-} = 9.7$, $S_3^{2-} = 7.5$, $S_4^{2-} = 6.3$, to S_5^{2-} 5.7 (*18*) confirms the observation that pH values of aqueous sulfide solutions gradually drop and the average sulfur chain length S_{avg} of the sulfidosilane, not unexpectedly of course, grows with the amount of elemental sulfur added. The operational range of pH has to be finally adjusted between around pH = 12 as the limitation of the stability of alkoxysilanes against condensation reaction and neutral conditions of around pH = 7 when formation and emission of hydrogen sulfide can be observed (pK_a (HS⁻) = 6.9). During the PTC reaction a constant drop of the pH value towards neutral conditions can be measured reflecting the ongoing consumption of the sulfide species.

The fine-tuning to the desired organic sulfane distribution is carried out by the adjustment of elemental sulfur amounts that are added to the aqueous sulfide solution. Figure 4 exhibits the strong dependence of the distribution on slightest changes in the starting ratio of sulfur to (hydrogen) sulfide. As shown, the weight percentage of bis(silylpropyl)disulfane S2 is rapidly and of trisulfane S3 slowly decreasing when the sulfur/sulfide ratio is increased from 2.6 to 3.0. At the same time, the tetrasulfane S4 content remains nearly constant while higher sulfanes are increasingly formed. It is also noted that as the sulfur / sulfide ratio increases unincorporated sulfur begins to be observed in the final product distribution. It is obvious that average sulfur chain lengths higher than S_{avg} = 4.5 - 5.0 will hardly be achieved within aqueous phase chemistry.

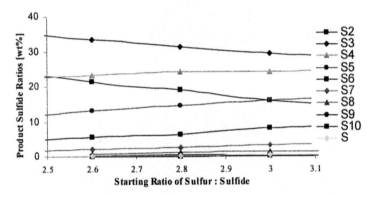

Figure 4. Sulfane distribution in 1 in dependence of sulfur /sulfide ratio.

HPLC Analysis

Figure 5 displays a typical HPLC chromatogram of bis(triethoxysilyl-propyl)tetrasulfane 1 prepared via PTC route. Using the described parameter the peaks are well separated, and sulfanes up to the dodecasulfane S12 have been identified. There is minimal evidence of condensed siloxanes at retention times of around 30 minutes typically detected in higher amounts in material prepared under far more drastic anhydrous conditions. Elemental sulfur is well separated and elutes between the disulfane S2 and trisulfane S3.

Applying the response factors of Table I onto the peak integrals an average sulfane chain length of $S_{avg.}$ = 3.75 has been calculated. The specific weight percentages of each sulfane species of the analysed mixture of 1 are listed in Table II. Elemental sulfur has been calculated to a level of about 0.6 wt%.

Table II. Typical sulfane distribution in 1

S_x	S2	S3	S4	S5	S6	S7	S8	S9	S10	$S_{avg.}$
$wt\%$	16.9	30.1	24.0	15.4	7.7	3.3	1.6	0.8	0.2	3.75

Equal experimental conditions have been used for the reaction of sodium polysulfide with chloropropyldiethoxymethyl-, -dimethylethoxy- and –trimethyl-silane to prepare the sulfidosilanes **2, 3**,and **4**. The HPLC analysis of the ethoxy substituted silanes showed a high similarity to the chromatogram of **1** with very close retention times and nearly identical peak integrals and appearances. As displayed in Figure 6, the peaks of the bis(trimethylsilylpropyl)tetrasulfane **4**, however, are shifted to longer retention times due to the reduced interaction of the non polar silane with the polar solvent mixture of the mobile phase. In the same way a peak broadening, especially of the signals of the shorter sulfanes as S2 and S3, is observed that leads to a false impression about the peak integral ratios. In fact, they are still very close to the chromatogram of silane **1**.

In order to further evaluate the feasibility of using the known response factors of the sulfanes of TESPT **1** as a basic system also for the determination of the sulfane distribution and the average sulfane chain lengths of silanes **2, 3**, and **4**, the UV absorption spectrum of each individual sulfane species has been recorded via diode array detector. The normalized UV absorption spectra of S2 to S8 of silane **1** are shown in Figure 7.

It is necessary to remark that further quantitative investigations of the UV spectra require pure material of each sulfane species. This would allow a correct determination of the molar extinction coefficients for final comparison of the absorption strength of the silanes and ultimately accurate response factors.

In Figure 7, S2 and S4 exhibit unique behaviour. S2 has a very strong and sharp absorption maximum in the quartz UV border region at $\lambda_{max} = 202$ nm and then only another weaker local maximum at $\lambda_{max} = 252$ nm. S4, however, as the only sulfane surprisingly does not show a maximum near $\lambda_{max} = 200$ nm, but the strongest absorption at $\lambda_{max} = 210$ nm and another well defined local maximum at 299 nm. For all other sulfanes maxima appear at around $\lambda_{max} = 200$ nm and, with a stepwise bathochromic shift, from 214 (S3) to 224 (S8) nm. They also exhibit a broad plateau towards the longer wavelengths side with onset points at $\lambda = 330 - 340$ nm for the longer sulfane chains. The bathochromic shift with every additional sulfur unit is correlating with an increasing conjugative effect of the free electrons at the sulfur atoms. This UV absorption behaviour has been also reported for the homologous sulfane series H_2S_n (*24*) and other organically endcapped polysulfanes. (*25*) Surprisingly, the correspondence of the individual absorption maxima is extremely good for all sulfanes of the same chain lengths. An excellent example is the absorption maximum of di-n-butyldisulfane that has been found at $\lambda_{max} = 251.5$ nm. This is very close to the absorption maxima of the disulfanes of **1, 2, 3**, and **4** analyzed at $\lambda_{max} = 252$ nm. This gives rise to the suggestion that the absorption in the lower energy region is rarely influenced by

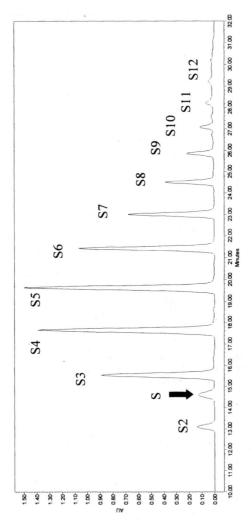

Figure 5. Typical HPLC chromatogram of TESPT 1.

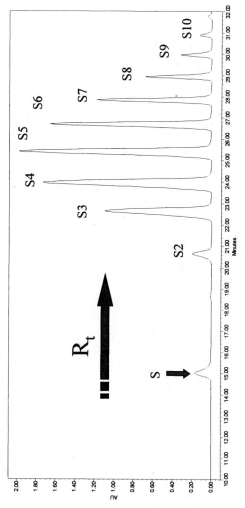

Figure 6. HPLC chromatogram of TMSPT 4.

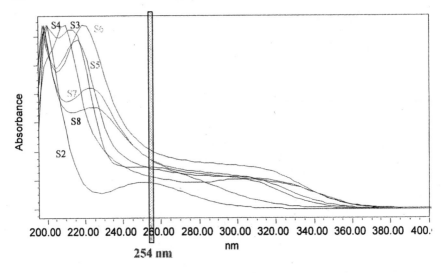

254 nm

Figure 7. UV-Absorption spectra of bis(triethoxysilylpropyl)sulfanes S2-S8.

the substituents if they are not carrying inductive or unsaturated groups that can interact themselves with the sulfane electrons. Figure 8 displays an overlay of the absorption spectra of the pentasulfanes S5 of the sulfidosilanes 1 – 4. In the for HPLC analysis relevant UV-region of λ = 240 – 260 nm (standard detector wavelength λ = 254 nm) all silanes demonstrate the same absorption behaviour.

In the far-UV region, of course, the influence of the substituents e.g. oxygen, is made visible by the different absorption strength of each silane. As mentioned earlier, it would be possible with pure sulfane species to determine the actual molar extinction coefficients at each individual wavelength, especially at λ = 254 nm. This would finally prove the accuracy of the increasing response factors with respect to the huge hyperchromic effect of every additional sulfur unit in the chain (*24*).

The UV absorption spectrum of elemental sulfur (usually orthorhombic α-Sulfur S_8 (*26*)) has also been taken and qualitatively compared to the linear octasulfane S8 of 1 (see Figure 9). Surprisingly, both molecules show the same maxima positions in the far-UV region at λ = 200 and 224. However, the elemental sulfur exhibits a local maximum of λ = 272 nm. This is close to the maximum normally found for the trisulfanes of 1 - 4. The ring strengths of the elemental sulfur corona (<SSS = 108.2°, <SSSS = 98.8° (*26*)) does obviously not allow a good conjugation of more than three sulfur atoms, while the open twisted chain of the octasulfane S8 (calculated <SSS = 107 - 107.5°, <SSSS = 78.1 – 84.7° (*27*)) leads to a good interaction of all sulfur electrons. Nevertheless, the population of the activated "S3" subunit state in elemental sulfur is very high leading to a high molar extinction coefficient even at λ = 254 nm requiring a far higher response factor in the HPLC analysis.

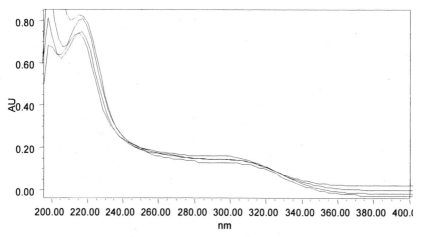

Figure 8 UV-Absorption spectra of pentasulfanes S5 of 1 - 4.

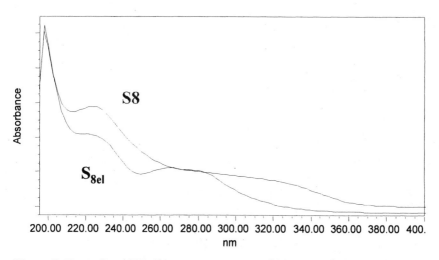

Figure 9. Normalized UV-Absorption spectrum of S8 in 1 and elemental sulfur.

NMR Analysis

As already described in literature (*17,18*) spectroscopic analysis of tetrasulfidosilanes via Nuclear Magnetic Resonance leads to very complex spectra for proton and carbon. Only the signals for the methylene group ($-S_x-C^\gamma H_2-$) next to the sulfur show a reasonable resolution due to effects from the poly-sulfur clusters and allow a further, however difficult quantification. As displayed in Figure 10, peaks for $-C^\gamma H_2-$ protons of all the silanes **1 – 4** are spread over a range of about 0.4 ppm (about 160 Hz) from $\delta = 2.6$ to 3.0 ppm. The peaks for

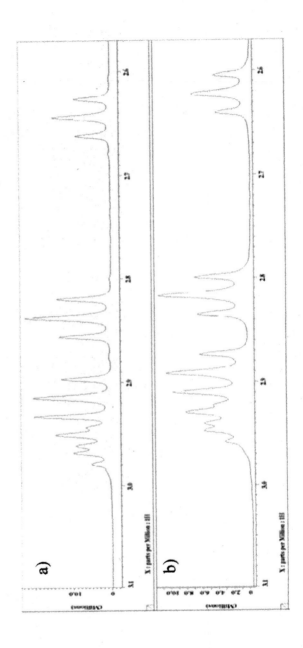

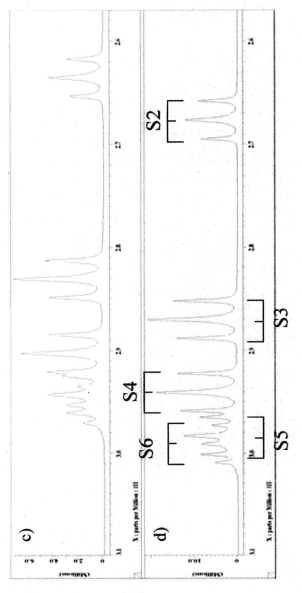

Figure 10. [1]H nmr spectra of sulfidosilanes a) 1, b) 2, c) 3, and d) 4.

S2 (δ = 2.62 – 2.68 ppm; $^3J_{H,H}$ = 7 Hz, similar for all -CH_2- proton peaks), and S_3 (2.82 – 2.87 ppm) are quantifiable and are used for comparison with HPLC results. The signals for S4 – S6/S8 are overlapping in the range of δ = 2.85 to 3.03 ppm and cannot be deconvoluted. In general, the integral ratio of S3/S2 is pretty much identical for all the four analysed silanes **1 – 4** confirming the excellent reproducibility of the phase transfer catalysis process.

The observation that the signals for the -$C^\gamma H_2$- protons of the trimethylsilyl substituted sulfanes of **4** are slightly shifted to lower field compared to the signals of the alkoxysustituted sulfanes is also made for the according carbon signals of these methylene groups. The ^{13}C NMR spectra of the -$C^\gamma H_2$- carbon atoms are displayed in Figure 11.

The spectra exhibit two striking features: (i) 7 peaks for the –Sx–$C^\gamma H_2$- carbons (x = 2 - 8) are well resolved at δ = 41.5 – 43.4 ppm while there are no signals for S9 and S10 detectable due to very low intensity, and (ii) the isotropic shifts of the peaks do not follow the regular order from S2 to S8 as observed in ^1H NMR. Analysis of the peak integrals proves again the nearly identical composition of all four silanes and confirms the HPLC results to a very close proximity. A curiosity is the observed migration of the signals of S2 and S7 towards the signals of S4 and S6, respectively, in case of the trimethyl substituted silane **4**.

The ^{29}Si NMR spectra demonstrate even better resolutions than the ^{13}C NMR with fine features for S_x with x even up to 9. This is a quite surprising observation considering the fact that the Si atoms are located at least four bonds away from the S_x- moiety. However, integrals of S9 cannot be obtained despite this level of resolution and the signals have been therefore not assigned in the spectra displayed in Figure 12. The ^{29}Si NMR analysis shows the expected iso-tropic shift ranges for trialkoxyalkylsilanes (**1**, δ = -45.88 - -46.33 ppm), dialk-oxydialkyl silanes (**2**, δ = -5.45 - -5.83 ppm), alkoxytrialkylsilanes (**3**, δ = 16.40 – 16.53 ppm), and tetraalkylsilanes (**4**, δ = 1.55 – 1.73 ppm) according to the different shielding power of the substituents. While the intensity profiles of the peaks are in sequential order for S2 to S8 from left to right for the signals of silane **1**, the signals of S2 and S7 in silane **2** are starting to migrate in direction towards the signals of S3 and S6. In the spectrum of silane **3** these signals are even overlapping. A full set of signals for S2 to S8, however, surprisingly in reverse order, is observed for the signals of silane **4**. Together with the general convergence of the signals ($\Delta\delta$(**1**) = 0.55 ppm, $\Delta\delta$(**3**) = 0.13 ppm), this is an indicator that the positive polarization of the silicon by the alkoxy substituents is recognized by the electronegative sulfur chain via some sort of backbonding through the space. On the other hand, an intermolecular interaction of the silicon atom with neighbouring sulfane chains cannot be completely ruled out either. The comparison of the integral ratios of resolved peaks confirms again the high similarity of the sulfane composition of the four silanes.

Conclusions

Four phase transfer catalysis experiments have been carried out preparing bis(silylpropyl)oligosulfanes with increasing number of ethoxy substituents being replaced by methyl groups at the silicon atom. NMR analyis of the generated mixtures demonstrated an excellent similarity of the molar distribution of the individual sulfane species and the corresponding average sulfane chain lengths. HPLC analysis also exhibited the same integral ratios of the sulfane species within each silane representing molecules of different sulfur content with respect of the molecular structure. The average formula of the silanes 1 – 4, the corresponding Molecular Weight and the consequently increasing theoretical sulfur content are listed in Table III.

Table III. Formula, Molecular Weights, and wt% Sulfur of 1 -4

Molecule	Formula	Mw	wt% S
TESPT 1	$C_{18}H_{42}O_6Si_2S_{3.75}$	530.95	22.65
DEMSPT 2	$C_{16}H_{38}O_4Si_2S_{3.75}$	470.89	25.54
DMESPT 3	$C_{14}H_{34}O_2Si_2S_{3.75}$	410.84	29.27
TMSPT 4	$C_{12}H_{30}Si_2S_{3.75}$	350.79	34.28

As a first result the following response factors are proposed for the HPLC analysis of the silanes 2 – 4 as listed in Table IV. The response factors are so far based purely on the ratio of the molecular weights of the specific sulfane Sx (2,3,4) to Sx (1) multiplied with the according response factor RF 1 of Table I. Better UV-absorption analysis will certainly lead to more accurate adjustments.

Table IV. Proposed HPLC Response Factors of 2, 3, and 4

RF/S_x	S2	S3	S4	S5	S6	S7	S8	S9	S10	S_{el}
RF 2	1.14	3.99	7.19	10.93	14.48	19.21	22.93	28.53	34.10	37.26
RF 3	1.34	4.61	8.22	12.39	16.28	21.45	25.45	31.49	37.45	37.26
RF 4	1.61	5.46	9.60	14.29	18.60	24.28	28.59	35.13	41.53	37.26

Acknowledgments

The authors would like to thank T. Materne, L. Stelandre, S.K. Mealey, S.C. Winchell, G.R. Haines, and D.J. Bunge for fruitful discussions. The help by H. Yue, T.J. Rivard, and M.R. Reiter (all Dow Corning) in the development of the analytical methods is gratefully acknowledged.

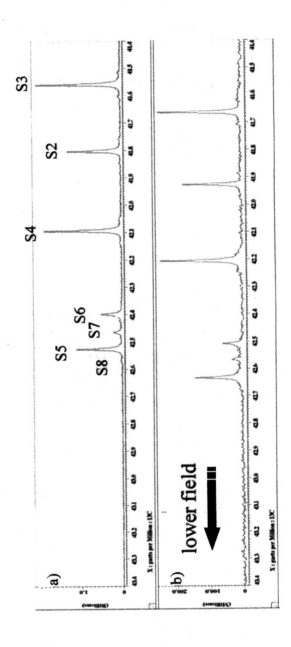

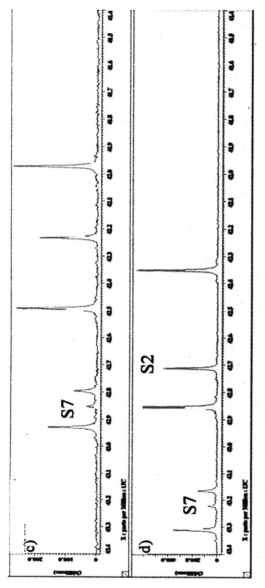

Figure 11. ^{13}C nmr spectra of sulfidosilanes a) 1, b) 2, c) 3, and d) 4.

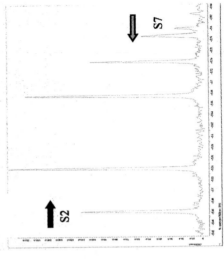

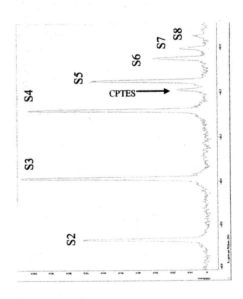

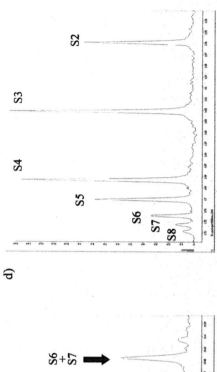

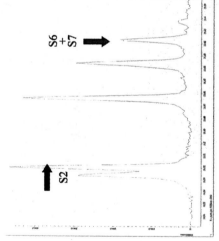

Figure 12. ²⁹Si nmr spectra of sulfidosilanes a) 1, b) 2, c) 3, and d) 4.

References

1. Plueddemann, E.P. *Silane Coupling Agents*, Plenum Press, New York, **1982**.
2. Agostini, G,; Bergh, J.; Materne, T.; *New Compound Technology,* paper presented at the Akron Rubber Group Meeting, October **1994**, Akron (OH).
3. R. Rauline, Compagnie Generale des Etablissements Michelin, Michelin & Cie, EP 0501127 **1991**, US 5227425 **1993**.
4. Braendstroem, A, *Principles of Phase Transfer Catalysis by Quarternary Ammonium Salts* in *Advances in Physical Organic Chemistry*, Gold, V.; Bethell, D., Eds., Academic Press, London **1977**, p. 267.
5. Starks, C.M.; Liotta, C.L.; Halpern, M.E., Eds. *Phase Transfer Catalysis – Fundamentals, Applications, and Industrial Perspectives*, Chapman & Hall, New York, **1994**.
6. The Goodyear Tire & Rubber Co., US 5,405,985, **1995**.
7. The Goodyear Tire & Rubber Co., US 5,468,893, **1995**.
8. The Goodyear Tire & Rubber Co., US 5,583,245, **1996**.
9. The Goodyear Tire & Rubber Co., US 5,663,396, **1997**.
10. Dow Corning Corporation, US 6,384,255, **2002**.
11. Dow Corning Corporation, US 6,384,256, **2002**.
12. Dow Corning Corporation, US 6,448,426, **2002**.
13. Dow Corning Corporation, US 6,534,668, **2003**.
14. Backer, M.W.; Buesing, C.A.; Gohndrone, J.M.; Maki, W.C.; Rivard, T.J.; Yue, H. *Polymer Preprints* **2004**, *45*, 660.
15. Luginsland, H.-D. *Reactivity of the Sulfur Functions of the Disulfane Silane TESPD and Tetrasulfane Silane TESPT*, paper presented at the ACS Meeting, April **1999**, Chicago (IL).
16. ASTM D6844 *Standard Test Method for Silanes Used in Rubber Formulations (bis-(triethoxysilylpropyl)sulfanes): Characterization by High Performance Liquid Chromatography (HPLC)*, **2002**.
17. Backer M.W.; Buesing, C.A.; Dingman, J.L., Gohndrone, J.M.; Hu, S.; Maki, W.C.; Mealey, S.K.; Thomas, B.; Yue, H. *Polymer Preprints* **2003**, *44(1)*, 245.
18. Backer M.W.; Buesing, C.A.; Dingman, J.L., Gohndrone, J.M.; Hu, S.; Maki, W.C.; Mealey, S.K.; Thomas, B.; Yue, H. *Use of LC, NMR and MS for Sulfidosilane Material Characterization*, presented by S.K. Mealey at the Fourth International Symposium on Silanes and other Coupling Agents, June **2003**, Orlando (FL).
19. Teder, A. *Acta Chem. Scand.* **1971**, *25*, 1722.
20. Schwarzenbach, G.; Fisher, A. *Helv. Chim. Acta* **1960**, *49*, 1365.
21. Teder, A. *Ark. for Kemi*, **1969**, *31*, 173.
22. Giggenbach, W. *Inorg. Chem.* **1972**, *11*, 1201.
23. Giggenbach, W. *Inorg. Chem.* **1974**, *13*, 1724.
24. Feher, F.; Münzner, H. *Chem. Ber.* **1963**, *96*, 1131.
25. Steudel, R. *Chem. Rev.* **2002**, *102*, 3905.
26. Wiberg N., Ed. *Hollemann-Wiberg, Lehrbuch der Anorganischen Chemie*, 91.-100. verbesserte Auflage, Walter de Gruyter, Berlin, **1985**, pp. 471-525.
27. Papamokos, G.V.; Demetropoulos, I.N. *J. Phys. Chem. A* **2002**, *106*, 1661.

Chapter 6

Catalysts for Silane and Silanol Activation

J. B. Baruah

Department of Chemistry, Indian Institute of Technology, Guwahati
781039, India (email: juba@iitg.ernet.in)

Dehydrogenative coupling reactions of various silanes with alcohols,
reductive coupling reactions of silanes with carbonyl compounds as
well as dehydration of silanols are studied for silicon oxygen bond
formation reactions under catalytic conditions. The dehydrogenative
coupling reactions are catalysed by amines and metal catalysts
derived from copper(II), gold(III), silver(I) palladium(II), rhodium(I),
platinum(II) complexes, whereas reductive coupling reactions
between quinone and different silanes are done by rhodium(I) and
palladium(II) catalysts. Under near neutral condition dehydration of
silanols can be caused by amines such as triethylenetetramine.

Introduction

Silicon-oxygen bond containing inorganic compounds constitutes a major portion of
the Earth's crust. [1] Thus, transformations leading to silicon oxygen bonds by reagents
under ambient condition are of great value. Such transformations have close relevance
towards the goal to convert SiO_2 to useful organic or inorganic materials.[2] For achieving
such a goal, it is a prime necessity to understand the reactivity of the silicon oxygen bond
starting from its formation to its cleavage. The process of developing new reagents
should also complement green chemical paths. We have been interested in the generation
of new Si-O bond forming reactions through dehydrogenative coupling[3], reductive
coupling of carbonyl compounds[4] with silanes as illustrated in scheme 1. The
dehydrogenative reaction would leave hydrogen as the side product, whereas in the case
of substitution of Si-OH group by an alcohol would lead to water as the side product. The
reductive coupling reaction will have excellent atom economy. We have studied these
reactions that are catalysed by various transition metal catalysts[5] and the important
observations that provide advantages of using a particular reagent are presented here.

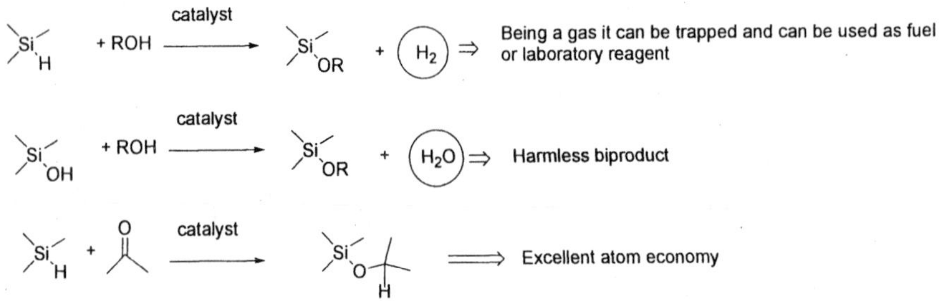

Scheme 1

Experimental

The synthetic aspects of most of the work presented here have been published elsewhere.[5]

Platinum catalysed Si-O bond forming reaction

In a typical experiment 1,3-diphenyl-2-propene-1-one (2.1g, 1mmol) was dissolved in 10ml of toluene in a Schlenck tube. To this K_2PtCl_4 (0.004g, 0.01mmol) was added make a homogeneous solution. The triethylsilane (0.47g, 3mmol) was added by a micro syringe to this solution. The reaction was continued for 6hrs at 80°C by constant stirring. The reaction was monitored by TLC from time to time and after 6hrs the stirring and heating was stopped and filtered. From the filtrate the solvent was removed under reduced pressure. The crude product obtained was further purified by column chromatography [silica gel with hexane/ethylacetate (5%)]. The products were further characterized by comparing their proton nmr, IR with authentic samples. The ratio of the isomers in each case was evaluated from the proton nmr spectra.

Results and Discussion

Palladium catalysed Si-H bond activation

Palladium complex catalysed silicon oxygen bond formation is well known.[6] Palladium complexes having phosphine ligands are generally used[6b] for such reactions but are expensive and toxic. Thus, we felt that the use of nitrogen containing ligands might reduce toxicity as well as ease the manipulation of air sensitive reagents. We found that the tetraethylethylenediamine (TEEDA), tetramethylethylenediamine (TMEDA) complexes of $PdCl_2$ are good catalysts for dehydrogenative coupling reaction between silane and alcohol. Varieties of silanes were reacted with different phenolic and alcoholic compounds to obtain the corresponding silyl ethers in excellent yield[5d]. In order to know the structure of the catalyst we have determined the structure of one of the catalyst Pd(TEEDA)Cl_2 and the structure is shown in Figure 1. The structure is a distorted square planar structure. Some of the important bond distances are N1-Pd, N2-Pd, Cl1-Pd, Cl2-Pd, 2.104, 2.078, 2.301 and 2.318Å respectively. Similarly the <N1-Pd-N2, <Cl1-Pd-N2,

<Cl2-Pd-N2 and <Cl1-Pd-Cl2 are 85.95, 92.83, 90.34, 90.87° respectively. The structure of the Pd(TMEDA)Cl$_2$ is reported recently and observed similar structural feature in it also.[7]

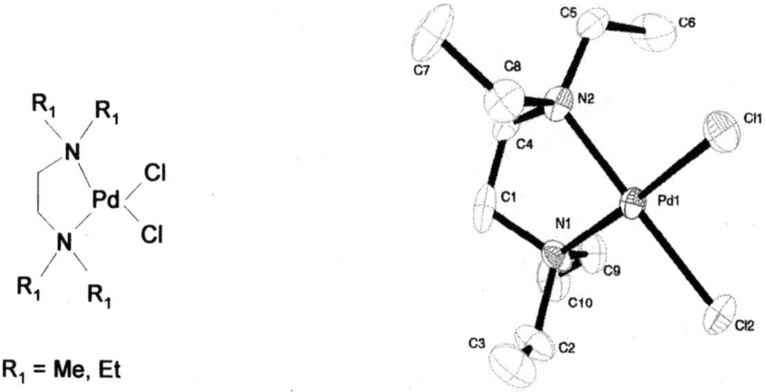

Figure 1. Left: Catalyst for Si-H bond activation; Right: The structure of Pd(TEEDA)Cl$_2$ (50% thermal ellipsoid, the hydrogen atoms are omitted for clarity).

The silicon-oxygen bond formation can be extended for oligomeric siloxane formation by reductive coupling of 1,4-benzoquinone with dihydro and trihydrosilanes. The advantage of using these catalyst systems is the versatility and mildness. It is to be noted that the dehydrogenative coupling reactions of silane with an alcohol to a lesser extent can be caused by the TMEDA also and but the coupling of silane with quinonic compounds require a palladium catalyst.

Copper(II) catalysed Si-H bond activation

It is reported in the literature that metallic copper[8]and copper (I) hydride[9] complex such as [Ph$_3$PCuH]$_6$ catalyse silicon-oxygen bond formation in the reactions between substituted silanes and alcohols. Although, metallic copper powder can be used for silicon-oxygen bond formation of silanes the reaction is limited to substrates containing phenyl groups attached to silicon and also requires drastic reaction conditions[10]. We have found that tetrachlorocuprates having anilinium, p-methoxyanilinium, p-methylanilinium, and pyridinium cations are fairly good catalysts for alcoholysis of silanes. The complexes are capable of converting 10-equivalent of the substrates with respect to one mole of the catalyst. The reactions proceeds at room temperature slowly but can be completed within 3-6hrs when carried out at 40-70°C. The tetrachlorocuprate catalyzed silicon-oxygen bond forming reactions can be represented by the equation 3 and 4.

These reactions can be performed under neutral and mild conditions. Since these reactions give only hydrogen gas as by-product separation is not difficult. But in the case of incomplete reactions and the reactions in which side products are formed, the purification of the products by column chromatography leads to lower yield. This happens as silica gel can degrade the silicon-oxygen bond, thereby decreasing the isolated yield after column chromatography.[11]

$$R_2\underset{R_1}{\overset{R_3}{\underset{|}{\overset{|}{Si}}}}H + ROH \xrightarrow{\text{copper(II) catalyst}} R_2\underset{R_1}{\overset{R_3}{\underset{|}{\overset{|}{Si}}}}OR + H_2$$

$R_1 = R_2 = R_3 = C_6H_5\text{-}$

$R_1 = R_2 = R_3 = \text{-CH}_2\text{CH}_3$

Equation 3

$$R_2\underset{H}{\overset{R_1}{\underset{|}{\overset{|}{Si}}}}H + ROH \xrightarrow{\text{Copper(II) catalyst}} R_2\underset{OR}{\overset{R_1}{\underset{|}{\overset{|}{Si}}}}OR + H_2$$

$R_1 = R_2 = C_6H_5\text{-}$

$R_1 = \text{-CH}_3, R_2 = C_6H_5\text{-}$

Equation 4

The reactions with various primary and secondary alcohols such as methanol, ethanol, isopropanol and allylalcohol gives the corresponding silylethers. The reaction is not applicable to tertiary butylalcohol (eqn 5). For example the reaction of triphenylsilane with tertiary butylalcohol in the presence of *bis*-pyridinium tetrachlorocuprate as a catalyst led to the formation of the 1,1,1,3,3,3-hexaphenylsiloxane only. In this reaction all the diphenylsilane is consumed. This suggests that the water present in the reaction competes with the alcohol. The allyl alcohol reacts with triphenylsilane to give silylether. In this case the double bond of the allyl group is not reduced. The phenols are found to be not suitable for these catalytic reactions. This is attributed to the fact that the copper catalysts that are under consideration can react with phenol under ambient condition either to give coupled product or to give phenylene ethers.[12] The reaction is very effective to dihydrosilanes, in which exclusively disubstituted silylethers are formed. The monohydrosilylethers are not observed in these reaction mixtures. Different alcohols such as methanol, ethanol, and isopropanol react with diphenylsilane and methyl-p-henylsilane in the presence of catalytic amount of *bis*-pyridinium tetrachlorocuprate. *Bis*-pyridinium tetrachlorocuprate has the advantage that it has a stable ligand, but *bis*-anilinium tetrachlorocuprate complex has a deficiency that it undergoes oxidation when left under aerial condition. The relative rates of different tetrachlorocuprate catalysts are determined by monitoring the alcoholysis reaction between diphenylsilane and ethanol with help of GC. Taking out solution from the concerned reaction mixtures using a micro syringe, followed by injection into a GC, does this. The products were compared with authentic samples. From the plot of change of concentration of diphenylsilane *vs* time it is observed that about 80% of the diphenylsilane reacts with alcohol and forms the corresponding silyl ether within half an hour. The reactions then slow down and some amount of diphenylsilane remains unreacted even after two and a half hours. Comparison of the percentage of conversion by the different tetrachlorcuprate catalysts suggests that the counter cation used in this investigation such as anilinium, *p*-methoxyanilinium, *p*-methylanilinium, and pyridinium does not effect the rate of the reaction in a significant manner.[5e]

$$R_3Si\text{-}H \xrightarrow[\text{t-butylalcohol}]{[PyH]_2[CuCl_4]} R_3Si\text{-}O\text{-}SiR_3$$

$$R = \text{-}C_6H_5\text{, -}C_2H_5$$

Equation 5

The possible involvement of a copper(I) species in the reaction of the copper catalysed silicon-oxygen bond forming reactions, can be seen from UV-Visible spectroscopy also. It is found that all these reactions initially pass through a decolorisation of copper(II) solution suggesting a reduction of copper(II) to copper(I) complex. The regeneration of the catalyst in the course of these reactions is very clear from visible spectroscopy. For example, the pyridinium tetrachlorocuprate has a well-defined absorption peak at 467nm. The absorption maximum decreases on reaction with diphenylsilane in the presence of ethanol (Figure 2 the graph marked as **a**). However, after the consumption of the diphenylsilane the absorption peak at 467nm reappears, suggesting that the Cu (II) is being regenerated (figure 2 marked as **b**) once all the diphenylsilane is consumed.

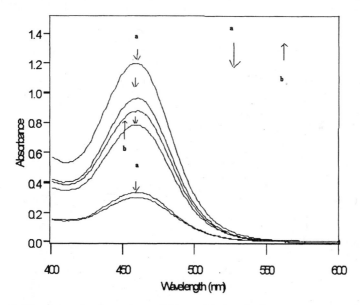

Figure 2. The change in absorption at 467 nm with time (at three minute intervals) of a solution of bis-pyridinium tetrachlorocuprate in ethanol on addition of diphenylsilane. (Downward arrows marked with a show the decrease in absorbance to show the conversion of copper(II) to copper(I) and the upward arrows marked with b shows the regeneration of the copper(II) species after consumption of diphenylsilane)

Performing the reaction at different concentrations of the *bis*-pyridinium tetrachlorocuprate complex, the relative rate of the decrease in the absorption at 467nm is studied during its reaction with diphenylsilane and ethanol. It is observed that the time taken to reach the minimum absorption is dependent on the catalyst concentration. Visible spectra show that as the concentration of the copper (II) catalyst is increased, the time required for a decrease in the absorption at 467nm is increased. For example, the reaction of diphenylsilane (0.1mmol), *bis*-pyridinium tetrachlorocuprate (0.011mmol) with ethanol (3.4 mmol) in acetonitrile (4 ml) reaches a minimum absorption at 467 nm after 15 min but the reaction of diphenylsilane (0.1mmol) *bis*-pyridinium tetrachlorocuprate (0.02mmol) with ethanol (3.4mmol) in acetonitrile (4ml) attains minimum absorption after 20 min. In the latter case, the regeneration of the catalyst is slower, because a higher amount of the catalyst is used than in the former case.

These observations suggests that once a substituted silane is added to the solution of *bis*-pyridinium tetrachlorocuprates it reduced the copper(II) to copper(I); this thereby decolorises the solution. As the reaction proceeds the silanes gets consumed and the copper(II) state is regenerated, such observation implies an interchange of copper(II) and copper(I) during the reaction. We have tested the reactivity of the anhydrous cupric chloride for silicon-oxygen bond forming reactions. It was found to be not suitable for such catalytic reactions. For example the reaction of triphenylsilane in methanol in the presence of cupric chloride gives 1,1,1,3,3,3-hexaphenylsiloxane. The result is attributed to the fact that the silicon-oxygen bonds are acid sensitive[13] and if at all a methylsilylether is formed in this reaction it gets hydrolysed to corresponding silanol, which on further reaction gives the 1,1,1,3,3,3-hexaphenylsiloxane.

As discussed earlier, the dihydrosilanes are good substrates that get difunctionalised on reaction with alcohol, quantitative products are formed through catalytic reaction. Thus, the tetrachlorocuprate complexes should be a good catalyst for the synthesis of silicon-oxygen bonded oligomers from polyhydric alcohol. It is found that the silicon-oxygen bond formation could be extended to synthesize oligomers from the reaction of dihydrosilanes (eqn. 6) and trihydrosilanes with polyhydroxyalcohols (eqn. 7).

HO-X-OH + $\underset{R_2}{\overset{R_1}{>}}Si\underset{H}{\overset{H}{<}}$ $\xrightarrow{\text{Cu(II) catalyst}}$ $*\left[\underset{R_2}{\overset{R_1}{|}}Si_O \, [CH_2]_m \, O \right]_n *$

$R_1 = R_2 = Ph, \; X = CH_2CH_2$
$R_1 = R_2 = Ph, \; X = CH_2CH_2CH_2$
$R_1 = R_2 = Ph, \; X = CH_2CH=CHCH_2$
$R_1 = Ph, R_2 = Me, X= CH_2CH= CHCH_2$
$R_1 = Ph, R_2 = Me, X= CH_2CH_2$
$R_1 = Ph, R_2 = Me, X= CH_2CH_2CH_2$

Equation 6

HO-X-OH + $\underset{R_1}{\overset{H}{>}}Si\underset{H}{\overset{H}{<}}$ $\xrightarrow{\text{Cu(II) catalyst}}$ $*\left[\underset{}{\overset{R_1}{|}}Si_O \, [CH_2]_m \, O \right]_n *$

$R_1 = Ph, \; X = CH_2CH_2$

Equation 7

The chromatograms of the oligomers are simple and unimodal suggesting that a majority of the oligomeric species are having similar chain length. The GPC profiles indicate that they are low molecular weight oligomer comprising of 8-10 repeated monomeric units. The polydispersity of this oligomer is again very close to unity showing it to have narrow distribution of molecular weight and uniform chain length.

In order to increase the molecular weight of the oligomer the reactions were carried out for prolonged time as well as increasing the catalyst concentration. Both the attempts were not successful and it is observed that prolonged heating and more catalysts lead to lower yield. This suggests that silicon-oxygen bond formed initially might be degraded by the same catalyst system. Similar observations on Si-O bond cleavage are found in Pd-catalysed reactions recently.[6d]

Rhodium catalysed reactions

The rhodium catalysed Si-O bond formation reactions are widely studied.[14] The reductive coupling reactions between quinone and hydrosilane by $RhCl(PPh_3)_3$ is an example of such reactions that is being well documented.[15] However, from our studies we could clearly see that monohydrosilane derivative is formed in reductive coupling recations of 1,4-naphthoquinone with diphenylsilane when catlaysed by $RhCl(PPh_3)_3$ (eqn 8) making the reaction more versatile The formation of such monohydrosilane derivative can be understood from experiment conducted in nmr tube of mixtures of diphenylsilane and 1,4-naphthoquinone in different ratios in the presence of $RhCl(PPh_3)_3$. The ^1HNMR spectra of the reaction carried out with diphenylsilane and 1,4-naphthoquinone in a ratio of 2:1 shows only the signals from the compound **A** as shown in eqn. 8.

Equation 8

Platinum catalysed reactions

So far we have shown that the catalytic activity and selectivity for reactions varies with catalysts. Platinum is very effective for Si-H bond forming reactions. Platinum catalyst are useful for reductive coupling recations of carbonyl compounds leading to Si-O bond formation.[16] However when an α,β-unsaturated carbonyl compound is under consideration we have to consider the chemoselectivity of such reactions as described in eqn. 9.

Rhodium complexes are used for selective hydrosilylation of varities of α,β-unsaturated carbonyl compounds[17]and the reactions are very useful for chemoselective transformations. On the other hand, in the platinum complex catalysed reaction of α,β-unsaturated carbonyl compounds the hydride of the silane is attacked from the γ-side.[16]

$$\underset{|}{\overset{|}{Si}}-H \;+\; R_1\diagdown\diagup\overset{O}{\underset{R_2}{\diagdown}} \longrightarrow R_1\diagdown\diagup\overset{\overset{|}{\underset{|}{Si}}-}{\underset{HR_2}{\overset{O}{|}}} \;+\; R_1\diagdown\diagup\overset{R_2}{\underset{O}{\diagup}}\overset{|/}{\underset{Si}{\diagdown}}-$$

α- attack   γ - attack

I   II

Equation 9

However, the question remains whether in a platinum catalysed reaction of silane on α,β-unsaturated carbonyl compound having phenyl substituent at the γ-position can decide the site of hydride attack or not (eqn 9). This question occurs because of the γ-hydride attack on a α,β-unsaturated carbonyl compound would lead to a thermodynamically less stable product.[18] With this point in mind we have studied the hydrosilylation reaction of a few α,β-unsaturated carbonyl compounds by triethylsilane with platinum catalyst. We have observed that the reaction of triethylsilane with various α-β unsaturated carbonyl compounds can be performed with triethylsilane in the presence of a catalytic amount of $K_2[PtCl_4]$. These reactions gave quantitative siloxyether as shown in eqn 9. In these reactions the ratios of hydride attack at α and γ positions are greatly influenced by electronic factor of the group present at α and γ position (product I and II in eqn 9). The triethylsilane was chosen as there are literature data available with this silane in similar reactions.[19]

The ratio of the isomeric products and yield in some of the reactions studied are summarized in Table 1. Several α-β-unsaturated carbonyl compounds are converted to the corresponding silylethers (Table 1) and selective substrates are studied as we intended to understand the γ-selectivity in platinum catalysed hydrosilylation when a phenyl ring is in conjugation to a carbonyl group though an intervening double bond. From Table 1 it is obvious that the ratio of α and γ-attack is decided by the electronic effect of the substituents. Exclusive γ-attack of the hydride of the triethylsilane on 1,3-diphenyl-2-propene-1-one was observed. This is also confirmed from the ^1H NMR spectra of the crude product obtained from K_2PtCl_4 catalysed reaction. From Table 1 it is also evident that the substituent at both the α, γ -position of the carbonyl compound has a role in deciding the hydride attack. It is clear from Table 1 that when there is an aromatic ring at the γ-position with an aromatic ring in the α-position then the γ-selectivity is enhanced (entry 2,3,7 of Table 1). Whereas the presence of an alkoxy group on the aromatic ring at γ-position reduces the selectivity. It is also to be noted that a hydrogen or alkyl group at the α-position reduces the selectivity of the products (entry 1, 8, 9,10 of table 1).

These results show that presence of an aromatic ring at the α-position leads to high selectivity in the hydrosilylation of unsaturated α,β-unsaturated carbonyl compounds containing phenyl group at the γ-position. But, such selectivity is effected if an electron donating group is placed at the aromatic ring at γ-position (entry 4, 6 of Table 1). This suggests that there is interaction of the aromatic ring at the α-position and the double bond of the α,β-unsaturated carbonyl compounds with the platinum(0). It is already reported in the literature that hydrosilylation by platinum(II) complexes involves colloidal platinum that is generated *in-situ* by the reductive reaction of silane on platinum (II). However, in our case we have not used a supporting agent to stabilize the metallic colloid if at all they are formed, so initially in our case the silane reduces the Pt(II) to

Table 1[#]

Sl No	R_1	R_2	Ratio of γ and α attack	Isolated Yield
1	C_6H_5	H	55 :45	96
2	C_6H_5	C_6H_5	100: 0	98
3	$p\text{-}CH_3C_6H_4$	C_6H_5	100: 0	95
4	$p\text{-}CH_3O\text{-}C_6H_4$	$C_6H_5\text{-}$	50 :50	92
5	$C_{10}H_7\text{-}$	C_6H_5	57 : 43	96
6	$p\text{-}EtOC_6H_4$	C_6H_5	52: 48	94
7	$p\text{-}EtC_6H_4$	C_6H_5	100:0	95
8	C_6H_5	CH_3	50:50	90
9	$p\text{-}CH_3C_6H_4$	CH_3	40: 60	92
10	$p\text{-}CH_3O\text{-}C_6H_4$	CH_3	55:45	86

[#] In all cases the reactions were carried out at 80°C in toluene with 1 mole% of K_2PtCl_4 catalyst[20], significance of R_1 and R_2 as per equation 9

Pt(0) to provide the precursor for the catalytic reaction.[21] The platinum(0) formed will be very reactive towards oxidative addition with the Si-H bond to form a H-Pt-Si(Et)$_3$ type of units.[22] The platinum(0) may also independently form complex with an unsaturated double bond.[23] The participation of an aromatic ring in the complexation with platinum in the selective hydroxylation by platinum complexes is already reported.[24]

Gold(III) and silver (I) catalysed reactions

The gold(III) chloride or auric acid can cause Si-O bond formation from silane (eqn 10).[25] Similar reactions can be caused by silver nitrate. The reactions results in

$$3Ph_3SiH + 3EtOH + H[AuCl_4] \rightarrow 3Ph_3SiOEt + H_2 + Au + 4HCl$$

Equation 10

formation of the respective metallic colloids which participate further in the catalysis. Although these reactions have a high conversion rate of silanes, they result in the liberation of acid as a side product. This results in low isolated yield of the ethers. Similar results on the cleavage of the Si-O bond by catalysts after formation are observed in palladium catalysts also.[6d]

Amine catalysed Si-H bond activation

There are numerous examples, which elucidates the catalytic role of amines in the dissolution of silica in aqueous and non-aqueous media.[2] However, the acid base chemistry of silanols is not studied in detail. Laine *et al* have demonstrated the role of different amines in solublising silica in ethyleneglycol[2a]. We have found that octaphenylcyclotetrasiloxane is formed from the reaction of diphenylsilanol with triethylenetetramine (eqn 11).

Equation 11

It is already known that octaphenylcyclotetrasiloxane can be prepared from the reaction of diphenylsilanediol in a sealed tube with a primary amine in the presence of ethylbromide[26]. But, in our case, the reaction is a catalytic reaction; it does not require a drastic procedure and the products can be separated very easily. We have found that the reaction also works well with amine catalysts such as ethylenediamine, N,N,N',N'-tetramethylethylenediamine, 1,10-phenanthroline. However, the reaction is not caused by amines such as aniline, triethylamine, diethylamine, dimethylaniline, pyridine under ordinary conditions. This clearly indicates that chelate effects may be responsible for the dehydration. It is important to note that the secondary amines such as diethylamine reacts with dihydrosilanes in the presence of a catalyst such as copper(I)chloride[27] or $Rh(PPh_3)_3Cl^{28}$ to give silazanes. Other silanes also undergoes dehydrogenative coupling reactions with alcohols in the presence of a catalytic amount of triethylenetetramine. The dehydrogenative coupling reaction by amine can be carried out in the presence of a metal catalyst leading to the formation of metallic colloids[29].

Among the silanes, the reaction of phenylsilane in aqueous acetone involves the most vigorous effervescence of hydrogen gas. This reaction leads siloxane oligomer as white solid within 15 minutes at room temperature with 6mole % of **I** (ref eqn. 11) with respect to silane. Polymerisation reactions of phenylsilane with different diols by catalytic amount of triethylenetetramine are found to give good yields under solvent free conditions.

Conclusions

In conclusion the dehydrogenative coupling reactions of different silanes can be caused under varieties of reaction conditions by different transition metal catalysts. Each of the metal catalysts has its own advantages depending on which one can perform specific Si-O bond forming reactions. Based on the selectivity each of these methods these findings should make impact in the synthesis of nano-materials.

Acknowledgements

The author thanks Mr. Kaustavmoni Deka, Dr. Arup Purkayshtha and Mr. Bani Kanta Sarma who contributed to this research.

References

1. Greenwood, N.N., Earnshaw, A., *Chemistry of Elements*, Butterworth-Heinmann, Oxford, 1997, 2nd Eds.
2. (a) Cheng, H., Tamaki, R., Laine, R. M., Babonneau, F., Chujo, Y., Treadwell, D. R., *J. Am. Chem. Soc.* **2000**, *122*, 10063-10072 (b) Kinrade, S. D., Del N. J. W., Schach, A. S., Sloan, T. A., Wilson, K. L., Knight, C. T. G, *Science* **1999**, *285*, 1542 (c) Boudin, A., Cerveau, G., Chuit, C., Corriu, R. J. P., Reye, C., *Angew. Chem. Int. Ed. Eng.* **1986**, *25*, 473 (d) Corriu, R. J. P., Perz, R., Reye, C., *Organometallics*, **1988**, *7*, 1165
3. (a) Luo, X-L., Crabtree, R.H., *J. Am. Chem. Soc.* **1989**, *111*, 2527-2535 (b) Steffen, K-D., *Die. Angew. Makromol. Chem.* **1972**, *24*, 1-20 (c) Bideau, F.L., Coradin, T., Hanique, J., Samuel, E., *Chem. Commun.* **2001**, 1408-1409 (d) Schubert, U., *Angew. Chem. Int. Ed. Eng.* **1994**, *33*, 419-421 (e) Gregg, B.T., Cutler, A.R., *Organometallics* **1994**, *13*, 1039-1043 (f) Marciniec, B., Gulinski, J., Nowicka T., **1999**, 3 pp. CODEN: POXXA7 PL 176036 B1 19990331, Patent written in Polish. Application: PL 94-303693 19940601 (g) Marciniec, B., *Pol. Roczniki Chemii* **1975**, *49*, 1565-1576 (h) Lukevics, E., Dzintara, M., *J. Organometal. Chem.* **1985**, *295*, 265-315
4. (a) Tilly, T.D., in The Silicon-heteroatom bond, Eds Patai S., Rappoport, Z., John Wiley, New York, **1991**, ch. 9, pp 246-307 (b) Lappert, M. F., Maskell, R. K., *J. Organomet. Chem.* **1984**, *264*, 217-228
5. (a) Purkayshtha, A., Baruah, J. B., *React. & Funct. Polymers* **2005**, *63*, 177-183 (b) Deka, K., Sarma, R. J., Baruah, J. B., *Inorg. Chem. Commun.* **2005**, *8*, 1082-1084 (c) Purkayastha, A., Baruah, J. B., *Appl. Organomet. Chem.* **2004**, *18*, 166-175 (d) Purkayshtha, A., Baruah, J. B., *J. Mol. Catal. A: Chemical* **2003**, *198*, 47-55 (e) Purkayshtha, A., Baruah, J. B., *Silicon Chemistry* **2002**, *1*, 229-232 (f) Purkayastha, A., Baruah, J. B., *Phosphorus, Sulfur and Silicon and the Related Elements* **2001**, 168-169, 333-338 (h) Purkayastha, A., Baruah, J. B., *Appl. Organomet.Chem.* **2001**, *15*, 693-698 (i) Purkayastha, A., Baruah, J. B., *Appl. Organomet. Chem.* **2000**, *14*, 477-483 (j) Baruah, J.B., Osakada, K., *Main group metal Chemistry* **1997**, *20*, 661
6. (a) Iwakura, Y., Uno, K., Toda, F., Hattori ,K., Abe, ,M., *Bull. Chem. Soc. Jpn.* **1971**, *44*, 1400 (b) Reddy, P. N., Chauhan, B. P. S., Hayashi, T., Tanaka, M., *Chem. Lett.* **2000**, 250 (c) Kim, B-H., Woo, H-G, *Advances in Organomet.Chem.* **2005**, *52* 143-174 (d) Mirza-Aghayan, M., Boukherroub, R., Bolourtchian, M., *J. Organomet. Chem.* **2005**, *690*, 2372-2375
7. Boyle, R. C., Mague, J.T., Fink, M. J., *Acta Crystal., Sec. E: Structure Reports Online* **2004**, *E60*, m40-m41
8. (a) Miller, W. S., Peake, J. S., Nebergall, W. H., *J. Am. Chem. Soc.* **1957**, *79*, 5604 (b) Sternbach, B., MacDiarmid, A.G., *J. Am. Chem. Soc.* **1959**, *81*, 5109
9. Lorenz, C., Schubert, U., *Chem. Ber.* **1995**, *128*, 1267
10. (a) Dolgov, B.N., Kharitonov, N. P., Voronkov, M. G, *Zh. Obshch. Khim.* **1954**, *24*, 1178 (b) Sternback, B., MacDiarmid, A.G., *J. Am. Chem. Soc.* **1959**, *81*, 5109 (c) Reikhsfel'd, V.O., Prokhorova, V.A., *Zh. Obshch. Khim.* **1961**, *31*, 2613
11. Corey, E.J., Venkateswarlu, A., *J. Am. Chem. Soc.* **1972**, *94*, 6190

80

12. Barooah N., Sharma S., Sarma B.C., Baruah J.B., *Appl. Organomet. Chem.* **2004**, *18*, 440-445

13. (a) Pilcher, A.S., DeShong, P., *J. Org.Chem.* **1993**, *58*, 5130-5134 (b) Babak, K., Daryoush, Z., *Tetrahedron Lett.* **2005**, *46*, 4661-4665 (b) Higashibayashi, S., Shinko K., Ishizu, T., Hashimoto, K., Shirahama, H., Nakata ,M. , *Synlett* **2000**, 1306-1308

14. (a) Corriu, R.J.P., Moreau, J.J.E., *J. Organomet. Chem.* **1977**, *127*, 7 (b) Corriu, R.J.P., Moreau, J.J.E., *J. Organomet. Chem.* **1976**, *120*, 337

15. Bakola-Christanopoulou M.N., *Appl. Organomet. Chem.* **2001**, *15*, 889

16. Zheng, G. Z., Chan, T.H., *Organometallics* **1995**, *14*, 70

17. Mori, A., Kato, T., *Synlett.* **2002**, 1167

18. DeWolf R.H., Young, W.G., *Chem Rev.* **1956**, *56*, 753

19. Bourhis, R., Frainnet, E. Moulines, F., *J. Organomet. Chem.* **1977**, *141*, 157

20. Unpublished results presented at National Symposium on Modern Trend in Inorganic Chemistry held at Indian Institute of Technology, Bombay, December 2003.

21. (a) Lewis, L.N., Lewis, N., *Chem. Mater.* **1989**, 1, 106 (b) Lewis, L. N., *Chem. Rev.* **1993**, *93*, 2693

22. Gulinski J., Klosin J., Marciniec, B., *Appl. Organomet.Chem.* **1994**, *8*, 409-414

23. Huber, C., Kokil, A., Caseri, W.R., Weder, C., *Organometallics* **2002**, *21*, 3817

24. Marsella, A., Agapakis, S., Pinna, F., Strukul, G., *Organometallics* **1992**, *11*, 3579

25. Prasad, B.L.V., Stoeva, S.I., Sorensen, C.M., Zaikovski, V., Klanblude, K.J., *J. Am. Chem. Soc.* **2003**, *125*, 10488

26. Shigemi, K., *Nippon Kagaku Zasshi* **1963**, *84*, 422

27. Liu, H. Q., Harrod, J. F., *Canadian J. Chem.* **1992**, *70*, 107

28. (a) Nagai, Y., Ojima, I., Kono, H., *Jpn. Kokai Tokkyo Koho* **1974**, 4 pp. CODEN: JKXXAF JP 49108024 19741014 (b) Ojima, I., Inaba, S.I., Kogure, T., Matsumoto, M., Matsumoto, H., Watanabe, H., Nagai, Y., *J. Organomet. Chem.* **1973**, *55*, C4

29. (a) Purkayastha, A., Baruah, J.B., Polymer Preprints (American Chemical Society, Division of Polymer Chemistry) **2004**, *45*, 666-667 (b) Chauhan, B.P.S., Rathore, J.S., Chauhan, M., Krawicz, A., *J. Am. Chem. Soc.* **2003**, *125*, 2876-2877

Copolymers

Chapter 7

Copolymers Based on Dimethylsiloxane and Diphenylsiloxane Units

Thomas M. Gädda and William P. Weber

D. P. and K. B. Loker Hydrocarbon Research Institute, Department of Chemistry, University of Southern California, Los Angeles, CA 90089–1661

A number of novel copoly(dimethylsiloxanes/ diphenylsiloxanes)s have been prepared and characterized. These copolymers have oligodiphenylsiloxane blocks located either alternatively in the polysiloxane main chain or at the chain ends of polydimethylsiloxane (PDMS). Alternatively, alkylheptaphenylcyclotetrasiloxanes were distributed either randomly as grafts along the PDMS backbone or at the chain ends of PDMS of different molecular weights. These copolymers were characterized by 1H, 13C and 29Si NMR, IR, and UV spectroscopy. Molecular weights were determined by GPC as well as by NMR end group analysis. Thermal transition temperatures and stabilities of copolymers were analyzed by DSC and TGA, respectively.

Introduction

A variety of copoly(dimethylsiloxanes/diphenylsiloxanes)s have been prepared and can be found in the literature.[1] These copolymers have been prepared to improve the thermal stability of polydimethylsiloxane (PDMS), alter its viscosity or to reduce its crystallinity. Triblock diphenylsiloxane-dimethylsiloxane-diphenylsiloxane polymers, on the other hand, have been prepared in the hope that they would exhibit thermoplastic elastomeric properties analogous to those of polystyrene-polybutadiene-polystyrene (PS-PB-PS). Siloxane polymers, compared to polyolefins, possess higher thermal stability due to the strength of the siloxane bond, which does not undergo thermolysis easily.[1] Thus, polysiloxanes are well known to undergo degradation by inter- and intramolecular cyclization reactions to lose volatile cyclosiloxanes. Polydiphenylsiloxane (PDPS) has been proven to exhibit very high thermal stability and does not decompose before 500 °C.[2-4] The bulky phenyl groups of PDPS may hinder depolymerization reactions. Thus, the thermal stability of PDPS is roughly 250 °C higher than PDMS in inert atmospheres. For this reason, aromatic or arylsiloxy units may be incorporated into PDMS to improve its thermal stability. Despite high thermal stability, PDPS has found little commercial use. This may be due to the difficult preparation of PDPS which results from its unfavorable ring-chain equilibria.[5] In addition, PDPS requires very high temperatures for isotropic melt formation, which in turn results in serious problems related to its melt processing.[3,4]

Besides good thermal stability, polydialkylsiloxanes are also well known for their low temperature flexibility. This predominately results from the characteristics of the siloxane bond.[1] However, polysiloxanes are semicrystalline and the formation of crystals has limited the low temperature use to their melting temperatures. Hence, arylsiloxane units have been incorporated into PDMS and polydiethylsiloxane to distrupt crystal packing and to prevent the undersirable crystallization.[6-9] These modified materials may retain rubbery properties at temperatures below their T_m and hence show significantly improved low temperature flexibilities.

The desirable properties of the first thermoplastic elastomers, i.e. polystyrene-polybutadiene (or polyisoprene)-polystyrene triblock copolymers, have provided encouragement to prepare analogous siloxane based materials. PDPS-PDMS-PDPS triblock copolymers have been prepared by sequential anionic ring-opening polymerization (AROP) of hexamethylcyclotrisiloxane (D_3) and hexaphenylcyclotrisiloxane (P_3) at high temperatures.[10,11] Elastomers may be obtained from PDPS-PDMS-polyvinyl-methylsiloxane triblock copolymers after peroxide initiated radical coupling.[12] Diblock copolymers have also been prepared and their thermal properties thoroughly investigated.[13] While these materials demonstrate the expected thermal transitions by DSC, no spectral data in support of their structures have been reported.

Herein, we report our preparation of copoly(dimethylsiloxanes/diphenylsiloxanes)s. The length of the diphenylsiloxane blocks were kept short to avoid the difficulties associated with higher molecular weight PDPS and its preparation. Never- the-less, we hoped that the copolymers would exhibit phase separation to yield new elastomers. While high molecular weight PDPS is practically insoluble, oligodiphenylsiloxanes with low degrees of polymerization (DP ≤ 11-18) are sufficiently soluble in common organic solvents at room temperature for analysis.[14,15] In addition, low molecular weight PDPS does not form liquid crystalline mesophases, which limits the processability of PDPS.

Hydrosilylation copolymerizations – PDPS-PDMS multiblock copolymers

A series of copolycarbosiloxanes were prepared by Pt-catalyzed hydrosilylation co-polymerization of 1,9-divinyldecaphenylpentasiloxane (I, Scheme 1), 1,7-divinylocta-phenyltetrasiloxane (II), 1,5-divinylhexaphenyltrisiloxane (III) and 1,3-divinyltetraphenyldisiloxane (IV) with α,ω-dihydridodecamethylpentasiloxane (V).[16] Oligomers I-III were prepared by treating a α,ω-dihydroxydiphenyloligosiloxane with butyl lithium and vinyldiphenylchlorosilane. A similar series of copolycarbosiloxanes was prepared by copolymerization of I with α,ω-dihydridooligodimethylsiloxanes.[17]

The α,ω-divinyloligodiphenylsiloxanes were synthesized in good yields. The oligomers and the copolymers were highly soluble in common organic solvents at room temperature. The majority of diphenylsiloxanes and silanes described in the literature are solids with melting points above room temperature. The oligodiphenylsiloxane described herein, however, were viscous liquids at room temperature, except for IV which displayed a melting point at 80 °C. Never-the-less, melting like transitions between -35 °C and -14 °C were measured by DSC for oligomers I-III. The temperatures of these transi-ions increased with increasing oligomer molecular weight. A similar trend between the copolymer T_g's was observed and a linear relationship was found between copolymer T_g and their diphenylsiloxy content (Figure 1a). Each copolymer exhibited only a single thermal transition by DSC. This is consistent with the lack of phase separation in the copolymers. These results and the low melting temperatures of oligomers I-III suggested that a longer oligodiphenylsiloxane segment may be required to achieve phase separation.

The NMR spectra of oligomers and copolymers confirmed their structures. However, the absence of end groups in the ^{1}H or ^{29}Si NMR spectra did not permit calculation of accurate number average molecular weights. In addition, no

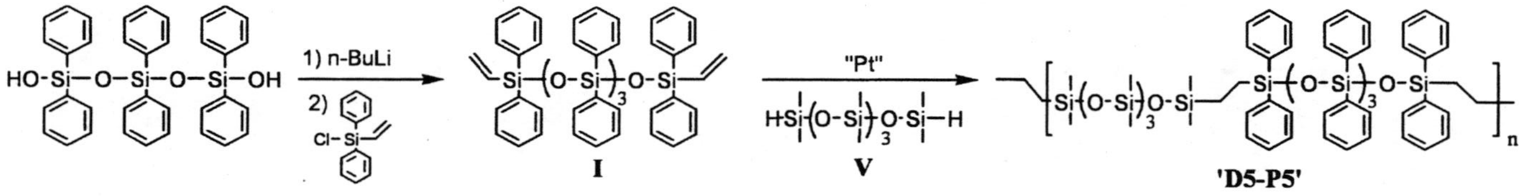

Scheme 1. Copolymerization of 1,9-divinyldecaphenylpentasiloxane and α,ω-dihydridodecamethylsiloxane.

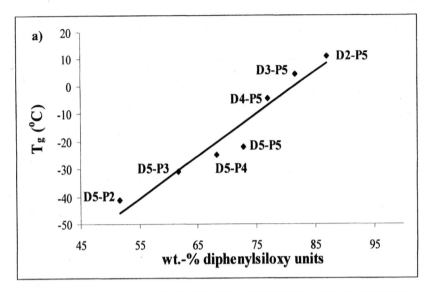

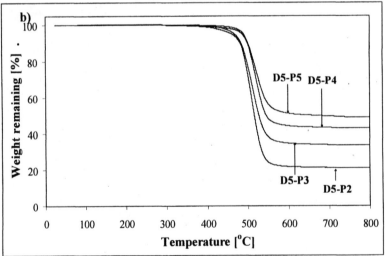

Figure 1. a) Glass transition as a function of diphenylsiloxy content and b) thermal stabilities of alt-copoly(dimethylsiloxane/diphenylsiloxanes). Dx-Py where x is the number of dimethylsiloxanes units (D) and y is the number of diphenylsiloxanes units (P).

Si-H signals could be observed in the IR spectra. All of the copolymers displayed good thermal stability in N_2 as expected for a copolymer with diphenylsiloxane units in the main chain (at least to 400 °C, Figure 1b). Above this temperature a single weight loss process occurs. Clear differences between the copolymers were observed for the onset and the quantity of char residue when the TGA analysis was carried out in N_2. Longer oligodiphenylsiloxane segments lead to a higher onset temperature for the thermal decomposition as well as to higher char yields.

PDMS with terminal and pendant 2-(heptaphenylcyclotetrasiloxanyl)ethyl groups

In general, there has been considerable interest in modified polymers and their preparation.[18-20] Recently, α,ω-bis(pyrene)PDMS was reported to exhibit thermoplastic elastomeric properties and reduced crystallinity. [21-23] Similarly, we have modified PDMS by end terminating and randomly modifying it with heptaphenylcyclotetrasiloxy moieties (Ph_7D_4-moieties, Scheme 2).[24] Vinylheptaphenylcyclotetrasiloxane (VI), prepared from vinylphenyldichloro-silane (VII) and hexaphenyltrisiloxane-1,5-diol (VIII), was introduced to various Si-H functional PDMS polymers by Pt-catalyzed hydrosilylation. The chemical modification significantly altered the physical properties of the material.

End group analysis by ^1H and ^{29}Si NMR was used to calculate molecular weights and polymer compositions. The molecular weights, obtained from the ratios of the integrals of repeating dimethylsiloxy units to the integral of the – $SiH(CH_3)_2$ or $-Si(CH_3)_2(CH_2)_2Ph_7D_4$ units, calculated before and after modification were in very good agreement with each other (Table 1). Also, the molecular weights obtained by GPC are in reasonable agreement with NMR end group analysis. These results indicated that a possible side reaction involving the Pt catalyzed oxidation of Si-H to Si-OH by water, followed by condensation of the Si-OH did not occur. Thus, the PDMS block before and after modification is constant and changes in the material properties are due to the chemical modification.

The measured T_g's for the modified polymers were very close to that of PDMS (Table 1). On the other hand, a significant reduction in crystallinity of the PDMS was observed for all modified samples when compared to their unmodified counterparts. In fact, some samples were completely amorphous materials and showed no signs of melting or cold crystallization, while higher molecular weight samples displayed reduced crystallinities (Figure 2a). Crystal formation was inhibited, when the bulky Ph_7D_4 unit is introduced at the ends of a low molecular weight PDMS or randomly as grafts to the PDMS main chain. No melting transition from a Ph_7D_4 phase was observed, consistent with a homogenous material that does not undergo phase separation.

88

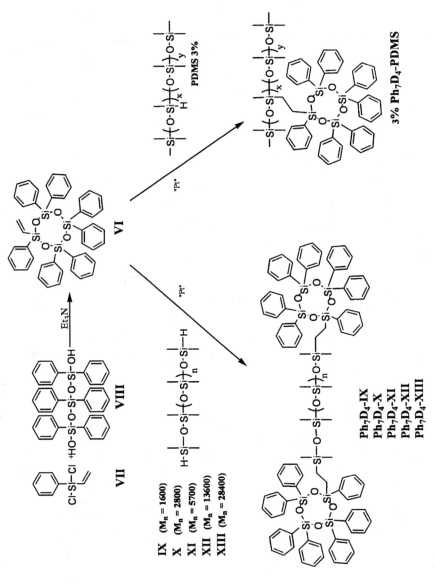

Scheme 2. Preparation of vinylheptaphenylcyclotetrasiloxane (VI) and modified PDMS.

VI

PDMS 3%

3% Ph₇D₄-PDMS

"Pt"

"Pt"

VIII

VII

IX   (Mₙ = 1600)
X   (Mₙ = 2800)
XI   (Mₙ = 5700)
XII (Mₙ = 13600)
XIII (Mₙ = 28400)

Ph₇D₄-IX
Ph₇D₄-X
Ph₇D₄-XI
Ph₇D₄-XII
Ph₇D₄-XIII

Thermal and thermo-oxidative stabilities of these modified materials were comparable or higher than those of α,ω-bis(trimethylsiloxy)PDMS (M_w/M_n = 17400/8300 by GPC). The modified polymers started to thermally degrade in N_2 at ~350 °C. While the onset of thermal degradation was nearly the same for all modified materials, significant increase in char yield was observed as the ratio of dimethylsiloxane units to diphenylsiloxane units decreased. This is consistent with a different degradation mechanism for phenyl containing siloxanes compared to PDMS. Higher char yields for PDPS has been attributed to formation of cross linked networks by way of radical coupling.[1,2] Although phase separation was not detected, relatively large increases in melt viscosities were measured when the viscosities of unmodified α,ω-bis(hydrido)PDMS is compared to the corresponding viscosities of modified materials (Table 1). Since the average molecular weights of the polymers have not drastically changed, the observed increases in viscosities may be due to intermolecular attactive interactions of the terminal and pendant Ph_7D_4 groups.

Table 1. Molecular weights, viscosities and crystallinities of materials before and after modification.

	Si-H functional PDMS			*Ph₇D₄ modified PDMS*			
	^{29}Si NMR M_n (g/mol)	Viscosity (cPs)	α_{cryst} (%)	^{29}Si NMR M_n (g/mol)	Viscosity (cPs)	T_g (°C)	α_{cryst} (%)
IX	1570	10	79	2620	1920	-57	0
X	3000	34	60	4750	720	-119*	0
XI	4300	80	64	7500	611	-121**	0
XII	14800	463	56	15400	1330	-123	40
XIII	28300	773	52	30500	2140	-123	44
3% SiH-PDMS	13300	234	---+	15300	3870	-123	0
CROP of XIV	---	---	---	13700***	-++	4	0
AROP of XIV	---	---	---	10500***	-++	-1	0

*T_g = -120 °C and **T_g = 123 °C after 2 h equilibration at -60°C. ***measured by ^1H NMR. +$\Delta H_f°$ unknown
++not measured

a)

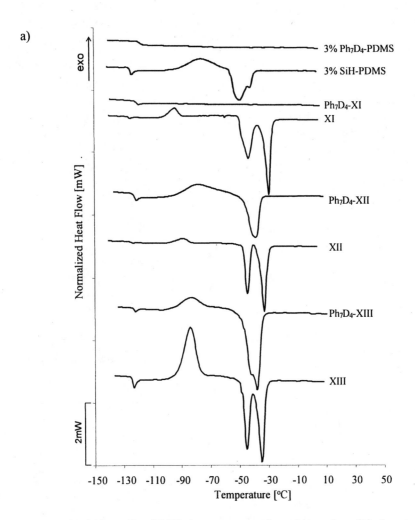

Figure 2. a) Normalized DSC thermograms of starting and modified materials; b) thermal decomposition profiles in N_2 of modified polymers and a PDMS standard.

b)

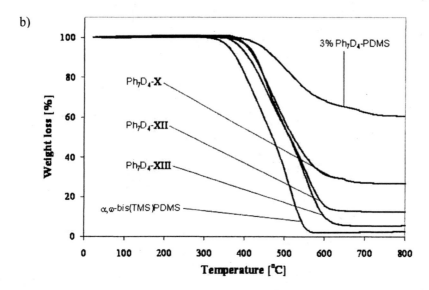

Figure 2. Continued.

Attempts to carry out similar Pt-catalyzed hydrosilylation of copoly(hydridomethylsiloxane/PDMS) materials with Si-H content higher than 3 mol-% gave crosslinked materials. Such crosslinking may be due to the undesired Pt-catalyzed oxidation of Si-H groups by water followed by condensation, if reactions are not carried out under strictly anhydrous conditions. To avoid crosslinking and to allow preparation of copoly-mers with higher percent of modification, we prepared 1-[2'-(heptaphenylcyclotetrasilo-xanyl)ethyl]-1,3,3,5,5-pentamethylcyclotrisiloxane (XIV) and studied its ring-opening polymerizations.[25] Analogous bicyclic siloxanes and a number of 1,2-bis(cyclotri- and cyclotetrasiloxanyl) substituted ethanes to XIV have been prepared.[26-31] Attempts to selectively polymerize these by ring opening of the more reactive cyclotrisiloxane to yield linear materials have been unsuccessful. To our knowledge, ROP of XIV is the first example of a regioselective cationic ring opening of a cyclotrisiloxane in the presence of a cyclotetrasiloxane (Scheme 3). The polymer properties of such a material may be determined by the bulky heptaphenylcyclotetrasiloxanyl group which is attached to every third silicon in the polymer back-bone.

Both acid (CROP) and base (AROP) catalyzed ring-opening polymerization of XIV gives highly viscous, transparent polymers. Based on ^{29}Si NMR spectroscopy, the polymers appear to result exclusively from ring opening of the cyclotrisiloxane ring. No evidence for ring-opening of the cyclotetrasiloxane ring was observed. T_gs of 4°C and -1 °C were measured for the polymers prepared by CROP and AROP, respectively. This is significantly higher than the T_g observed

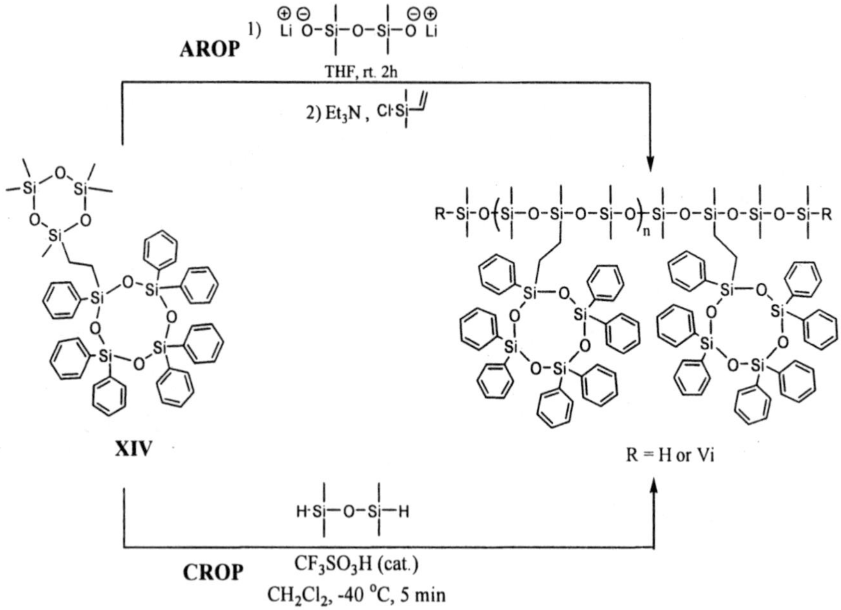

Scheme 3. AROP and CROP of XIV.

for the 3 mol-% randomly modified PDMS reported earlier, which had a T_g of -123 °C. This suggests that bulky side groups reduce the free segmental mobility of the flexible Si-O bond. However, evidence for phase separation as a result of increased amounts of the pendant heptaphenylcyclotetrasiloxane groups was not observed.

PDPS-PDMS-PDPS triblock copolymers

PDPS-PDMS-PDPS triblock copolymers were prepared by sequential monomer addition of hexamethylcyclotrisiloxane (D_3) and hexaphenylcyclotrisiloxane (P_3) to a difunctional anionic initiator, or by a convergent method (Scheme 4). The convergent method uses Pt-catalyzed hydrosilylation to couple known blocks. This was used to confirm the properties and structure of the polymers prepared by sequential monomer addition. Starting materials are shown in Table 2. Diphenyl ether has been the usual solvent used for anionic polymerization of P_3. We found that *o*-dichlorobenzene is a suitable solvent for AROP of P_3. This and the use of lower reaction temperatures allowed us to prepare well defined triblock copolymers. As previously mentioned, PDPS-PDMS block copolymers have been prepared but no spectral evidence in support of their structures has been provided.

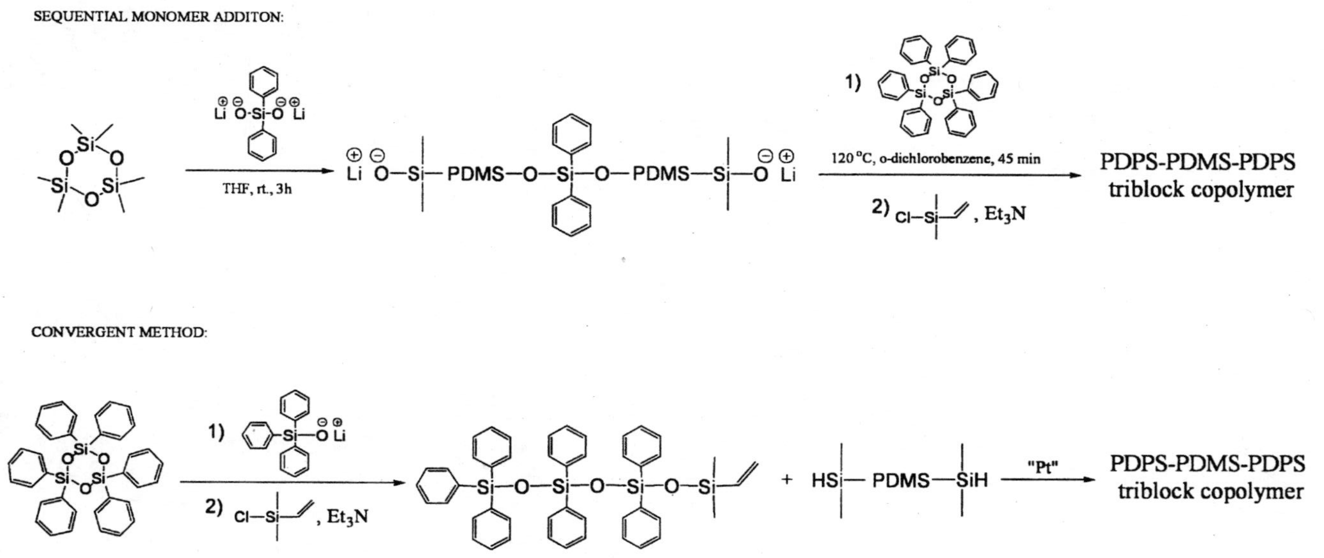

Scheme 4. Sequential monomer addition and convergent route to PDPS-PDMS-PDPS triblock copolymers.

The PDPS-PDMS-PDPS copolymers (XV-XVII) prepared by sequential monomer addition were obtained in relatively low yields and molecular weights. This is may be largely due to the nature of the ring-chain equilibria for diphenylsiloxanes.[5] It is well known that the equilibrium for siloxanes bearing bulky substituents are strongly shifted towards cyclics as the conformational freedom for the linear polymer is reduced when compared to dimethylsiloxanes. Despite this, they behaved like chewing gum or hard, slightly elastic waxes at room temperature. Polymers prepared by the convergent method were similar in physical appearance. As expected for triblock materials thermal transitions for both phases can be observed for all copolymers (Figure 3, Table 3). A T_g (~-120 °C) and a T_m (-50 to -40 °C) for the PDMS phase was seen, while the PDPS phase of the triblock copolymers displayed two melting endotherms, T_{m1} and T_{m2} (160–190 °C), as observed by DSC. T_{m1}'s and T_{m2}'s measured are in the temperature range in which melting transitions of low molecular weight diphenylsiloxane oligomers have been reported to occur. However, these melting transition of the PDPS phase are somewhat lower than those reported.[14,15,32] On the other hand, the T_g of PDPS (40 °C) of the highly crystalline PDPS blocks could not detected.[33] XVII displayed a partially random structure and hence its melting temperatures were lower than for the other copolymer prepared. The PDPS oligomer used in the preparation of copolymers XVIII and XIX displayed its highest melting peak at ~200 °C. Heats of fusion obtained for PDMS and PDPS crystal melting in copolymers XV and XVI were determined. XV displayed a $\Delta H_{f,PDMS}$ = 9.3 J/g and a $\Delta H_{f,PDPS}$ = 3.5 J/g while corresponding values for XVI were $\Delta H_{f,PDMS}$ = 10.0 J/g and $\Delta H_{f,PDPS}$ = 3.7 J/g. The $\Delta H_{f,PDPS}$ is the total heat of fusion over the whole melting transition and is close to values previously reported.[15] On the other hand the enthalpies measured for the PDMS phases were lower than those of PDMS homopolymer.[34] This suggests that the PDPS blocks restrict mobility of PDMS chains and in this way reduces their ability to form crystals. Degrees of crystallinities were 15 % and 16 % for XV and XVI, respectively, using $\Delta H_f°$ = 61,19 J/g in calculations for a fully crystalline PDMS.[35] Thermal transition temperatures were confirmed and mechanical properties of the materials were evaluated by DMA. Similar storage and loss moduli profiles were obtained for all copolymers.

The regular structure of our triblock copolymers was confirmed by ^{1}H, ^{13}C, ^{29}Si NMR and DSC. In the ^{1}H NMR spectra, a complex phenyl region was observed. The triblock structure can be confirmed form the ^{1}H and ^{13}C NMR spectra since only a sharp singlet for the -Si(CH$_3$)$_2$O- group can be observed in the dimethylsiloxy region of the spectra (Figure 4a). If randomization occurs, several additional peaks will be visible in this -(CH$_3$)$_2$SiO- region (Figure 4b).

Similarly, for a regular triblock copolymer, a single peak at ~-21.8 ppm should be observed in the -(CH$_3$)$_2$SiO- region and at ~-46.5 ppm in the -Ph$_2$SiO- region of the ^{29}Si NMR spectra. In the in the -(CH$_3$)$_2$SiO- region a single peak is observed (Figure 5a). However, two relatively broad peaks are observed in the

Table 2. Starting materials used and molecular weights of copolymers.

Sample	Feed			Theoretical Mn:s			GPC Molecular Weights		
	D_3 g	P_3 g	Initiator mmol	PDMS kg/mol	PDPS kg/mol	$M_{n\,total}$ kg/mol	M_w kg/mol	M_n kg/mol	M_w/M_n
XV	0.38	0.6	0.017	22.5	36	58.6	15	8.8	1.7
XVI	0.5	0.6	0.017	30	36	66	18	10.3	1.8
XVII	0.5	0.5	0.023	22	22	44	16.1	11.8	1.4
XVIII*	1	0.62	-	14	3.8	21.6	36.4	24.9	1.5
XIX*	1	0.31	-	28	3.8	35.6	52.7	31.3	1.7

*Feed in grams of polymer.

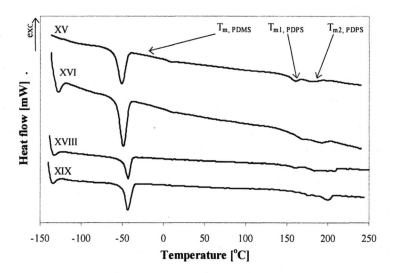

Figure 3. DSC thermograms of PDPS-PDMS-PDPS triblock copolymers.

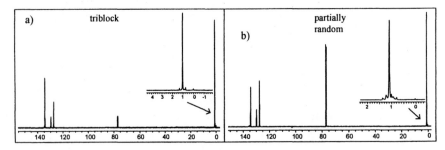

Figure 4. ^{13}C NMR of a PDPS-PDMS-PDPS triblock copolymer XV (a) and partially random copolymer XVII (b).

-Ph$_2$SiO- region in the ^{29}Si NMR spectra instead of one signal as expected (Figure 5a). In addition to the signal at ~-46.5 ppm, a signal of lower intensity at ~-47.5 ppm in the -Ph$_2$SiO- region is observed.[15,36] This signal results from -Ph$_2$SiO- units that are adjacent to -(CH$_3$)$_2$SiO- unit.

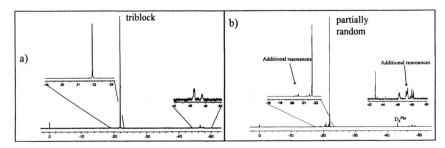

Figure 5. ^{29}Si NMR of a PDPS-PDMS-PDPS triblock copolymer XV (a) and partially random copolymer XVII (b).

Although the ^{29}Si NMR confirms distinct blocks for PDMS and PDPS, the complete analysis of the oligodiphenylsiloxane blocks by ^{29}Si NMR is problematic. Integration of the -(CH$_3$)$_2$SiO- and -Ph$_2$SiO- regions should give information on the composition of the copolymers. However, the integration results from ^{1}H and ^{29}Si NMR are different and ^{1}H NMR gives higher -Ph$_2$SiO- content than that obtained from ^{29}Si NMR integration (Table 3). This discrepancy probably results from different relaxation times that ^{29}Si nuclei have depending on their chemical environment. In general, ^{29}Si NMR spectra are acquired with a similar pulse sequence used to acquire ^{13}C NMR for which it is known that adjacent protons enhance the relaxation of the nuclei during NMR experiments.[37,38] Thus, the lower signal intensities for -Ph$_2$SiO- units compared to -(CH$_3$)$_2$SiO- units, is probably due to the fact that the closest proton is three bonds away in a phenyl substituted Si atom, whereas it is only two bonds away in a methyl substituted Si atom. Although integration of ^{13}C spectra is often thought to be impossible, scientists appear to believe that ^{29}Si NMR can be integrated.[36,39] In our experience, this is not always fully possible. To solve similar integration problems paramagnetic impurities have traditionally been added to NMR samples. These provide a different relaxation mechanism for the [29]Si nuclei.[37,38,40] Addition of Cr(acac)$_3$ increases the -Ph$_2$SiO- [29]Si signal intensity by ~30 % relative to the -(CH$_3$)$_2$SiO- signal. For instance, SiMe$_2$O/SiPh$_2$O ratio for copolymer XVII decreased from 7.07 to 4.89. However, attempts to achieve comparable [29]Si integration results to ^{1}H NMR or to sharpen the broad -Ph$_2$SiO- peaks by variable temperature NMR were unsuccessful. Similar inconsistencies in the integration results of ^{1}H NMR and ^{29}Si NMR of copolymers XVIII and XIX to those of copolymers XV-XVII were obtained.

Table 3. Polymer yields, NMR integration and thermal properties of copolymers.

Sample	Polymer yield [%]	Theoretical Integration SiCH$_3$/SiPh ^{1}H NMR^a / ^{29}Si NMR	Integration Ratio SiCH$_3$/SiPh ^{1}H- NMR^a	^{29}Si NMR^b	DSC Thermal Properties T_g [°C]	$T_{m, PDMS}$ [°C]	$T_{m, PDPS}$ [°C]
XV	51	1.67	3.06	3.79	-121	-51	161, 181
XVI	58	2.23	3.43	5.34	-117	-47	176, 193
XVII	51	2.68	3.23	4.89	-123	-37	116
XVIII*	93	4.97	4.79	6.78	-120	-43	161, 206
XIX*	94	9.95	6.01	9.12	-119	-43	184, 199

^{a}ratio = $(I_{Si-Me}/6)/(I_{Si-Ph}/10)$ ^{b}ratio = I_{Si-Me}/I_{Si-Ph}

The presence of 2,2-dimethyl-4,4,6,6,8,8-hexaphenylcyclotetrasiloxane (P$_4^{Me2}$) in supernatant liquids from polymer precipitations confirmed that intramolecular backbiting (reversion) accompanies the polymerization of P$_3$ even at the very beginning of the reaction.

Conclusions

Block copolymers of AB type were prepared from oligodiphenylsiloxy and oligodimethylsiloxy segments by hydrosilylation. A single glass transition was observed indicating materials that do not phase separate. These materials were highly thermally stable, as expected. In a similar manner, PDMS polymers bearing alkylheptaphenylcyclotetrasiloxane moieties did not undergo phase separation, but displayed significantly increased viscosities and somewhat improved thermal stability compared to PDMS. These results suggest that longer diphenylsiloxy segments are required to prepare thermoplastic elastomers. Triblock copolymers were prepared by sequential addition of monomers and by a convergent approach. Spectroscopic analysis by NMR confirms their triblock structure. These block copolymers showed the expected phase separation.

Acknowledgement

T. M. Gädda thanks fellowship support from the Foundation of Technology and the Swedish Academy of Engineering Sciences in Finland.

References

1. Noll, W. Chemistry and Technology of Silicones, Academic Press: New York, 1968, pp. 437-530.
2. Dvornic, P. R. In Silicon Containing Polymers; Jones, R. G.; Ando, W.; Chojnowski, J. Eds.; Kluwer Academic Publishers, Dordrecht Netherlands, 2000, pp. 185-212.
3. Papkov, V. S.; Buzin, M. I.; Gerasimov, M. V.; Obolonkova, E. S. *Macromolecules* **2002**, *35*, 1079.
4. Harkness, B.R.; Tachikawa, M., Yue, H.; Mita, I. *Chem. Mater.* **1998**, *10*, 1700.
5. Wright, P. V.; Semlyen, J. A. *Polymer* **1970**, *11*, 462.
6. Polmanteer, K. E.; Hunter, M. J. *J. Appl. Polym. Sci.* **1959**, *1*, 3.
7. Brewer, J. R.; Tsuchihara, K.; Morita, R.; Jones, J. R.; Bloxsidge, J. P.; Kagao, S.; Otsuki, T.; Fujishige, S. *Polymer* **1994**, *35*, 5109.
8. Brewer, J. R.; Tsuchihara, K.; Morita, R.; Jones, J. R.; Bloxsidge, J. P.; Kagao, S.; Otsuki, T.; Fujishige, S. *Polymer* **1994**, *35*, 5118.
9. Liu, L.; Yang, S.; Zhang, Z.; Wang, Q.; Xie. J. *J. Polym. Sci. Part A: Polym. Chem.* **2003**, *41*, 2722.
10. Bostick E. E. *U.S. Patent* 3,337,497, **1967**.
11. Bostick E. E. *Polym. Prepr.* **1969**, *10*, 877.
12. Ibemesi J.; Gvozdic, N.; Keumin, M.; Lynch, M. J.; Meier, D. J.; *Polym. Prepr.* **1985**, *26*, 18.
13. Yang, M.-H.; Huang, W.-J.; Chien, T.-C.; Chen, C.-M.; Chang, H.-Y.; Chang, Y.-S.; Chou, C. *Polymer* **2001**, *42*, 8841.
14. Harkness, B. R.; Tachikawa, M.; Itaru, M. *Macromolecules* **1995**, *28*, 1323.

99

15. Harkness, B. R.; Tachikawa, M.; Itaru, M. *Macromolecules* **1995**, *28*, 8136.
16. Gädda, T. M.; Weber, W. P. *J. Polym. Sci. Part A: Polym. Chem.* **2005**, *43*, 2155.
17. Lee, N. S.; Gädda, T. M.; Weber, W. P. *J. Polym. Sci. Part A: Polym. Chem.* **2005**, *43*, 6146.
18. Carraher, C. E. Jr.; Moore, J. A. In Modification of Polymers, Plenum Press, New York 1983, 420 p.
19. Benham, J. L.; Kinstle, J. F.; In Chemical Reactions on Polymers, ACS Symp. Series 363, American Chemical Society, Washington D. C. 1988.
20. Guo, H.; Tapsak, M. A.; Weber, W. P. *Polym. Bull.* **1994**, *33*, 417.
21. Kim, S. D.; Torkelson, J. M. *Macromolecules* **2002**, *35*, 5943.
22. Jones, B. A.; Torkelson, J. M. *ANTEC Conference Proceedings*, Vol. III, Chicago IL, **2004**, 4162.
23. Jones, B. A.; Torkelson, J. M. *J. Polym. Sci. Part B: Polym. Phys.* **2004**, *42*, 3470.
24. Gädda, T. M.; Weber, W. P. *J. Polym. Sci. Part A: Polym. Chem.* **2005**, *43*, 5007.
25. Gädda, T. M.; Weber, W. P. *J. Polym. Sci. Part A: Polym. Chem.* **2006**, *44*, 137.
26. Andrianov, K. A.; Zachernyuk, A. B. *Appl. Polym. Symp.* **1975**, *26*, 123
27. Andrianov, K. A.; Zachernyuk, A. B. *Pure & Appl. Chem.* **1976**, *48*, 251.
28. Nitzsche, S.; Burkhardt, J.; Wegehaupt, K.-H. *German Pat.* 1,794,219 (Wacker-Chemie G.m.b.H., 1968)
29. Zhdanov, A. A.; Kotov, V. M.; Pryakhina, T. A. *Izv. Akad. Nauk. SSSR Ser. Khim.* **1986**, 241.
30. Zhdanov, A. A.; Astapova, T. V.; Lavrukhin, B. D. *Izv. Akad. Nauk. SSSR Ser. Khim.* **1988**, 657.
31. Cheng, P.-S.; Hughes, T. S.; Zhang, Y.; Webster, G. R.; Poczynok, D.; Buese, M. A. *J. Polym. Sci. Part A: Polym. Chem.* **1993**, *31*, 891.
32. Li, L.-J.; Yang, M.-H. *Polymer* **1998**, *39*, 689.
33. Lee, C. L.; Marko, O. W. *Polym. Prepr.* **1978**, *19*, 250.
34. Wang, B.; Krause, S. *Macromolecules* **1987**, *20*, 2201.
35. Lebedev, B. V.; Mukhina, N. N.; Kulagina, T. G. *Vysokomol. Soedin. Ser. A* **1978**, *20*, 1458.
36. Cypryk, M.; Kazmierski, K.; Foruinak, W.; Chojnowski, J. *Macromolecules* **2000**, *33*, 1536.
37. Crews, P.; Rodriguez, J.; Jaspars, M. In Organic Structure Analysis; Oxford University Press: New York NY, 1998; pp. 36-38.
38. Günther, H. In NMR Spectroscopy; John Wiley & Sons: Chichester England, 1995; pp. 221-238.
39. Babu, G. N.; Christopher, S. S.; Newmark, R. A. *Macromolecules* **1987**, *20*, 2654.
40. Brook, M. A. In Silicon in Organic, Organometallic, and Polymer Chemistry; John Wiley & Sons: New York NY, 2000; pp. 14-26.

Chapter 8

Silicone–Urea Copolymers Modified with Polyethers

Iskender Yilgor and Emel Yilgor

Chemistry Department, Koc University, Sariyer 34450, Istanbul, Turkey

Silicone-urea copolymers were modified by the chemical incorporation of polyethers (PEO and PPO) into the polymer backbone. Resulting materials were characterized by dynamic mechanical analysis, tensile tests, water absorption studies and refractive index measurements. Depending on the type, average molecular weight and amount of the polyether incorporation and urea hard segment content of the copolymer it was possible to modify the microphase morphologies and physicochemical properties of silicone-urea copolymers. We believe using this novel approach and simple, one-pot reaction procedure it is possible to tailor design and prepare silicone-urea copolymers with controlled hydrophobic/hydrophilic balance, improved tensile properties and controlled refractive indices.

Physicochemical and mechanical properties of multiphase copolymers are closely related to their complex, heterogeneous supramolecular structures or microphase morphologies (1-3). Large number of parameters including chemical (bond strength, hydrogen bonding) and structural factors (crystallinity and defects), processing conditions and thermal history play major roles in determining the microphase morphology of copolymers. In addition to these, an especially important parameter is the nature and properties of the interface (or interphase) between the microphase separated hard and soft segment phases. A sharp interface between two phases with dissimilar properties could create a weak point in a block copolymer. In biological systems (e. g. proteins) this problem is solved by creating a smooth gradient rather than a sharp transition between phases (4). Interestingly, major aim of synthetic polymer chemists has always been the preparation of block copolymers with well-defined hard and soft segment structures, block molecular weights, molecular weight distributions and architectures (1-3). Such copolymers would preferentially display well phase separated microphase morphologies with relatively sharp interfaces between hard and soft phases. Investigation of the structure-morphology-property relations in such well-defined block and segmented copolymers has been one of the most active areas of research for over 40 years.

We have been investigating the synthesis; characterization and structure-morphology-property behavior of polydimethylsiloxane (PDMS) based urea and urethane type segmented copolymers for about two decades (5-7). Due to the substantial differences between the solubility parameters of PDMS [15.6 $(J/cm^3)^{1/2}$], urethane [37.2 $(J/cm^3)^{1/2}$] and urea [45.6 $(J/cm^3)^{1/2}$] segments (8), it is possible to prepare PDMS-urea and PDMS-urethane type copolymers with very good phase separation and a sharp interface between continuous soft segment matrix and hard segment domains (7, 9).

As shown in Table 1, it is interesting to note that when compared with their polyether-based homologs, silicone-urea copolymers display much higher modulus but lower ultimate tensile strength and elongation at break values. Sample codes used in Table 1 are designed as follows: PSU indicates a PDMS (M_n=2000 g/mol) based polyurea, whereas PEU indicates a poly(ethylene oxide) (PEO) (M_n=2000 g/mol) based polyurea. Second group of letters indicate the diamine chain extender used (ED=ethylene diamine, HM=hexamethylene diamine and DY=2-methyl-1,5-diaminopentane). Two digit numbers that follow indicate the hard segment content of the polymer in weight percent. These high molecular weight copolymers as indicated by the intrinsic viscosity values, formed clear and strong films, when cast from IPA solution.

It is interesting to note from Table 1 that PDMS-urea copolymers display much higher Young's modulus when compared to PEO-urea copolymers. On the other hand, PEO-urea copolymers show much higher tensile strengths and elongation at break values than the PDMS-urea copolymers. This behavior may be due to several factors, which include; (i) the lack of stress-induced crystallization in PDMS at room temperature, (ii) the inherent mechanical weakness of the PDMS chains, since at room temperature they are about 150°C above their Tg values, or (iii) the presence of a very sharp interface between continuous PDMS matrix and urea domains, leading to a poor energy transfer mechanism between the PDMS and urea phases and thus poor tensile properties.

Table 1. Tensile properties of HMDI based homologous silicone-urea and polyether-urea copolymers

Sample	Soft segment Type	Mn (g/mole)	Urea (wt %)	Mod. (MPa)	Tens. St. (MPa)	Elong. (%)
PSU-DY-20	PDMS	2500	20	20.6	7.90	205
PSU-HM-20	PDMS	2500	20	18.9	8.10	195
PSU-ED-19	PDMS	2500	19	21.3	8.30	180
PEU-DY-20	PEO	2000	20	4.30	25.4	1320
PEU-HM-20	PEO	2000	20	4.20	25.8	1325
PEU-ED-19	PEO	2000	19	4.50	26.5	1450

Phenomena described in (i) and (ii) is inherent to PDMS. However, it is possible to controllably modify the structure and nature of the interface (interphase) between PDMS and urea phases by using reactive polyether oligomers, such as poly(ethylene oxide) (PEO), poly(propylene oxide) (PPO) or poly(tetramethylene oxide) (PTMO). There are three major reasons for choosing the polyether oligomers. One is the availability of telechelic, reactive oligomers with number average molecular weights ranging from 200 to 4,000 g/mol. The second important reason is the solubility parameters of PEO, and PPO which are 24.5 $(J/cm^3)^{1/2}$ and 23.5 $(J/cm^3)^{1/2}$, in between that of PDMS and urea. The third reason is strong hydrogen bonding interaction between ether and urea groups (10), leading to the formation of a gradient interphase in these copolymers. Critical parameters, which may influence the microphase morphology and the nature or structure of the interface are; (i) type of the polyether, (PEO, PPO or PTMO), (ii) average molecular weight of the polyether, (iii) urea hard segment content of the polymer, and (iv) average molecular weight of the PDMS oligomer in the copolymer.

In this study preparation and various physicochemical properties of silicone-urea copolymers that contain a second soft segment component, namely poly(ethylene oxide) (PEO) or poly(propylene oxide) (PPO), which is chemically incorporated into the polymer backbone in a controlled manner was investigated. Since polyethers can form hydrogen bonding with the urea groups, we believe a gradient type interphase (instead of a sharp interface) forms between PDMS matrix and urea hard segments as demonstrated in our earlier publication (11). In addition, incorporation of polyether segments, suct as PEO will also provide hydrophilicity to otherwise totally hydrophobic materials.

Experimental

Materials. α,ω-Aminopropyl terminated PDMS oligomers with $<M_n>$ of 3,200 g/mol were obtained from Wacker Chemie, Munich, Germany. Amine terminated PEO and PPO oligomers with $<M_n>$ between 240 and 2000 g/mol were obtained from Huntsman Chemical Corp. $<M_n>$ values of amine-terminated oligomers were determined by titration of the end groups with standard hydrochloric acid. Bis(4-isocyanatocyclohexyl)methane (HMDI) with a purity better than 99.5 % was kindly provied by Bayer AG. 2-Methyl-1,5-diaminopentane (Dytek A) (DY) was provided by Du Pont. Reagent grade isopropanol (IPA) was purchased form Carlo Erba and used as the reaction solvent without further purification.

Polymerization procedure. All polymerization reactions were carried out in IPA, at room temperature in a three-neck, round bottom flask equipped with an overhead stirrer, nitrogen inlet and an addition funnel. As shown in Reaction Scheme I, polymers were prepared using a three-step procedure. Calculated amount of HMDI was weighed into the reaction flask and dissolved in IPA. Calculated amounts of amine terminated PDMS and PPO oligomers were separately weighed into the Erlenmeyer flasks and dissolved in IPA. These solutions were sequentially introduced into the addition funnel and added dropwise into the reactor containing the HMDI solution, under strong agitation, to prepare the prepolymer. Finally, stoichiometric amount of diamine chain extender was dissolved in IPA and added drop-wise into the reactor, through the addition funnel. Progress and completion of the reactions was monitored by FT-IR spectroscopy, following the disappearance of the strong isocyanate peak at 2270 cm^{-1} and formation of urea (N-H) and (C=O) carbonyl peaks around 3300 and 1650 cm^{-1} respectively. Reaction mixtures were homogeneous and clear throughout; no precipitation was observed. The polymers films (0.3-0.5 mm thick) utilized for analysis were cast from IPA solution into Teflon molds, dried at room temperature overnight, and dried at 65°C until a constant weight was reached.

Instrumentation. FTIR spectra of thin films cast on KBr disks from IPA solutions were obtained on a Nicolet Impact 400D spectrometer, with a resolution of 2 cm^{-1}. Thermal properties of the polymers were studied on a TA DSC Q100 differential scanning calorimeter. DSC scans were obtained between -150 and 200 °C, under nitrogen atmosphere, at a heating rate of 10 °C per minute. Dynamic mechanical analysis (DMA) of the polymers were obtained on a TA DMA Q800 instrument. Measurements were made in tensile mode, between -150 and 200 °C, under nitrogen atmosphere, at a heating rate of 3 °C per minute and a frequency of 1 Hz. Stress-strain tests were carried out on an Instron Model 4411 Universal Tester, at room temperature, with a crosshead speed of 25 mm/min. Dog-bone samples were punched out of thin copolymer films using a standard die (ASTM D 1708). Intrinsic viscosities were determined in Ubbelohde viscometers at 25°C in IPA. Refractive indices of thin polymer films were determined by using an Abbe refractometer at 25°C. Films were cast from solution on glass Petri dishes cleaned by successive washings with toluene, acetone, isopropanol and triple distilled water. Solvent was evaporated by keeping the samples at room temperature overnight and then in a vacuum oven at 60 °C, until a constant weight is reached. Refractive index values reported are the average of 10 different readings. Water absorption of polymer films was determined gravimetrically at room temperature. Three rectangular polymer films with dimensions of 3.0x3.0x0.1 cm were immersed into three different covered jars containing distilled water at 25°C. They were intermittently removed from water bath, surfaces were dried by using a tissue paper and weight gains were determined gravimetrically by using an analytical balance. Results reported are the average of three measurements.

Results and Discussion

Novel polydimethylsiloxane-urea copolymers with a polyether gradient between PDMS matrix and the urea hard segments were prepared and characterized. Using this novel approach and a simple, one-pot reaction procedure a large number of interesting and versatile polymers with unique combination of properties can be designed and synthesized. The properties that can be tailored include hydrophilic/hydrophobic balance, tensile strength, modulus, refractive index and water absorption. These novel polymers may find applications as biomaterials, coatings, adhesives, elastomers and textile fibers.

We have recently published a detailed report on the structure-morphology-property behavior of silicone-urea copolymers modified with amine terminated PPO oligomers with molecular weight ranging from 400 to 2000 g/mole (11). All of the materials displayed microphase morphology as determined from dynamic mechanical analysis (DMA) and small angle X-ray scattering (SAXS) studies. DMA and SAXS results suggested that the ability of the PPO segments to hydrogen bond with the urea segments results in a limited inter-segmental mixing which leads to the formation of a gradient interphase, especially in the PPO-2000 containing copolymers. DMA also demonstrated that the polyureas based on only PDMS possessed remarkably broad and nearly temperature insensitive rubbery plateaus that extended up to ca. 175 °C. Incorporation of PPO resulted in more temperature sensitive rubbery plateaus. A distinct improvement in the Young's modulus, tensile strength, and elongation at break in the PPO-2000 containing copolymers was observed due to inter-segmental hydrogen bonding and presumably the formation of a gradient interphase.

In the present study in addition to the hydrophobic PPO, we also utilized hydrophilic PEO oligomers as modifiers for silicone-urea copolymers. Although our main aim was the preparation of silicone-urea copolymers with much improved mechanical properties, other physicochemical characteristics, such as equilibrium water absorption and refractive indices of the resultant materials were also investigated. All polymers were prepared by using HMDI as the diisocyanate and PDMS with $<M_n>$ = 3200 g/mol as the soft segment. Dytek A was used as the chain extender. Polyether modifiers were either PPO or PEO oligomers, with molecular weights ranging from 230 to 2000 g/mol. Table 2 provides the chemical compositions (PDMS, polyether and hard segment (HS) contents) and tensile properties (Young's modulus (Mod), ultimate tensile strength (TS) and elongation at break (E) values) of all silicone-urea and polyether modified silicone-urea copolymers prepared in this study. All of the copolymers synthesized were of high molecular weight and formed strong films, as also indicated by the tensile tests, which are discussed in detailed below.

Nomenclature or coding of the polymers were done as follows: The coding of the first four polymers, which are based on PDMS, HMDI and Dytek A chain extender and do not contain any polyethers was as follows: The first two digit number gives the weight percent of PDMS-3200 in the copolymer, followed by

Table 2. Chemical compositions and tensile properties of silicone-urea copolymers and polyether modified silicone-urea copolymers

Sample code	PDMS (wt %)	Polyether (wt %)	HS (wt %)	Mod (MPa)	TS (MPa)	E (%)
83-U-17	83.3	--	16.7	16.4	9.70	265
76-U-24	75.8	--	24.2	22.3	13.2	205
70-U-30	69.6	--	30.4	34.4	17.9	160
64-U-36	64.4	--	35.6	52.9	22.0	150
72-P240-11-U-28	71.6	10.7	17.7	38.5	14.0	280
61-P240-9-U-30	61.3	9.2	29.5	80.8	17.8	130
59-P240-18-U-24	58.5	17.5	24.0	90.5	14.0	125
57-P450-16-U-27	56.7	16.0	27.3	92.0	15.8	205
34-P2000-42-U-24	33.7	42.1	24.2	18.8	17.2	580
32-P2000-41-U-27	32.4	40.5	27.1	29.6	19.8	525
42-P2000-26-U-32	42.0	26.2	31.8	45.6	17.7	375
53-E630-21-U-26	53.3	21.0	25.7	31.7	17.0	365
49-E960-14-U-37	49.1	13.8	37.1	81.0	20.2	170
52-E960-15-U-33	52.1	14.7	33.2	63.8	15.5	200
44-E960-25-U-31	43.8	24.7	31.5	41.0	17.5	340
49-E960-28-U-24	48.9	27.5	23.6	9.3	11.5	600
49-S960-28-U-24	48.9	27.5	23.6	105	11.8	205
33-E2000-40-U-27	32.6	40.4	27.1	15.7	10.7	830

letter (U) indicating urea hard segments (HMDI+DY). The last two digit number provides the wight percent of the urea hard segments in the copolymer. As for the polyether modified materials, where (P) indicates PPO and (E) indicates PEO type modifiers, the coding was done as follows:

PDMS (wt %)—Polyether type, MW and weight %—Urea content (wt %)

For example; 34-P2000-42-U-24 describes a copolymer with 34% by weight PDMS-3200, 42% by weight PPO-2000 and a urea (HMDI+DY) hard segment content of 24% by weight.

Dynamic mechanical behavior

In order to understand the influence of polyether modification on the modulus-temperature behavior and more importantly on the microphase morphologies of silicone-urea copolymers DMA analysis were performed. Figure 1 gives the modulus-temperature and tanδ-temperature curves for the silicone-urea copolymer containing 24% by weight hard urea segment (76-U-24) and two copolymers modified with PEO-2000 (33-E2000-40-U-27) and PEO-630 (53-E630-21-U-26). As expected the silicone-urea copolymer shows a well defined and sharp PDMS glass transition between -120 and -100 °C, followed by a very long and temperature insensitive rubbery plateau extending to +100 °C. Strong hydrogen bonding in the urea hard segments slowly starts breaking up above 100 °C (rubbery flow) followed by an accelerated breakage (viscous flow) above 200 °C, which can clearly be followed in both modulus-temperature and tanδ-temperature curves. This is a typical behavior for silicone-urea copolymers (7,9). Silicone-urea copolymer modified with 40% by weight of PEO-2000 (33-E2000-40-U-27) shows two low temperature glass transitions, one for the PDMS, between -120 and -100 °C and the other for PEO, between -60 and -40 °C, followed by a very short rubbery plateau extending to just about to room temperature. Interestingly, although the urea hard segment content of this copolymer is slightly higher than that of 76-U-24, it becomes very weak and shows flow at around 100 °C, much earlier than that of pure silicone-urea copolymer. This is most probably due to the weakening (plasticization) of the urea hard segments through mixing with ethylene oxide units in PEO (10, 12, 13). These results indicate the presence of three regions in the microphase morphology of this polymer, which are: (i) PDMS soft segment phase, (ii) PEO soft segment phase, and (iii) PEO mixed urea hard segment phase. Interestingly, as shown in Figure 1, the copolymer modified with the shorter polyether oligomer (PEO-630) shows only one very well defined low temperature glass transition, which is due to PDMS. It is followed by a rubbery plateau extending from -100 to about room temperature. The material softens and flows before reaching 100 °C. This we believe indicates a two phase morphology consisting

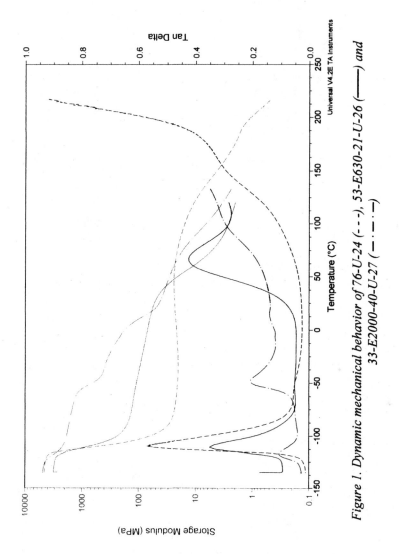

Figure 1. Dynamic mechanical behavior of 76-U-24 (- -), 53-E630-21-U-26 (——) and 33-E2000-40-U-27 (— · — · —)

of (i) PDMS phase and (ii) mixed PEO-urea phase. As discussed below, these interesting microphase morphologies clearly manifest themselves in unusual tensile properties and water absorption behavior.

Figure 2 provides the modulus-temperature and tanδ-temperature curves for unmodified silicone-urea copolymer (76-U-24) and copolymers modified with PDMS and PEO oligomers of identical molecular weights (M_n=960 g/mol). Modified systems also have identical urea hard segment contents of 24% by weight. As a result, the DMA data presented here directly compares the influence of the chemical structure (and solubility parameter) of the oligomeric modifier on the microphase morphology of the new system. PDMS-960 modified material (49-S960-28-U-24) displays a fairly sharp glass transition between -120 and -90 °C, slightly, higher than that of unmodified silicone-urea copolymer (76-U-24), followed by a temperature insensitive rubbery plateau until about +100 °C, followed by a flow, between 100 and 180 °C due to breakage of strong hydrogen bonding in the urea hard segment domains. Modulus of the rubbery plateau is about an order of magnitude higher than that of pure silicone-urea, indicating the formation of a well phase separated and stiffer elastomer upon modification with PDMS-960, which is also supported by the tensile data as discussed below.

Modulus-temperature behavior of the PEO-960 modified homologous material (49-E960-28-U-24) is quite different. It also shows a single glass transition between -120 and -90 °C, followed by an extremely short plateau between -90 and -50 °C. After -50 °C a continuous drop in the storage modulus is observed until about 100 °C, where the polymer fails. These results clearly indicate extensive mixing between PEO and urea groups, which was also observed in model blends (10,12-14). As a result of this mixing urea hard domains are plasticized and softened by PEO. As shown in Table 2, this material (49-E960-28-U-24) has a much lower Young's modulus than unmodified and PDMS modified silicone-urea copolymers.

Tensile properties

As can be seen from Table 2, silicone-urea copolymers or polyether modified silicone-urea copolymers with very high urea hard segment contents (up to almost 40% by weight) can be prepared in isopropanol (15). When tensile properties of silicone-urea copolymers, with hard segment contents ranging from 17 to 36% by weight are compared (the first four samples on Table 2), gradual increases in the Young's modulus and ultimate tensile strengths and a gradual decrease in the elongation at break values is observed with increasing HS content. Silicone-urea copolymer with HS content of 36% by weight has an ultimate tensile strength of 22.0 MPa, which is much higher than a crosslinked

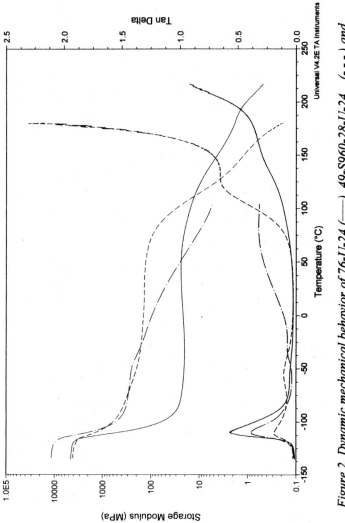

Figure 2. Dynamic mechanical behavior of 76-U-24 (——), 49-S960-28-U-24 (- - -) and 49-E960-28-U-24 (—·—·—)

and silica filled PDMS elastomer, which has a tensile strength of about 10-12 MPa (16). However, the elongation at break value is only 150%, much lower than that of a filled PDMS elastomer, which is generally above 500%. As can be seen both from Table 1 and Table 2, silicone-urea copolymers with chain extended urea hard segments behave more like toughened plastics than elastomers at room temperature. Based on the comparative tensile data on polyether-ureas (Table 1), chemical incorporation of polyether chains into the polymer backbone seemed to be a feasible approach to provide elastomer-like properties to silicone-urea copolymers. For this purpose silicone-urea copolymers were modified by the chemical incorporation of different amounts of amine terminated PPO (M_n 230, 450, 2000 g/mol) and PEO (Mn 630, 960, 2000 g/mol) oligomers into their backbone structures. As can be seen in Table 2, low molecular weight PPO based modifiers (P230 and P450) behaved almost like simple diamine chain extenders (e. g. Dytek A) and did not influence the tensile properties much, when compared to unmodified silicone-urea copolymers. As a result they did not provide any improvement in the elongation at break values. On the other hand when high molecular weight PPO (P2000) was used as the modifier, dramatic improvements in the elongation at break values were obtained, with no negative effect on the modulus or ultimate tensile strengths of the polymers, when compared with the unmodified silicone-urea copolymers. Elongation at break values increased with increasing PPO incorporation from about 200% for the unmodified silicone-urea to about 580% for a copolymer modified with 42% by weight of P2000. A similar behavior is also observed for PEO modified silicone-urea copolymers, regarding the tensile properties. Incorporation of 21% by weight of low molecular weight PEO oligomer (E630) improved the elongation at break value of a silicone urea copolymer with 26% by weight HS from less than 200% to 365%, without affecting the modulus or ultimate tensile strength values. Chemical incorporation of higher molecular weight PEO (E960) provided higher levels of improvement in the elongation at break values, up to 600%, which, as expected dependent on the amount of polyether incorporation. Again there was no dramatic change in the modulus and ultimate tensile values when compared with unmodified silicone-urea copolymers of similar urea HS content. On the other hand when 41% by weight PEO-2000 was used as the modifier in a copolymer with 27% HS, although the elongation at break value reached to 830%, modulus and tensile strength values were substantially lower than that of a similar copolymer modified with P2000 (32-P2000-41-U-27 versus 32-E2000-41-U-27). These results indicate that polyethers can successfully be used as modifiers in improving the tensile properties of silicone-urea copolymers. Best results are obtained by using oligomers with average molecular weights around 2000 g/mol. Based on the limited data provided on Table 2, PPO (P2000) modification seems to provide better overall improvement in tensile properties when compared with PEO (E2000).

When the stress-strain data on silicone-urea and polyether-urea copolymers provided in Table 1and the stress-strain results presented in Table 2 are compared, it is interesting to note the synergistic effect of the polyether modification on tensile properties. Polyether incorporation into the silicone-urea copolymers does not average out the tensile properties, but improves the inferior one (elongation at break) without any noticeable (negative) influence on the modulus or ultimate strength at break. As we reported in our earlier publication (7) and also discussed above in DMA analysis, we believe this is due the modification of the microphase morphology of the silicone-urea copolymer, especially the nature of the urea hard segments. As a result of the miscibility between urea and ether groups in PEO and PPO, we believe the *sharp interface* (transition) between PDMS matrix and the urea HS is modified and becomes a *diffuse* or a *gradient interface.*

Finally, an interesting comparison, regarding the influence of the structure of the modifier on the *nature of the interface* between the PDMS matrix and the urea domains and eventually on the tensile properties of the system, is provided by samples 49-E960-28-U-24 and 49-S960-28-U-24. These two materials have identical compositions except the chemical structure of the modifier. Both consist of 49% by weight PDMS-3200, 24% by weight urea hard segments (HMDI+DY) and 28% by weight modifier which has an $<M_n>$ value of 960 g/mol. In 49-E960-28-U-24, the modifier is PEO in 49-S960-28-U-24, the modifier is PDMS. Both of the samples have very similar ultimate tensile strengths of 11.5 and 11.8 MPa, however, substantially different modulus and elongation at break values. PDMS modified copolymer displays a very high modulus of 105 MPa and a fairly low elongation at break value of 205% indicating the formation of a stiff thermoplastic. The tensile strength and elongation of 49-S960-28-U-24 is very similar to that of the unmodified silicone-urea copolymer with 24% HS content (76-U-24, Table 1). Interestingly, a five-fold increase is observed in the modulus (22.3 versus 105 MPa) when it is modified with PDMS-960. On the other hand PEO modified copolymer displays a fairly low modulus of 9.3 MPa and elongation at break value of 600%, clearly indicating the formation of a soft, elastomeric material. We believe this is due to the change in the microphase morphology and the nature of the interface (or formation of a new interphase) between PDMS matrix and the urea domains in the system by the incorporation of PEO, since PEO is miscible with urea groups (10, 12, 13). As discussed before, strong support for this behavior in PEO modified material is provided by the dynamic mechanical analysis (DMA).

Water absorption

Another very interesting effect of the polyether modification on the physicochemical properties of silicone-urea copolymers is on the water absorption behavior. Equilibrium water uptake data for unmodified, PPO and PEO modified silicone-urea copolymers is provided in Table 2.

Table 2. Equilibrium water uptake (50 days at 23 °C)

Composition	PDMS (wt %)	Polyether (wt %)	HS (wt %)	Water uptake (wt %)
70-U-30	69.6	--	30.4	0.64
42-P2000-26-U-32	42.0	26.2	31.8	1.18
34-P2000-42-U-24	33.7	42.1	24.2	1.69
49-E960-14-U-37	49.1	13.8	37.1	10.8
49-E960-28-U-24	48.9	27.5	23.6	29.3
51-E2000-29-U-20	51.2	29.2	19.6	61.5
33-E2000-40-U-27	32.6	40.4	27.0	75.1

Due to very hydrophobic nature of PDMS, unmodified silicone-urea copolymer with 70% PDMS and 30% urea hards segments absorbs only 0.64% by weight of water at equilibrium, which is expected. When PPO is used as the modifier, since it is also hydrophobic in nature, a very small increase in the equilibrium water absorption is observed. Silicone-urea copolymer modified with 42% by weight of PPO-2000 shows an equilibrium water absorption value of only 1.69% by weight. On the other hand when hydrophilic PEO is used as the modifier, as expected, a dramatic increase in the level of equilibrium water absorption is achieved. As can be seen in Table 2, the level of water absorption is dependent both on the average molecular weight of the PEO modifier and also on the amount incorporated into the system. Silicone-urea copolymer modified with 13.8 and 27.5% by weight of PEO-960 shows equilibrium water absorption values of 10.8 and 29.3% by weight respectively, much higher than the unmodified or PPO modified materials. When higher molecular weight PEO (E2000) is used as the modifier, the level of water absorption gets even higher. As can be seen in Table 2, at similar levels of modifier incorporation, higher molecular weight PEO containing silicone-urea copolymer shows twice as much higher level of water absorption. Sample 51-E2000-29-U-20, which contains 29% by weight PEO-2000 shows 61.5% by weight of water absorption, while the sample 49-E960-28-U-24, containing 29% by weight of PEO-960, absorbs only 29.3% by weight of water. We believe the difference is related to the structure of the microphase morphology of the copolymer upon polyether modification. Our previous studies by DMA and SAXS on PPO modified silicone-urea copolymers (7) indicate the formation of a well-defined polyether phase in the system, with PPO-2000 incorporation. On the other when lower molecular weight polyether modifiers (e. g. P400) are used, they do not form a separate phase. Instead they mix with the urea hard segments and plasticize them (7, 10).

Refractive indices

One of the interesting applications of silicone polymers is their use in contact lenses and intraocular lenses (IOL). IOLs are surgically implanted into the human eye to replace the damaged or diseased natural lense of the eye. For optimum performance in IOL application a unique set of optical and mechanical properties, such as; high optical clarity, a refractive index higher than 1.44, a low durometer hardness (Shore 30-35A), reasonable tear (~ 20 pli) and tensile strength (~ 3 MPa) and good flexibility (17). At the present time crosslinked and silica filled dimethyl-diphenyl silicone copolymers are used for IOL applications (18-20). Thermoplastic silicone copolymers with tailor designed properties may provide added benefits in IOL applications, such as improved processability, high tensile strengths to enable production of thin IOLs and better control of overall polymer properties and performance. With this in mind, we investigated the refractive indices of thin copolymer films as a function of copolymer composition. These results are provided in Table 3, together with the refractive indices of amine terminated PDMS oligomers as reference (21). Cyclic D_4 has a refractive index value of 1.3964. As for the amine terminated PDMS oligomers, refractive index decreases as the molecular weight increases. PDMS-960 has a refractive index value of 1.4218, compared to 1.4095 and 1.4050 for PDMS-3200 and PDMS-10000. Silicone-urea copolymers display considerably higher refractive index values as a function of their urea hard segment contents. As shown in Table 3, silicone-urea copolymers with hard segment contents of 17, 20 and 24 weight % have refractive index values of 1.4335, 1.4406 and 1.4880 respectively. Incorporation of polyethers (PEO or PPO) increases the refractive index values of these copolymers further. Depending on the type and amount of the polyether, it is possible to obtain silicone-urea copolymers with refractive indices up to about 1.46. As shown for sample 49-S960-28-U-24, both the air side (A) or glass side (G) of the films display very similar refractive index values.

Conclusions

Silicone-urea copolymers are modified through chemical incorporation of PEO and PPO oligomers into their backbones. Microphase morphologies of the resultant polymers strongly depended on the type (PEO versus PPO) amount and the average molecular weight of the modifier. Low molecular weight modifiers (e. g. $<M_n>$ in 200-450 g/mol range) behave as chain extenders. Medium molecular weight PEO modifiers (e. g. PEO-630 and PEO-960) are miscible with urea hard segments, and therefore act as plasticizers for the hard segments. High molecular weight oligomers PEO-2000 and PPO-2000 form separate polyether phases. Physical properties of modified silicone-urea copolymers, such

Table 3. Refractive indices of PDMS oligomers, silicone-urea copolymers and silicone-polyether-urea terpolymers

Polymer	PDMS (%)	Polyether (Mn)	Polyether (%)	HS (%)	RI
D_4	100				1.3964
PDMS-960	100				1.4218
PDMS-2500	100				1.4105
PDMS-3200	100				1.4095
PDMS-7000	100				1.4073
PDMS-10000	100				1.4050
83-U-17	83.3	--	--	16.7	1.4335
76-U-24	75.8	--	--	24.2	1.4406
30-U-70	69.6	--	--	30.4	1.4480
49-S960-28-U-24	76.5	--	--	23.5(A)	1.4490
49-S960-28-U-24	76.5	--	--	23.5(G)	1.4483
57-D460-16-U-27	56.7	450	16.0	27.3	1.4550
53-E650-21-U-26	53.3	630	21.0	25.7	1.4573
49-E900-U-28	48.9	900	27.5	23.6	1.4589
34-D2000-42-U-24	33.7	2000	42.1	24.2	1.4608

as tensile strengths, hydrophilic/hydrophobic balance, water absorption and refractive indices show strong dependence on the type, molecular weight and amount of the modifier used.

References

1. Noshay, A.; McGrath J.E. *Block Copolymers. Overview and Critical Survey*, Academic Press, NY (1977).
2. McGrath, J.E. (Editor) *Anionic Polymerization. Kinetics, Mechanisms and Synthesis*, ACS Symp. Ser. 166, Washington D.C. (1981)
3. Matyjaszewski, K. *Prog. Polym. Sci.,* **2005**, *30*, 858-875.
4. Alper, J. *Science* **2002**, *297*, 329.
5. Yilgor, I.; Sha'aban, A.K.; Steckle, Jr., W.P.; Tyagi, D. Wilkes, G.L.; McGrath, J.E. *Polymer,* **1984**, *25*, 1800-1806.
6. Yilgor, E.; Yilgor, I. *Polymer,* **2001**, *42*, 7953-7959.
7. Sheth, J.P.; Aneja, A.; Wilkes, G.L.; Yilgor, E.; Atilla, G.E.; Yilgor, I.; Beyer, F.L. *Polymer* **2004**, *45*, 6919-6932.
8. Van Krevelen, D.V. *Properties of Polymers,* Elsevier, Amsterdam, Netherlands (1990), Ch. 7.
9. Tyagi, D.; Yilgor, I.; McGrath, J.E.; Wilkes, G.L. *Polymer,* **1984**, *25*, 1807-1816.

10. Yilgor, E; Yurtsever, E.; Yilgor, I. *Polymer,* **2002**, *43(24)*, 6561-6568.
11. Sheth, J.P.; Yilgor, E.; Erenturk, B.; Ozhalici, H.; Yilgor, I.; Wilkes, G.L. *Polymer*, **2005**, *46(19)*, 8185-8193.
12. Coleman, M.M.; Painter, P.C. *Prog. Polym. Sci.,* **1995**, *20*, 1-59
13. Coleman, M.M.; Skrovanek, D.J. Hu, J.; Painter, P.C. *Macromolecules,* **1988**, 21(1):59-66.
14. Yilgor, E.; Yilgor, I.; Yurtsever, E. *Polymer*, **2002**, *43(24)*, 6551-6560.
15. Yilgor, E.; Atilla, G.E.; Ekin, A.; Kurt, P.; Yilgor I. *Polymer*, **2003**, *44(26)*, 7787-7793.
16. Butts, M.; Cella, J.; Word, C.D.; Gillette, G.; Kerboua, R.; Leman, J.; Lewis, L.; Rubinsztajn, S.; Schattenmann, F.; Stein, J.; Wicht, D.; Rajaraman, S.; Wengrovius, J. *Silicones*. In Kirk-Othmer Encycl. of Chem. Tech. DOI: 10.1002/0471238961.1909120918090308.a01.pub2
17. Christ R, Nash B. A. and Petraitis D. J. *US Patent* 5,236,970 (1993)
18. Travnicek, E.A. *US Patent* 3,996,187
19. Travnicek, E.A. *US Patent* 3,996,189
20. Koziol, J.E. *US Patent* 4,615,702
21. Ingebrigtson, D. N; Klimish, H.M; Smith R.C., in *"Analysis of Silicones"*, Ed. Smith A. L., Wiley Interscience, New York, 1974, Ch. 6.

Chapter 9

Application of Poly(styrene-*b*-methylphenylsiloxane) Diblock Copolymers for the Steric Stabilization of Organic Polymerizations in Carbon Dioxide

Abhijit V. Jadhav[1], Siddharth V. Patwardhan[1], Hyeon Woo Ahn[1], Craig E. Selby[1], James O. Stuart[1], Xiao Kang Zhang[1], Chung Mien Kuo[1], Ephraim A. Sheerin[1], Richard A. Vaia[2], Steven D. Smith[3], and Stephen J. Clarson[1,*]

[1]Department of Chemical and Materials Engineering, University of Cincinnati, Cincinnati, OH 45221–0012
[2]Materials and Manufacturing Directorate, AFRL, Wright-Patterson Air Force Base, OH 45433
[3]Proctor & Gamble, Miami Valley Innovation Center, 11810 East Miami River Road, Ross, OH 43061
*Corresponding author: Stephen.Clarson@UC.edu

The ability to isolate the individual stereochemical isomers of the cyclic methylphenylsiloxanes provides one with novel siloxane monomers for the preparation of block copolymers. For the case of the kinetically controlled anionic ring-opening polymerization of the cyclic trimers, one is able to synthesize poly(styrene-b-methylphenylsiloxane) (PS-b-PMPS) diblock copolymers of well defined composition and structure. Herein we describe an investigation of the application of PS-b-PMPS as a surfactant for the dispersion polymerization of styrene in supercritical carbon dioxide.

Introduction

The seventieth anniversary of the commercialization of silicones will soon be upon us and, fortunately, the number of applications and the quantity of sales of silicones continues to grow.[1-4] The workhorse of the silicones industry remains poly(dimethylsiloxane) (PDMS) $-[(CH_3)_2SiO]-$. There are, however, a number of products that are phenyl-containing silicone materials and these are based upon the methyphenylsiloxane repeat unit $-[(CH_3)(C_6H_5)SiO]-$, the diphenylsiloxane repeat unit $-[(C_6H_5)_2SiO]-$, or a variety of copolymers containing either of these repeat units. These phenyl-containing silicones or polysiloxanes find applications as fluids, elastomers and resins.

The investigations of phenyl-containing silicones in our laboratories that we have reported to date have been focused on linear, cyclic and network materials based upon the methyphenylsiloxane repeat unit $-[(CH_3)(C_6H_5)SiO]-$. We have described the synthesis of PMPS[5-8] and various physical characteristics of PMPS including solution[9-11], bulk[12-19], and surface properties.[20,21]

The conventional ring opening polymerization to prepare linear PMPS usually proceeds to equilibrium and hence the resulting linear polymer is typically stereochemically atactic[9-12] (see Figure 1) and has a most probable molar mass distribution ($M_w/M_n = 2$). Furthermore, an equilibrium distribution of cyclic MPS species is found and, following isolation, some of the properties of these cyclic species have been characterized and compared with the corresponding linear materials.[5,9,11,12]

The ability to isolate the individual stereochemical isomers of the cyclic methylphenylsiloxanes provides novel monomers for the preparation of poly(methylphenylsiloxane) with control of the stereochemical structure of the polymer. This is achieved by carrying out ring-opening polymerizations with kinetic control.[7,8] Furthermore, the living nature of such polymerizations allows one to prepare linear PMPS with molar mass distributions less than 1.1.

Living ring-opening polymerizations under kinetic control are also useful for the preparation of poly(methylphenylsiloxane)-polystyrene diblock copolymers. The addition of small amount of organic-siloxane block copolymer to the parent polymer can lead to improved fire resistance, dielectric properties, toughness, crack resistance and surface properties.[22-24]

Liquid and supercritical carbon dioxide (SCCO$_2$) have been utilized as a solvent for a wide range of chemical reactions and processing technologies. These include organic synthesis systems[25], enzymatic catalyzed systems[26,27] and polymerization reactions[28,29,30]. Of the commercial polymeric materials only fluoropolymers and polysiloxanes have been shown to exhibit appreciable solubility in SCCO$_2$. Due to this limited solubility of polymer systems in SCCO$_2$, polymerization reactions are typically carried out in dispersions or emulsions that are stabilized by polymeric surfactants. A variety of polymerization reactions have been carried out using diblock copolymer

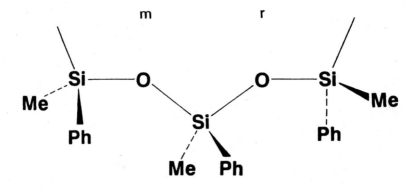

Figure 1: Configuration of atactic PMPS showing meso(m) and racemic(r) diads

surfactants based on poly(dimethylsiloxane).[28,30,31] For free radical styrene polymerizations carried out in the presence of poly(dimethylsiloxane) homopolymer the poor results obtained can be explained by the inefficient anchoring of the homopolymer onto the surfaces of the polystyrene particles.

The fractionation of various polysiloxane and siloxane-containing copolymers using supercritical fluids has been described by McHugh and Krukonis.[32] Although poly(dimethylsiloxane) (PDMS) based siloxanes[22,30] have been widely studied, there are very few studies of phenyl-containing siloxanes in supercritical fluids to date. One such study is where the use of carbon dioxide in a reaction/separation scheme has been described for the poly(dimethylsiloxane-co-diphenylsiloxane) system. As to the poly(methylphenylsiloxane) (PMPS) system, Barry, Ferioli and Hubball have reported extraction studies of linear PMPS with carbon dioxide at 40°C and at a pressure of 103 bar (1,500 psia).[33] McHugh and Krukonis[32] have studied linear PMPS of modest molar mass (M_n = 2,130 g mol^{-1}) in carbon dioxide, ethylene, propane and propylene.[34] In the case of supercritical ethylene the parent linear PMPS was successfully separated into a series if eight fractions ranging in molar mass from 1,510 g mol^{-1} up to 5,300 g mol^{-1}.

Herein we describe an investigation of the application of PS-b-PMPS as a surfactant for the dispersion polymerization of styrene in supercritical carbon dioxide. As far as we are aware, the work described below is the first study of PS-PMPS block copolymers in supercritical carbon dioxide.

Experimental

Materials:

The styrene monomer ($C_6H_5CH=CH_2$, 99+% from Aldrich, inhibited with 10-15 ppm 4-tert-butylcatechol) was purified prior to use by the removal of the inhibitor. This was achieved by passing it through a column of aluminium oxide. The column, filled with the alumina, was connected to the middle neck of a three-necked round bottom flask. One of the necks was sealed with a rubber septum and the third neck was equipped with a rubber septum to which a syringe was connected through which argon was passed to remove the oxygen from the styrene. The initiator 2,2-azobisisobutyronitrile (AIBN, Aldrich) was used as received. Carbon dioxide (SFC/SFE Grade) with or without a helium head pressure was purchased from Air Products and was used as received.

The MPS trimers were prepared by reacting methylphenyldichlorosilane (Aldrich) with zinc oxide, as we have described elsewhere.[7,8] Isolation of the MPS trimers was achieved via a combination of vacuum distillation, selective seeding using pure isomer crystals and fractional crystallization methods.[7,8]

Poly(styrene-b-methylphenylsiloxane) (PS-b-PMPS]) were synthesized in our laboratory as shown below by living anionic polymerization[22,23] carried out on a custom built vacuum line.

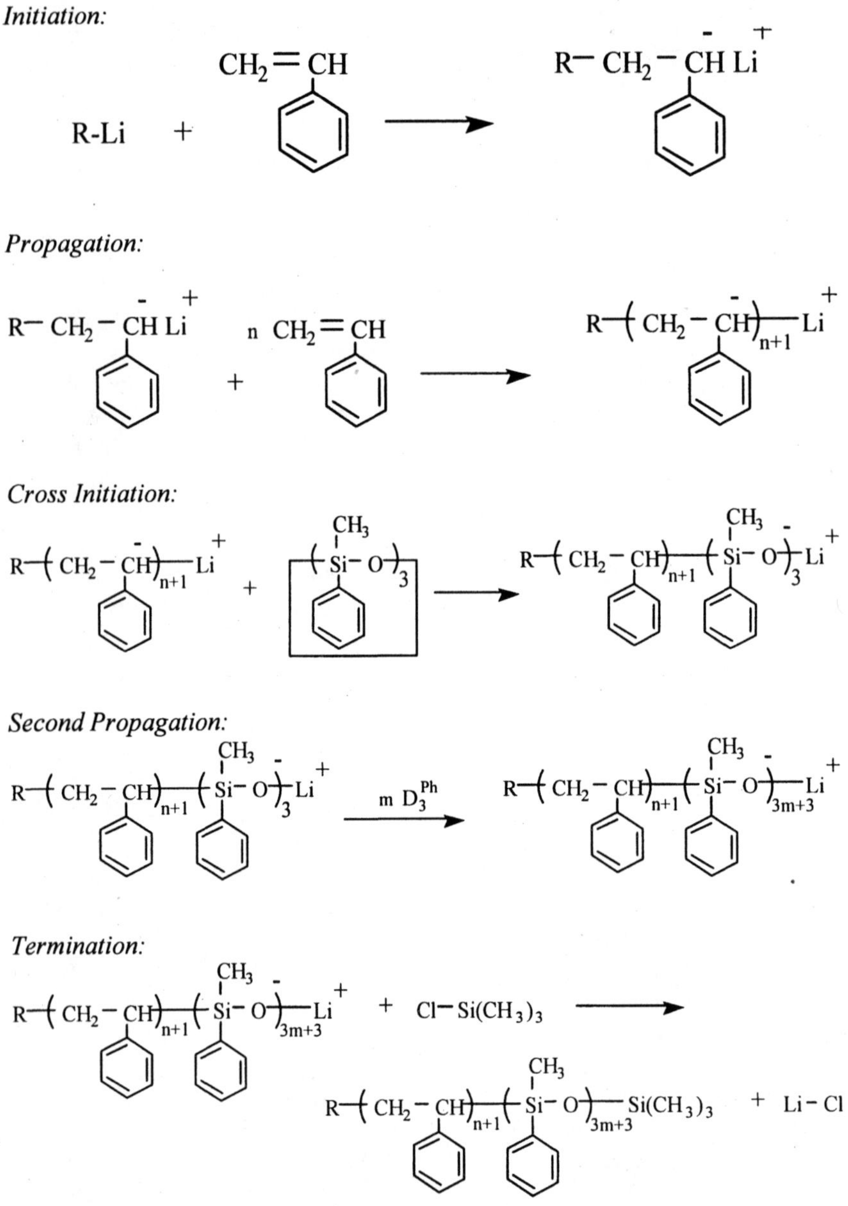

120

Initiation:

$$CH_2{=}CH \cdot C_6H_5 + R\text{-Li} \longrightarrow R{-}CH_2{-}\overset{-}{C}H\;\overset{+}{Li}\cdot C_6H_5$$

R-Li   +   CH$_2$=CH(Ph) $\longrightarrow$ R$-$CH$_2-$$\overset{-}{C}$H $\overset{+}{Li}$(Ph)

Propagation:

R$-$CH$_2-$$\overset{-}{C}$H $\overset{+}{Li}$(Ph)   +   n CH$_2$=CH(Ph) $\longrightarrow$ R$\left(\text{CH}_2-\text{CH(Ph)}\right)_{n+1}\overset{-}{}\overset{+}{Li}$

Cross Initiation:

R$\left(\text{CH}_2-\overset{-}{\text{C}}\text{H(Ph)}\right)_{n+1}\overset{+}{Li}$ + $\left(\overset{CH_3}{\underset{Ph}{Si}}-O\right)_3$ $\longrightarrow$ R$\left(\text{CH}_2-\text{CH(Ph)}\right)_{n+1}\left(\overset{CH_3}{\underset{Ph}{Si}}-O\right)_3\overset{-}{}\overset{+}{Li}$

Second Propagation:

R$\left(\text{CH}_2-\text{CH(Ph)}\right)_{n+1}\left(\overset{CH_3}{\underset{Ph}{Si}}-O\right)_3\overset{-}{}\overset{+}{Li}$ $\xrightarrow{\; m\, D_3^{Ph}\;}$ R$\left(\text{CH}_2-\text{CH(Ph)}\right)_{n+1}\left(\overset{CH_3}{\underset{Ph}{Si}}-O\right)_{3m+3}\overset{-}{}\overset{+}{Li}$

Termination:

R$\left(\text{CH}_2-\text{CH(Ph)}\right)_{n+1}\left(\overset{CH_3}{\underset{Ph}{Si}}-O\right)_{3m+3}\overset{-}{}\overset{+}{Li}$ +   Cl$-$Si(CH$_3$)$_3$ $\longrightarrow$

R$\left(\text{CH}_2-\text{CH(Ph)}\right)_{n+1}\left(\overset{CH_3}{\underset{Ph}{Si}}-O\right)_{3m+3}$Si(CH$_3$)$_3$ +   Li$-$Cl

Figure 2: Living anionic polymerization scheme for the preparation of the PMPS-b-PS diblock copolymers

The copolymer (CP) molar masses and compositions as determined by Gel Permeation Chromatography[19, 21] and ^{1}H and ^{13}C NMR are listed in Table 1.

An experimental apparatus for carrying out polymerization reactions in supercritical CO_2 was designed and built in our laboratory and is shown in Figure 3. The polymerizations were conducted in a reactor obtained from High Pressure Equipment Company (Model 62-6-10). The maximum capacity of this reactor was 45 ml. The temperature and pressure recorders were purchased from Omega (models DP460 and DP24-E, respectively) and were each calibrated before use.

Table 1 : Characterization of the PS-b-PMPS Block Copolymers.

Sample	Structure	$M_n / g \, mol^{-1}$	M_w / M_n	S / PS
CP1	s-Bu-PS-PMPS-OSi(CH₃)₃	3,300	1.18	1.7 / 1.6
CP2	s-Bu-PS-PMPS-OSi(CH₃)₃	9,400	1.10	3.5 / 5.9
CP3	s-Bu-PS-PMPS-OSi(CH₃)₃	12,700	1.06	4.1 / 8.6
CP4	s-Bu-PS-PMPS-OSi(CH₃)₃	15,700	1.05	5.4 / 10.3

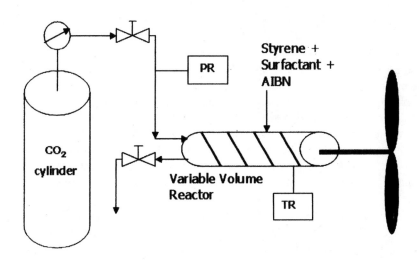

PR=Pressure Recorder, TR=Temperature Recorder

Figure 3: Illustration of the Experimental Equipment for the Dispersion Polymerization Reactions in SCCO₂.

Dispersion Polymerization Reactions Of Styrene In Supercritical CO_2 Using Ps-B-Pmps:

0.04 g of 2,2-azobisisobutyronitrile initiator was weighed and then added to 0.2 g PS-b-PMPS diblock copolymer (surfactant) and both of these were then charged to the reactor. 2 ml of purified styrene (monomer) was added using a syringe. The pressure input to the cylinder was 1200 psi. Later on it was increased to 2100 psi by reducing the volume of the reactor. It was further increased to 5000 psi by increasing the temperature to 60 °C. All the valves were then closed and the reaction mixture was kept for 24 hours at the above temperature and pressure and the reaction was allowed to proceed. After 24 hours the reactor was then depressurized by slowly venting the CO_2 and the vented CO_2 was bubbled through methanol in order to trap any polystyrene particles which sprayed out during the venting process. The reactor was then cooled in a refrigerator and the solid polymer product was removed from the reactor. The resulting polystyrene powder was then characterized as described below. Typical yields of polystyrene were around 60% based on styrene monomer.

Characterization:

Molar Mass Determination by Gel Permeation Chromatography

A Waters analytical Gel Permeation Chromatograph (GPC) with four Ultra-Styragel columns (500, 1000, 10,000 and 100,000 Å nominal porosities) was employed for molar mass characterization. A Waters 410 differential refractometer (RI detector) and a Viscotec Dual Detector (Light Scattering and Viscometer) was used along with Viscotec's Omnisec software to calculate absolute molecular weights and the molecular weight distributions of the polystyrene products from each reaction. Tetrahydrofuran (THF) was used as the mobile phase with a flow rate of 1.0 mL/min and the instrument was calibrated using PS standards.

Spectroscopic Analysis by NMR and FTIR:

^1H NMR spectra of the monomer, polymer and diblock copolymer surfactants were obtained using a Bruker WM250 spectrometer. The solvent used was deuterated chloroform.

A Perkin Elmer FTIR Spectrometer (Range 4000–400 cm^{-1}) was used to confirm the formation of polystyrene by solution Fourier Transform Infrared Spectroscopy in chloroform.

Structure and Morphology of the Dispersion Polymerized Styrene Particles by ESEM

The size, shape and morphologies of the resulting polymer particles were determined using a JEOL 6400 FE Environmental Scanning Electron Microscope (ESEM) in the Department of Chemistry, University of Cincinnati.

Results And Discussion

Spectroscopic Analysis

The ^{1}H NMR spectrum and the FTIR spectrum of dispersion polymerized polystyrene were different from that of the starting compound (styrene monomer). This can be seen from Figures 4, 6, 7, 9. The spectral characteristics of the polystyrene showed complete disappearance of the olefinic signal due to the H-8 proton at δ 5.2 ppm as a doublet, the H-7 proton at δ 5.7 ppm as doublet and the H-6 proton at 6.3 ppm as a quartet in styrene (see Figure 6). Instead, we observed a presence of two new signals at δ 1.8 ppm and δ 1.4 ppm as a broad singlet confirming that the olefinic moiety has undergone chemical modification during the course of the reaction. The proton at δ 1.8 ppm (broad singlet) can be assigned to the H-6 proton and δ 1.4 ppm (broad singlet) can be assigned to the H-7 and the H-8 proton of the product showing that styrene moiety has converted to polystyrene (see Figure 6). The product was confirmed by comparing the spectra of the polystyrene product with a polystyrene standard.

No significant amonut of the surfactant was detected. Utilizing Fourier Transform Infrared (FTIR) spectroscopy, for example, note there was no siloxane peak at 1083 cm^{-1}. This band was observed in the spectrum of the surfactant (PS-b-PMPS) (see Figure 8). Furthermore there was an absence of a triplet signal of a characteristic siloxane proton peak at around δ 0.09 ppm and the phenyl peaks around δ 7.26 ppm, 7.33 ppm and 7.39 ppm (see Figure 5), thus confirming the purity of the resulting polymer. In terms of side reactions, one possibility would be free radical chain transfer to the siloxane by hydrogen abstraction. Pelton, Osterroth and Brook[35] have reported that the use of AIBN does not lead to chain transfer to PDMS and hence we can reasonably neglect the possibility of chemical grafting to the PMPS in this study.

124

STY.001

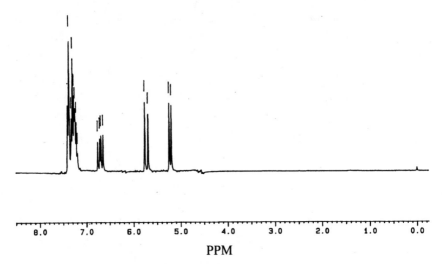

PPM

Figure 4: ^1H NMR spectrum of the styrene monomer

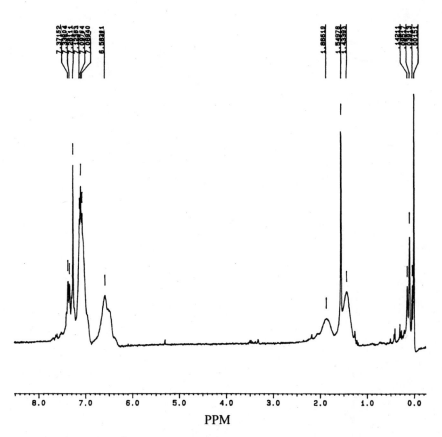

Figure 5: ^1H NMR spectrum of the PS-b-PMPS surfactant

126

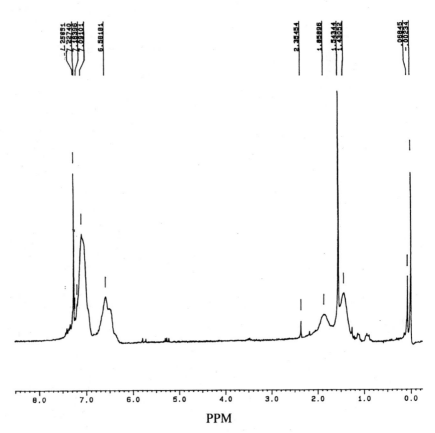

PPM

Figure 6: ¹H NMR spectrum of polystyrene synthesized in SCCO₂ with PS-b-PMPS as the surfactant

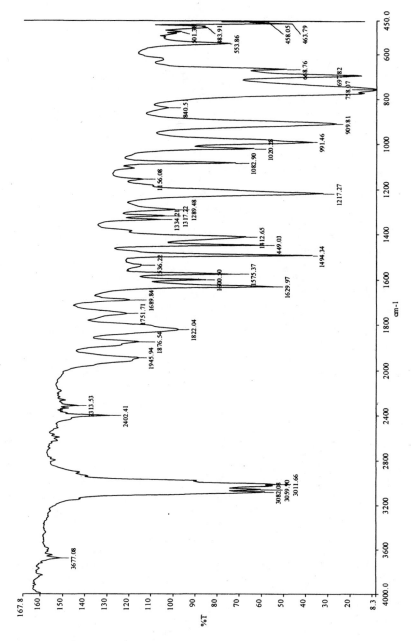

Figure 7: IR spectrum of the styrene monomer

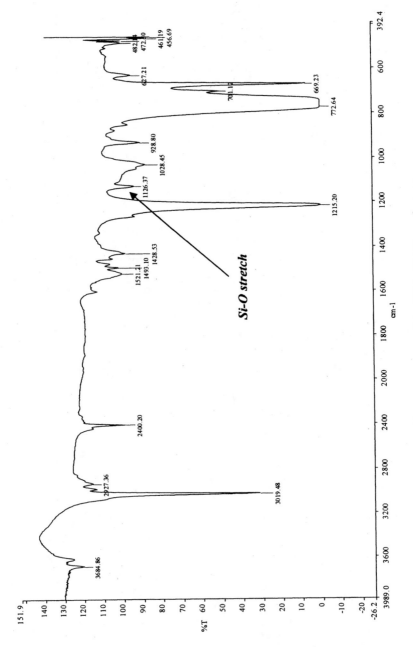

Figure 8: IR spectrum of the PS-b-PMPS surfactant

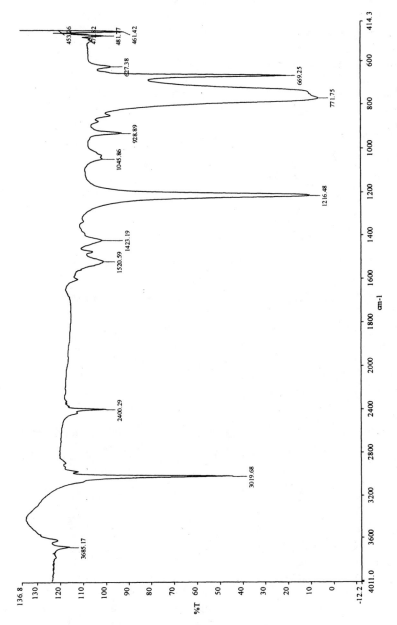

Figure 9: IR spectrum of polystyrene synthesized in SCCO₂ with PS-b-PMPS as the surfactant

130

Gel Permeation Chromatography

GPC analysis of the polystyrene samples produced by the PS-b-PMPS surfactant system were found to have number average molar masses in the range 5,000-30,000 g/mol. This is illustrated by the polystyrene sample shown in Figure 10, where a number-average molar mass $M_n \sim$ 29,100 g/mol and a polydispersity index $M_w/M_n \sim$ 2.04 was obtained. The three detectors used were a Light Scattering detector (LS), Refractive Index detector (RI) and an Intrinsic Viscosity detector (IV).

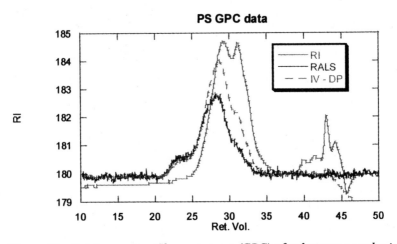

Figure 10: Gel Permeation Chromatogram (GPC) of polystyrene synthesized in SCCO$_2$ with PS-b-PMPS as the surfactant.

Table 2: GPC characterization of polystyrene samples synthesized in SCCO$_2$ with PS-b-PMPS as the surfactant.

Sample	M_n / g mol^{-1}	M_w / g mol^{-1}	M_w/M_n
PS1	6,750	16,150	2.39
PS2	29,100	59,200	2.04

Scanning Electron Microscopy

Figure 11 shows polystyrene particles produced in supercritical CO$_2$ using the methods described above. The polystyrene particles were in the range 0.5-5μm. Smaller and well defined polystyrene particles have been reported in the

131

Figure 11: ESEM of polystyrene produced by dispersion polymerization of styrene in SCCO₂ with PS-b-PMPS as the surfactant.

case of $SCCO_2$ using polystyrene-b-polydimethylsiloxane (PS-b-PDMS) block copolymers.[28,30,36]

It is clear that if we can transform this system from a dispersion polymerization to a 'classical' emulsion system[37] then we will be able to produce smaller and better defined PS particles and furthermore one can anticipate that the molar mass of the resulting PS will be higher.

Conclusions

Cyclic methylphenylsiloxane trimers when isolated with high purity undergo living anionic ring-opening polymerization to give PS-b-PMPS diblock copolymers of well defined structure ($M_w/M_n < 1.1$). Here we have shown that PS-b-PMPS is useful as a surfactant for the supercritical CO_2 polymerization of styrene. Our future work will focus on optimizing the process conditions for the PS-b-PMPS / CO_2 system.

Acknowledgements

The authors thank Professor J. M. DeSimone and his group members at the University of North Carolina at Chapel Hill for kindly hosting a research visit for SVP and for advice on polymerizations in $SCCO_2$.

References

1. *Silicones Under The Monogram*, H. A. Liebhafsky, Wiley-Interscience, New York, **1978**.
2. *Forty years of Firsts*, E. L. Warrick, McGraw Hill, New York, **1990**.
3. *Siloxane Polymers*, S. J. Clarson and J. A. Semlyen, Eds., Prentice Hall, Englewood Cliffs, New Jersey, **1993**.
4. *Silicon in Organic, Organometallic, and Polymer Chemistry*, M. A. Brook, Wiley-Interscience, New York, **1999**.
5. Clarson; S. J.; Semlyen; J. A. *Polymer* **1986**, *27*, 1633-1636.
6. Van Dyke, M. E.; Clarson, S. J. *J. Inorganic and Organometallic Polymers* **1998**, *8(2)*, 111-117.
7. Ahn, H. W.; Clarson, S. J. *Journal of Inorganic and Organometallic Polymers* **2001**, *11(4)*, 203-216.
8. Ahn, H. W.; Clarson, S. J. in *Synthesis and Properties of Silicones and Silicone-Modified Materials*, S. J. Clarson, J. J. Fitzgerald, M. J. Owen, S. D. Smith and M. E. Van Dyke, Eds. ACS Symposium Series, Vol. 838, Oxford University Press, New York, **2003**, pp. 40-49.
9. Clarson, S. J; Dodgson, K.; Semlyen, J. A. *Polymer* **1987**, *28*, 189-192.
10. Lima, C. J. C.; Macanita, A. L.; Dias, F. B.; Clarson, S. J.; Horta, A.; Pierola, I. F. *Macromolecules* **2000**, *33(13)*, 4772-4779.
11. Clarson, S. J., in *Polymer Data Handbook*, Oxford University Press, New York, **1999**, pp. 732-733.
12. Clarson, S. J.; Semlyen, J. A.; Dodgson, K. *Polymer* **1991**, *32*, 2823-2827.
13. Kuo, C. M.; Clarson, S. J. *Macromolecules* **1992**, *25*, 2192-2195.
14. Kuo, C. M.; Clarson, S. J; Semlyen, J.A. *Polymer* **1994**, *35*, 4623-4626.
15. Kuo, C. M.; Clarson, S. J. *European Polymer Journal* **1993**, *29*, 661-664.
16. Kuo, C. M.; Clarson, S. J. *Polymer* **2000**, *41*, 5993-6002.
17. Clarson, S. J.; Mark, J. E. *Polymer Communications* **1987**, *28*, 249-252.
18. Clarson, S. J.; Mark, J. E. *Polymer Communications* **1989**, *30*, 275-277.
19. Goodwin, A. A.; Beevers, M. S.; Clarson, S. J.; Semlyen, J. A. *Polymer* **1996**, *37(13)*, 2597-2602.
20. Kuzmenka, D. J.; Granick, S.; Clarson, S. J.; Semlyen, J. A. *Langmuir* **1989**, *5*, 144-147.
21. Kuzmenka, D. J.; Granick, S.; Clarson, S. J.; Semlyen, J. A. *Macromolecules* **1989**, *22*, 1878-1881.
22. Selby, C. E.; Stuart, J. O.; Clarson, S. J.; Smith, S. D.; Sabata, A.; van Ooij, W. J.; Cave, N. G. *J. Inorganic and Organometallic Polymers* **1994**, *4(1)*, 85-93.
23. Clarson, S. J.; Selby, C. E.; Stuart, J. O.; Sabata, A.; Smith, S. D.; Ashraf, A. *Macromolecules* **1995**, *28*, 674-677.
24. Ahn, H. W.; Jadhav, A. V.; Selby, C. E.; Stuart, J. O; Zhang, X. K.; Kuo, C. M.; Clarson, S . J. *Polymer Preprints* **2004**, *45(1)*, 595.

25. Jacobson, G. B.; Lee, C. T., Jr.; da Rocha, S. R. P.; Johnston, K. P. *J. Org. Chem.* **1999**, *64(4)*, 1207-1210.
26. Johnston, K. P.; Harrison, K. L.; Clarke, M. J.; Howdle, S. M.; Heitz, M. P.; Bright, F. V.; Carlier, C.; Randolph, T. W. *Science* **1996**, *271*, 624-626.
27. Holmes, J. D.; Steytler, D. C.; Rees, G. D.; Robinson, B. H. *Langmuir* **1998**, *14(22)*, 6371-6376.
28. Kendall, J. L.; Canelas, D. A.; Young, J. L.; DeSimone, J. M. *Chem. Rev.* **1999**, *99*, 543-563.
29. Cooper, A. I. *J. Mater. Chem.* **2000**, *10*, 207-234.
30. Folk, S. L.; DeSimone, J. M. In *Synthesis and Properties of Silicones and Silicone-Modified Materials*, S. J. Clarson, J. J. Fitzgerald, M. J. Owen, S. D. Smith, M. E. Van Dyke, Eds., ACS Symposium Series, Vol. 838, Oxford University Press, New York, **2003**, pp. 79-93.
31. Canelas D.A.; Betts, D. E.; Desimone, J. M. *Macromolecules* **1996**, 29, 2818-2821.
32. *Supercritical Fluid Extraction*, M. McHugh and V. Krukonis, Butterworth-Heinemann, Boston, **1994**, pp. 217-250.
33. Barry, E. F. B.; Ferioli, P.; Hubball, J. A. *J. High Resolution Chromatogr. Chromatogr. Commun.* **1983**, *6*, 172-177.
34. *Supercritical Fluid Extraction*, M. McHugh and V. Krukonis, Butterworth-Heinemann, Boston, **1994**, pp. 264-265.
35. Pelton, R. H.; Osterroth, A.; Brook, M. A. *J. Colloid Interface Sci.* **1990**, *137*, 120-127.
36. Canelas, D.A.; DeSimone, J. M. *Macromolecules* **1997**, *30*, 5673-5682.
37. *Physical Chemistry of Surfaces*, 5th Edition, A. W. Adamson, Wiley-Interscience, New York, **1990**, pp. 525-559.

Networks

Chapter 10

Large-Scale Structures in Some End-Linked Polysiloxane Networks: A Critical Review

Brent D. Viers and James E. Mark*

Department of Chemistry and the Polymer Research Center,
The University of Cincinnati, Cincinnati, OH 45221–0172

A variety of model end-linked poly(dimethylsiloxane) (PDMS) networks with a turbid, phase separated micro-structure are compared and contrasted. A critical review of the various PDMS end-linking techniques is presented to understand the phenomenology of the structure development. The necessary and sufficient requirements for developing such textures appear to be (i) large concentrations of short PDMS chains, and (ii) large amounts of the catalyst used in the end-linking or cross-linking reactions. Scattering and microscopic techniques demonstrate that the observed structures are likely high cross-link density clustered phases. The phase separation is not the result of side reactions or system-specific artifacts, and thus appears to be a general phenomenon in such network-forming systems. The formation of the phases does not appear to be directly linked with the speed of the reactions, which suggests that the phases are not the result of a precipitation, or a semi-stable colloidal structure, etc. Interesting phase patterns can be developed, including an interpenetrating spinodal-like structure and spheres trapped within the network mesh. These structures bear a close similarity to structures expected from classical thermodynamic phase separation mechanisms (spinodal decomposition and binodal nucleation and growth); furthermore, these structures persist even after attempts at dispersal/redissolution. These results suggest that the observed structures are the unique result of a reaction induced phase separation which is in turn mediated by and eventually trapped by the extensive cross linking. Hydrosilylation cross linking favors the interpenetrating spinodal texture, while condensation cross linking favors dispersed spheres. Bimodal formulations do not appear to affect the observed hydrosilylation cross-linked structure, whereas bimodal and/or hybrid reactions tend to favor smaller/polydisperse spheres for condensation cross linking.

Introduction

This volume focuses on the unique properties of silicones, mainly poly(dimethyl siloxane) (PDMS). PDMS is thought to be biologically inert and biocompatible, with exceptional thermo-oxidative stability, atomic oxygen resistance, and ultraviolet radiation resistance. PDMS surfaces have a low surface energy and are hydrophobic. PDMS has a high permeability to small molecules, is nonflammable, and has an extraordinarily high decomposition temperature. However, in common parlance, silicone is synonymous with silicone rubber, and accounts for applications ranging from drug delivery vehicles to the de rigueur fashion accessory for some Hollywood starlets.

Commonly available elastomeric PDMS rubber formulations take advantage of the fact that the cross-linking chemistry is particularly specific. The polymerization of PDMS oligomers can be controlled to give a polymer with functionalized ends. These telechelic PDMS chains can react with the complementary functionality on small cross-linking molecules. This particular reaction class is termed the room temperature vulcanization (RTV), since these reactions readily take place under ambient conditions.

Some Cross-linking Chemistries

Several network forming reactions used in this work are shown in Figure 1. One type of end-linking involves a hydrosilylation reaction (*1-4*), in which a vinyl-terminated PDMS reacts with a tetrafunctional hydrosilane, tetrakisdimethylsiloxy silane. The hydrosilylation reaction is very specific and, since it is an addition reaction, there are no byproducts. Generally, an excess of the cross-linking agent must be added for the closest approach to reaction completion, possibly because side reactions with atmospheric water consume potential reactive sites.

The corresponding condensation approach for making model networks utilizes hydroxyl-terminated PDMS reacting with a tetraalkoxysilane, with the

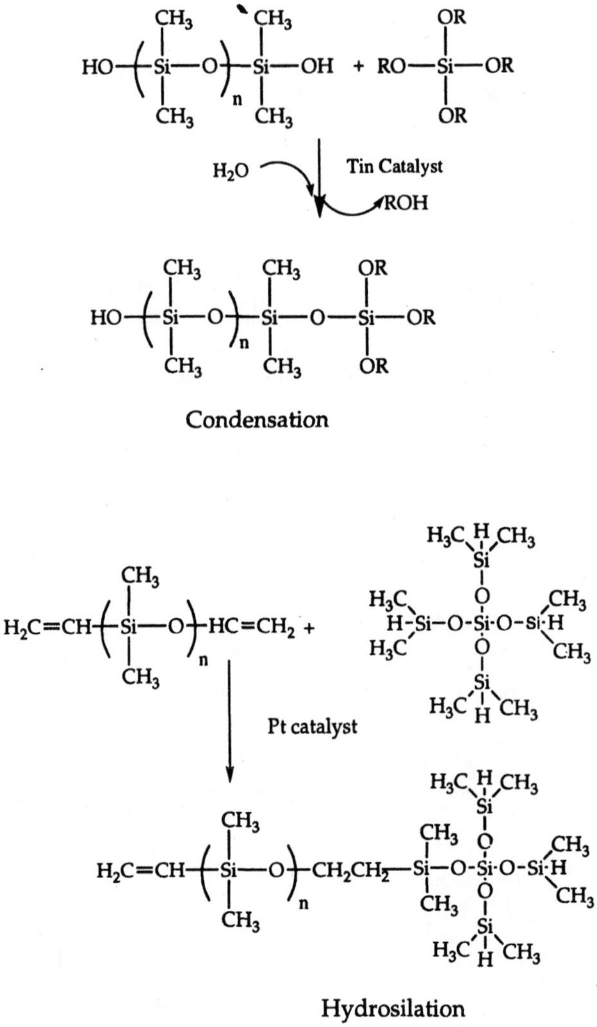

Condensation

Hydrosilation

Figure 1. PDMS End-linking Chemistries

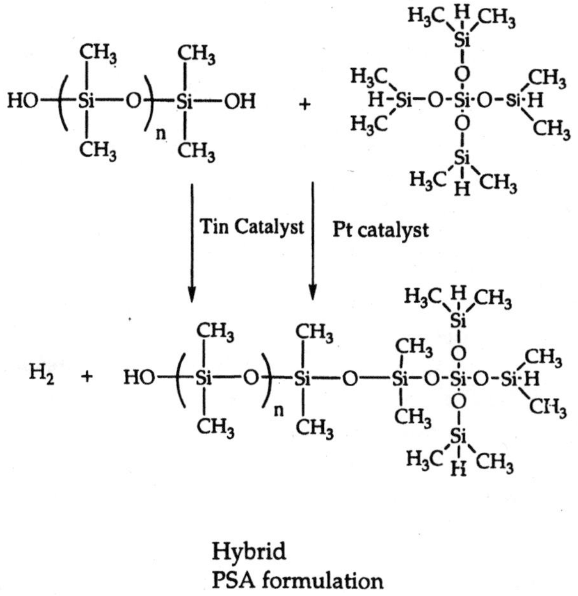

$$\underset{\substack{CH_3 \\ | \\ CH_3}}{HO-\left(Si-O\right)}\underset{\substack{CH_3 \\ | \\ CH_3}}{Si-OH} + \underset{\substack{H_3C\diagdown\overset{H}{|}\diagup CH_3 \\ Si \\ | \\ O \\ | \\ H_3C^{\diagup}\overset{|}{H}\diagdown CH_3}}{\overset{H_3C\diagdown}{\underset{H_3C\diagup}{H-Si}-O-Si\cdot O-Si\cdot H}\diagdown_{CH_3}^{CH_3}}$$

Tin Catalyst | Pt catalyst

$$H_2 + \underset{\substack{CH_3 \\ | \\ CH_3}}{HO-\left(Si-O\right)}\underset{\substack{CH_3 \\ | \\ CH_3}}{Si-O}-\underset{\substack{CH_3 \\ | \\ CH_3}}{Si-O-Si\cdot O-Si\cdot H}$$

Hybrid
PSA formulation

Figure 1. Continued.

evolution of an alcohol (*5-12*). Various metal salts catalyze the reaction, although tin (II) and tin (IV) carboxylates tend to be the most common, and were used in the present study. Whereas it has been suggested that catalyst levels as low as 0.1 weight percent are best for making model networks (*5*), it is common to use 1 - 2 weight percent (*13*). Competing reactions can occur during network formation with chain-chain and cross-link–cross-link condensations occurring at the same time as the desired chain-end–cross-link condensations (*13-16*). There have been suggestions that atmospheric moisture is a necessary component of the end-linking reaction (*17,18*).

Finally, infrequently used is a hybrid hydridosilane/silanol cross-linking reaction (*19*). In this type of cross linking, both the cross-linking agent and the chains are formed from dimethylsiloxane moieties, thus making the whole system essentially an athermal mixture. Also, the process could conceivably be carried out under conditions similar to those used in the hydrosilylation and/or condensation approaches, although the hybrid hydridosilane/silanol cross linking has not been studied in depth.

Formation of Model Networks

End-linked elastomeric PDMS networks are also of more fundamental interest since the molecular weight between cross links would simply be the molecular weight of the precursor PDMS chains. Their known molecular weight/mesh size have made these "model", "ideal", or "tailor-made" PDMS networks eminently suitable for studies of some of the molecular aspects of rubber-like elasticity (1-12). The end-linking method also makes possible the preparation of networks consisting of bimodal distributions of unusually short chains end-linked with the much longer chains more typical of elastomeric materials. Such bimodal elastomers are of particular interest because of their unusually good mechanical properties (7-12).

The present investigation used very short, well-characterized chains, since this would give the highest molar concentration of reactive groups, and would maximize the effect of changing conditions. The focus was exclusively on the "rational" case where there was a stoichiometric balance between the cross-linking sites and the chain end groups. Some bimodal elastomers containing an overwhelming proportion of short chains (ca. 99 mole percent) that nominally corresponded to our published reports were analyzed (20-22). Interestingly, it is often common practice to add large excess of the cross-linking agents to produce the tightest networks (least swollen at equilibrium) which are thought to be the best exemplars for comparison to rubberlike elasticity theories. And it seems self apparent that bimodal networks, while rationally cross linked, should certainly have a different micro/nanostructure when compared to a similar unimodal network. However, traditional rubberlike elasticity analyses do not address these structural details.

Nanostructure of Model Networks: Scattering Analyses

The existence of supramolecular structures in elastomeric networks systems has been argued based chiefly on the unique scattering of radiation observed for the networks in the swollen state (23-34). Small-angle neutron scattering profiles have been reported not only for model end-linked PDMS networks (25,27-30), but also for networks cross linked via high energy irradiation (36,37), as well as other non-silicone networks (25) and thus the scattering derived structure appears to be a general phenomenon. Small-angle scattering is generally expressed as a function of the intensity dependence on the scattering vector $q = 4\pi/\lambda \sin(\theta/2)$ where λ is the wavelength of the incident radiation. Here, θ is the scattering angle, and q has units of Å^{-1} and is thus a length-scale measure (31-34). A semi-dilute solution has a simple scattering pattern for the intensity I as a function of length scale, showing self-avoiding coil scaling at small length scales (high q) that varies as $I \sim q^{-5/3}$ which transitions to featureless (flat plateau) scattering at large length scales (low q) (25). The lack of low-q scattering has been described by Edwards (35) as a screening of long-range interactions, analogous to the well known screening of long-range Coulombic forces in electrolyte solutions upon addition of salt. Thus, a semi-dilute solution is thought to be structureless at large length scales. A swollen network of the same polymer volume fraction as the equivalent semi-dilute solution shows excess scattering at large length scales (low q) over the equivalent solution. This excess scattering ostensibly arises from contrast between regions of differing concentration that become descreened upon the swelling deformation.

A caveat should be noted when analyzing the scattering from swollen networks. Often, a large length scale (low q) plateau is approached (or assumed by extrapolation) in the small-angle neutron scattering of swollen gels. This is taken to be evidence for the lack of larger (micron) scale structure in analogy with the plateau of semidilute solutions. Similarly, analysis of the bulk unswollen gel is generally not undertaken since it is tacitly assumed that the long-range interactions would be screened. We have previously shown that these assumptions are not warranted and a network with known supramolecular structure can have small-angle scattering very similar to previously reported (and assumed structureless) spectra (20,22). Furthermore, a full fitting analysis of the scattering profile based on (assumed) lack of large-scale structure can be complex; fitting algorithms based on a generalized Ornstein-Zernicke (Lorentzian) or a combination of functions such as a Lorentzian and a squared Lorentzian, a Lorentzian and a Gaussian, a Lorentzian and a stretched exponential, or fractal regimes, etc. have been applied. We do not specifically address the fitting problem here, but interested readers are directed to Bastide (25). Instead, the focus here will be on the analysis of the scaling regimes for elastomers with demonstrable micron scale supramolecular structure.

Possible Micron-Sized Structures in Model PDMS Networks

Even if the prevalence of nanostructure in networks (as delineated by small-angle scattering) is accepted, the prevailing wisdom is that rubberlike networks are homogeneous on a coarse scale; in the case of silicone elastomers, this seems completely reasonable because they are often completely transparent. To date, most structures were viewed as random inhomogeneities. *A priori*, if true "phase separation", then the sizes of the phases can get quite large to minimize unfavorable interfacial interactions (this argument is developed further in the thermodynamics of phase development argument given below).

We have noted several instances when PDMS network formation shows a distinct phase separation. The liquid-like components are generally mixed into what appears to be a completely homogenous medium. At some later point during the reaction (indeed, even after the apparent gel point), a noticeable turbidity frequently forms. Such phase separation can be quite pronounced, with miniature "Liesegang"-like phenomena suggesting diffusion control in the system. The turbidity can continue to develop, in some cases spreading throughout a sample. Such separations would be expected to have significant effects on the physical properties of the elastomers, including their mechanical behavior. The present investigation was therefore undertaken in an attempt to establish the conditions that would lead to such phase separations in end-linked PDMS elastomers.

de Gennes (*38*), in considering network formation, argues that the introduction of cross links into a system tends to introduce segregation between chains and solvent (and in this case, the solvent could refer to unreacted but potentially reactive sol components). In this way, segregation could entail expulsion of the solvent, but not as pronounced as the usual case of "syneresis". Ostensibly, such a segregation and reorganization could be a very slow process in a gel system, and may be part of the diffusive aspect that was ascribed to the "Liesegang" phenomenon. Two regimes were envisioned (*38*): if the gelation proceeded relatively slowly, then there should be a smooth expulsion of the solvent from the cross-linked regions. However, if the gelation occurred more quickly, then gross reorganization couldn't occur, and the phase separation would occur at fixed concentration (*38*).

Experimental

Materials and Network Formation

All precursor chains, and cross-linking agents were obtained from Gelest, and were used as received. A sample of nearly monodisperse trimeric hydroxyl encapped PDMS was generously donated by Dr. Mark Buese of PCR Chemicals. The sample had a number-average molecular weight M_n = 300 g/mol, a polydispersity index M_w/M_n ~1.1, and was found to be essentially free of cyclics (as determined by gas chromatography on end-capped samples). The gelling solutions were formed by mixing the PDMS chains with the stoichiometric amount of cross-linking agent and catalyst at ambient conditions. Specific details and additional information on the components are given below, and elsewhere (*20-22*).

Molds for the samples were made from Bytac FEP/aluminum adhesive film (Fisher) fastened onto glass plates, with microscope slides then cyanoacrylate superglued onto the Bytac in the appropriate geometry. The liquid mixtures were poured into such molds having nominal dimensions 5 cm x 5 cm and 1 mm thick. Samples were allowed to cure for approximately 1 day under the chosen conditions.

Samples suitable for microscopy were prepared by placing 2 drops of the material onto a glass slide. One drop was immediately covered with a glass coverslip. This allowed for a uniform thickness film that was approximately 0.05 mm thick. The other drop was allowed to spread out unfettered to form a thin film that was approximately 0.1 mm thick. The gel points of the samples were estimated as the point at which the free samples were able to resist flow when overturned.

Two benchmark bulk morphological samples were used for comparison between the various scattering and microscopic techniques below. The micron scale spherical ("binodal") texture was from the hydridosilane/silanol hybrid network which was formed by end-linking α,ω-hydroxyl-functionalized short chains having M_n = 300 g/mol ($M_w/M_n \sim 1.1$) (PCR Chemicals) with tetrakisdimethylsiloxysilane (Gelest) and 10 wt% dibutyltin(IV)dilaurate catalyst (M&T Chemicals) at room temperature. The micron-scale interpenetrating phase ("spinodal") networks were formed by end-linking 99 mole% of 1,3 divinyl-1,1,3,3,-tetramethyldisiloxane with 1 mole % α,ω-divinyl PDMS of M_n 12,000 ($M_w/M_n \sim 2$) in the presence of a 1,000 ppm platinum complex catalyst diluted in divinyl siloxanes.

Light Microscopy

Light microscopy was carried out on a Nikon Labophot transmission optical microscope, as described elsewhere (22). The sample was illuminated with white light in Koehler focusing. In this manner, differing vertical sections could be sampled. The images were obtained with a 10x ocular and 400x objective lens. The images were captured via a Javelin CCD detector connected to a frame grabber on a PC. In no case did the image analysis tools change the observed morphology.

Atomic Force Microscopy

The present model elastomers were thought to be good candidates for atomic force microscope (AFM) studies (22), being both elastic to some degree and having a well-characterized bulk morphology. Such measurements provide a real-space topography map of the surface contours, and measurements of the torsion of a AFM's cantilever can characterize the frictional properties of a surface. The sample and the tip could have an in-phase periodic modulation applied, with the phase lag between the measured signal and the applied signal characterizing viscoelastic responses.

Relevant here are the facts that PDMS has a very low surface free energy, and as a result tends to migrate to sample surfaces (39,40). There have been some recent AFM studies on PDMS elastomeric networks (40-45) and many advocates of atomic force microscopy favor using PDMS networks for AFM method development due to the lack of hysteresis in the force curves (41,42). It should be emphasized that the focus of this effort was geared toward morphology characterization rather than absolute mechanical measurements per se. The phase separation in the model networks is large and thus might be visible even with coarse-grained imaging techniques

Atomic Force Microscopy studies were carried out on a Burleigh Metrix 2000, using a silicon nitride probe tip (tip radius ca 100 Å) mounted on a V-shape cantilever with spring constant 0.37 N/m. Contact topography and contact frictional force imaging were obtained on ca. 25 micron square sections (the limit of the instrumental resolution).

Some additional measurements were carried out on a Park Autoprobe CP, using a gold coated silicon tip mounted on a V-shape cantilever with a spring constant 13 N/m. Tapping mode topography, phase modulation, and vertical force studies were carried out by tapping the tip at its resonant frequency. There was good agreement in the results from the two different instruments utilizing different analysis conditions (22).

Small-angle Neutron Scattering

Small-angle neutron scattering (SANS) was conducted on the time-offlight SAND instrument at Argonne National Laboratories. Polychromaticradiation (1 – 14 Å) was impinged on the sample, and the time-resolved collection of the radiation occurred on a 40 x 40 cm^2 two-dimensional detector. The default instrument configuration allowed usage of a q range of $0.0035 < q(Å^{-1}) < 0.6$. The elastomeric samples were swollen to equilibrium in perdeuterated benzene and placed into 2 mm quartz cells. The data were corrected for background noise (dark current), empty cell scattering, solvent scattering, and the incoherent signal from the hydrogen in the PDMS. The scattered intensity was converted to an absolute scale using a silica standard, although arbitrary intensities are reported here. Further instrumental details can be found on the web page http://www.pns.anl.gov/instruments/sand/.

Small-angle X-ray Scattering

Small-angle X-ray scattering (SAXS) measurements were performed on the Proctor and Gamble Miami Valley SAXS. CuK_α radiation ($\lambda = 0.154$ nm) was generated with a Rigaku RU-300 rotating anode x-ray generator operating at 40 kV and 40 mA with a 0.2 x 0.2 mm focal size (a 0.2 x 2 mm filament run in point mode). The patterns were collected with the Siemens 2dimensional small-angle scattering system which consisted of the HI-STAR wire detector and Anton Parr HR-PHK collimation system. Collimation was achieved with a single 100 µm diameter pinhole positioned 490 mm from the focal spot. The size of the focal spot restricted beam divergence. A 300 µm guard pinhole was placed 650 mm from the focal spot, just in front of the sample. The detector was placed a distance of approximately 650 mm from the sample. A Ni filter was used to eliminate the K_β radiation. Because of the small beam size and large sample-to-detector distance, two-dimensional profiles (q_x, q_y) could be obtained with a minimum of instrumental smearing; thus no smearing corrections were employed. The two-dimensional spectra were reduced and reported as one-dimensional spectra.

In order to better compare the spectra, the observed intensities were arbitrarily scaled (e.g., the SAXS intensity scaled to overlap the absolute SANS intensities). Semi-dilute solution scattering from an equivalent solutions of uncross-linked PDMS in deuterated benzene confirms the general details of the SANS low-q plateau and high-q self-avoiding coil-blob scaling.

Results and Discussion

Phase Separation in Hydrosilylation Cross Linking

There are a number of reports on hydrosilylation cross-linked model networks (*46*). The effects of cross-link functionality for low functionality cross links (tri- and tetrafunctional) (*47*), very high functionality cross links (*46,48,49*), bimodal chain length distributions (*50*), and random vs. localized cross links (*50-53*) have been considered. The first catalyst noted to be effective in the hydrosilylation reaction was Speier's catalyst – a solution of hexachloroplatinic acid in isopropanol (*54*). The mode of activity of the Speier's catalyst is a matter of some debate, but the active species was thought to be $H[(C_3H_6)PtCl_3]$ (*55*). Unfortunately, the Speier's catalyst was not very soluble in the PDMS matrixes, and the cross-linking reaction occurred very slowly. Thus, various alternatives were attempted, including cross linking at higher temperature (*4,56*) or use of a platinum(IV) species complexed with organic ligands, and soluble in organic solvents such as toluene (*56*).

Another suitable hydrosilylation catalyst is a Pt(0) metal colloid activated by molecular oxygen (*57,58*), and was used here. Specifically, the catalyst was platinum-divinyltetramethyldisiloxane colloidal complex diluted either in divinyltetramethyldisiloxane (3 - 3.5% platinum, Gelest SIP6830) or diluted in xylene (3 - 3.5% platinum, Gelest SIP6831). Ojima has reviewed the various differences between the various (homogenous and heterogeneous) noble metal catalyzed hydrosilylation mechanisms (*59*).

A model network was formed from 1,3-divinyltetramethyldisiloxane (Gelest SID4613, M_n = 186 g/mol) with tetrakisdimethylsiloxysilane. The amount of catalyst needed to drive this reaction past the gel point was approximately 10 ppm Pt/Si; however, if 500 - 1000 ppm Pt/Si was used, thin films were able to develop the phase-separated texture, both constrained under the coverslip and free-standing thin films. No large-scale phase separation was observed to occur in bulk unimodal short chain samples. The time to gelation for these samples was relatively rapid (seconds for the slide samples to 1-2 hrs for the bulk sample). The development of phase separation appeared to proceed well after the gel point.

One problem was the very exothermic reaction, since high temperatures are known to affect the course of the hydrosilylation reaction. For example, Gorshov and coworkers have investigated the cross-linking reaction of a hydridosilane functional polymer with a PDMS sample having pendant vinyl groups (*60-63*). The catalyst was a platinum-ammonium complex, and the material was cured at high temperatures. Calorimetric studies indicated that the desired chemical crosslinking (vinyl-hydridosilane reaction) was initiated at approximately 130 - 140°C, and that the mechanism becomes diffusion controlled near the gel point. Quan also investigated high-temperature hydrosilylation cross linking in a novel manner (64). First, an α,ω-divinyl PDMS chain was reacted with a poly(hydromethyl-co-dimethylsiloxane) random copolymer. The catalyst was chloroplatinic acid suspended in vinylmethylcyclosiloxanes and curing occurred at room temperature over 24 hrs. ^{1}H NMR measurements indicated that all vinyl groups had reacted. The as-formed network was post-cured at high temperatures (137 °C), which in general increased the modulus, and thus indicated that additional cross links were generated. The modulus increase was lower at lower post-curing temperatures. A series of control experiments showed that deactivation of the platinum catalyst with ammonia reduced, but did not eliminate, the post-cure effect, while treatment with ethylene (which capped any residual Si–H) or post-curing in inert atmosphere eliminated the post-cure effect. This behavior highlights the role of Si–H side reactions with atmospheric water, which were favored due to the molar Si–H:Si–CH=CH$_2$ excesses involved.

It does not seem possible to reconcile the phase separation solely in terms of the side reactions (22). In fact, several researchers have noted that side reactions with atmospheric water are negligible during cross linking at room temperature (which was approximated in the cover slip constrained slide samples) (2,65). A possible side reaction that occurs at ambient temperatures could be a cross-linker redistribution reaction. Macosko and Saam (66) found that this mechanism applied to PDMS polymers having Si–H groups in the presence of a variety of noble metal complexes (67). The redistribution mechanism (in an ideal case) leads to higher functionality cross-linking agents. Furthermore, a possible redistribution product could be volatile dimethyl silane, and evolution of volatile products would destroy the stoichiometric balance. Our previous reports contrasting the "rational" cases of crosslinking suggest that a balance between the chain end group concentration and the crosslink site concentration is a prerequisite towards understanding (22, 68) the rubberlike elasticity, and thus calls into question the "common" practice (4) adding an excess of crosslinking agent.

One experiment focused on the molecular weight distribution, specifically through the use of a "rationally" cross-linked bimodal elastomer, having a stoichiometric balance beween the chain ends (both long and short chains) and crosslinking agent. It consisted of 99 mole % of 186 g/mol short PDMS chains with 1 mol% of a high molecular weight ca 12,000 g/mol vinyl end-capped PDMS. This class of bimodal elastomer was the basis of further light scattering study by Kulkarni and Beaucage (21). Bimodal elastomers do not have as great a volumetric concentration of endgroups as their short chain unimodal counterparts, and did not cure with a noticeable exotherm. The time for gelation was notably longer, ca. several hours to 1 day (22). Bimodal networks had the added benefit that they were reasonably tough (the "bimodal effect") (7). Phase separation occurred both in bulk samples and in thin films and pervaded the entire volumes of the bulk samples, which suggests that the exotherm was detrimental for the formation of phase separations. The observed structure in this bulk bimodal phase separated elastomer appeared to mimic the structure seen in the unimodal thin film in all morphological respects.

Figure 2 shows the small-angle scattering of the phase separated bulk bimodal elastomer. The SANS scattering of an equivalent semi-dilute solution shows the expected behavior, viz. self avoiding coil (blob) behavior with scaling

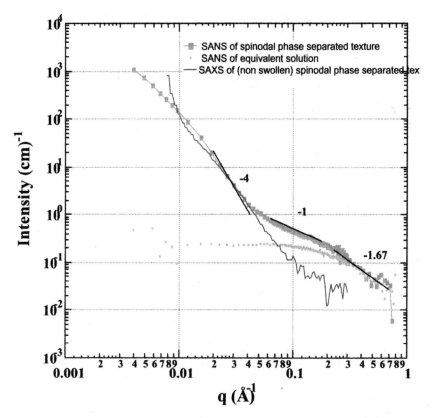

Figure 2 Small-angle Scattering Comparisons of Phase Separated Spinodal Texture. (SANS data from the SAD instrument at IPNS, Argonne National Laboratory.) Diamonds reflect the equivalent semi-dilute solution (50 volume % PDMS of M_n 20,000 g/mol in d6 benzene). Squares are associated with the spinodal texture swollen to equilibrium ($v_2 = 0.5$). The solid line gives the SAXS results for a non swollen "spinodal texture" network morphology.

$I\sim q^{-5/3}$ for $q > 0.2$ Å^{-1}. The semi-dilute solution scattering transitions to the screened plateau at low q. The bimodal elastomer scattering exactly tracks the solution scattering at $q > 0.3$ Å^{-1}. The median q scattering for the swollen elastomer shows a transition to a scaling regime of $I\sim q^{-1}$, which is the signature of a tensile blob regime as discussed by Beaucage (69). At $q < 0.04$ Å^{-1} q a strong upturn is present. This large overscattering is the characteristic indication of large-scale structure. The scaling for this low-q regime is seen to approach $I\sim q^{-4}$, which is the Porod regime that is indicative of a smooth interface for the high cross-link density domains (31-33). However, the scattering at the lowest q appears to show a transition, perhaps toward a semi-dilute solution like plateau. The small-angle X-ray sattering from the non-swollen bimodal elastomer (as shown by the solid line in Figure 2) clearly shows that this apparent approach to a low-q pseudoplateau, as indication of lack of coarse-grained microstructure, is not meaningful. The SAXS indicates essentially Porod $I\sim q^{-4}$ scaling persisting over all sampled q ranges. These results clearly connect the micron scale structure observed visually to the nanostructure from which the turbidity arises.

Comparisons with Condensation Cross-Linking

Condensation cross linking has been used to prepare a variety of model networks (70-74). As expected, in these networks the soluble fraction was small, and the modulus and swelling depended on the molecular weight of the precursor polymer. The effects of junction functionality have also been considered (74), as well as the molecular weight distributions (22). The condensation cross-linking mechanism provides an interesting contrast for hydrosilylation cross linking. First and foremost, it produces a very stable Si–O–Si bridge. This type of cross linking, however, is not as specific as hydrosilylation, and potential side reactions can occur such as chain extension (15,16) and formation of silica particles. The reaction necessarily entails evolution of a byproduct. There are many variable parameters, such as choice of the precursor PDMS molecular weight, alkoxy functional group on the cross-linking agent, choice of the catalyst, and curing conditions. These various effects on the formation of a phase-separated texture are considered below.

Effects of PDMS Precursor Molecular Weight

The primary prerequisite for forming a phase-separated structure appears to be using chains of low molecular weight. This fits in with the de Gennes model (38) of percolation gelation insofar as the network would likely be formed from bonds on adjacent "lattice" sites. The molecular weight range that exhibits a well-developed large-scale phase separated structure appears to be from about

300 g/mol (trimeric species) to 2,500 g/mol. It has not been possible to prepare model phase-separated networks from long chains alone. There were no significant differences regarding the development of the texture or reaction conditions between the 300 g/mol and 2,500 g/mol networks, and phase-separated thin films and bulk samples could be prepared. The degree of turbidity would range from a slight bluish haze to a white, milky texture. As might be expected, there were significant differences in the elastic properties (specifically the equilibrium degree of swelling, which was 0.43 for a unimodal 2,500 g/mol network and 0.51 for a unimodal 300 g/mol network) (22). The time to gel also varied significantly, from about 12 hrs for the short chain network of 300 g/mol to only 1 - 2 hrs for the 2,500 network. Also, the 300 network developed turbidity slowly after the gel point, while the 2,500 network was much faster.

When bimodal networks are formed by end-linking large molar proportions of the short chains (300 g/mol) with much longer chains (ca. 20,000 g/mol), the mechanical properties are distinctly improved. The number-average molecular weights of the common networks fall well below the 2,500 g/mol limit cited above. In most cases the networks were clear. This clarity may be accidental; in our previous publications (20-22,68,69) a series of cross-linked labelled networks were prepared (c.f. Effects of Cross-Linking Groups discussion below) which were opaque after staining treatment. A similar bimodal sample (formed from non-stainable tetraethoxysilane (see (20) for experimental details) did not show obvious texture. However, if this clear bimodal network was swollen with trivinylcyclohexane and then the swelling solvent was stained, the phase separated microstructure was emphasized (Figure 3). This morphology also corresponded very well with the tagged netork morphology we have previously described.(20) What is particularly interesting is that only the stained interfaces were emphasized when utilizing an "inverse stain technique," and furthermore the interfaces are seen to be smooth. Smooth-surfaced domains are expected for a thermodynamic instability phase separation (vide infra) and agrees well with the Porod scaling regime seen in Figure 2.

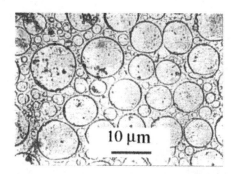

Figure 3. Inversely Stained Bimodal Network (Stained Swelling Solvent)

Nevertheless, several bimodal networks were observed to be turbid. For example, a model network from 97 mol% of a 300 g/mol short-chain polymer appeared to be phase separated spheres throughout its entire volume. Furthermore, this network was surprisingly tough, as is described elsewhere (22). The phase separation may be an intrinsic molecular weight distribution effect (viz., segregation of the short chains into high-cross-link density clusters) and thus may be facilitated by long chains being sequestered in the low-cross-link-density regions.

Effects of Catalyst

The other necessary condition for formation of the phase-separated texture is a relatively high level of catalyst (22). In the present studies, it was necessary to use rather high levels, around 1 wt %, to form networks that were not tacky, had reasonably low soluble fractions, maintained a uniform aspect ratio when swollen, etc. This is consistent with the levels of catalyst used in industrial formulations, which is on the order of 0.5 - 2 wt%. Kuo (13) has given an exhaustive account of the various effects of tin(II) and tin(IV) catalysts on the network formation and chain extension side reactions for PDMS networks from hydroxyl-functionalized chains and favors dibutyltin(IV)dilaurate as the model end-linking catalyst. For formation of both model networks and phase separation, both stannous octoate and dibutyltin(IV)dilaurate acted similarly.

At relatively high levels of catalyst (ca. 2%), a light, milky phase separation is obvious, albeit spotty and with incomplete coverage. At 5% catalyst loadings, the phase separation gives noticeable turbidity and pervades the sample. Microscopic investigation reveals that the size of the domains was reasonably constant for differing catalyst levels. It seems that the large level of catalyst allows for greater phase separation as a function of the depth from the surface. Several successive layers of the phase separation might be necessary to cause the opaque regions due to large-scale scattering, while any degree of phase separation will cause some scattering (a bluish tint) and possibly a haze.

Stannous octoate catalyzed networks tended to have a distinct yellowish cast (even after extraction). There was a significant difference in the molecular weights of the two standard catalysts (450 g/mol for stannous octoate vs. 631 g/mol for dibutyltin(IV)dilaurate). Thus, the tin(IV) catalyst has fewer catalytic centers per unit weight. Hence, dibutyltin(IV)dilaurate should be a less problematic catalyst on a given weight basis, and is the catalyst that was invariably used for further studies (13,22).

Various authors have considered the mechanism by which tin salts catalyze the condensation of silanol-terminated PDMS (13,16-18). The first step of the reaction appears to be a hydrolysis of the tin carboxylate bond to form a tin hydroxide group. The tin center coordinates with the silicon atom in the cross-linking agent, forming Sn–O–Si based stannasiloxane complexes. These bridges are subsequently hydrolyzed (75) and then the energetically favored Si–O–Si condensation occurs. The exact mechanism of insertion of a PDMS chain, and subsequent liberation of the stannasiloxane, is a matter of debate that is beyond the scope of the present study.

Effects of Cross-Linking Groups

The differences in reactivity of the various alkoxysilane substituents might be expected to have an effect on the cross-linking reactions. Specifically, a more reactive alkoxysilane should favor both the chain end-cross-link site reaction, as well as condensation of cross links to form silicate structures.

Furthermore, one would expect that a more active cross-linking reaction might mimic a higher level of catalyst. We used three cross-linking agents, listed in decreasing order of reactivity as tetramethoxysilane > tetraethoxysilane > vinyltriethoxysilane. One would expect that the most active, tetramethoxysilane, might not need co-reactants such as moisture to condense with the silanol on PDMS (76). Of course, the trifunctional cross linker is the lowest functionality that will allow for gel formation. It is not only the least reactive, but also has an unreactive site. Thus, it would be expected that this molecule could not form large-scale inhomogeneity (generally silicate particles), due to the fact that the surface has many non-hydrolyzeable and hydrophobic groups (22).

Mallam et al. (77) have prepared networks from α,ω-hydroxylfunctionalized PDMS with a large excess of an ethyl triacetoxysilane as both cross-linking agent and sol-gel silicate precursor. The ethyl triacetoxysilane was present at 4 wt%, which corresponds to a 10-fold molar excess of cross-linking groups based on the precursor PDMS molecular weight of ca. 40,000 g/mol. Acetoxysilane groups hydrolyze readily with atmospheric water and liberate volatile acetic acid. However, the silanols formed in this way can subsequently condense. It was noted that while discrete silicate particles do exist, they tended to be only around 800Å (77), while the present results suggest that the phase sizes tend to be several microns in size (ca. 10,000 - 50,000 Å) (22). All three cross-linking agents were able to form a phase-separated structure, and the present study did not indicate any significant differences in the nature of the phase separation for a given molecular weight and catalyst loading (22).

Phase Separation in Silanol-Hydridosilane Networks

The formation of the energetically favored Si–O–Si bond is the basis of the condensation approach. This bridge can also be formed by the reaction of a silanol with a hydridosilane as in Figure 1. This reaction might be considered as a hybrid of the condensation and hydrosilylation reaction, and the catalysts used to drive this reaction could be the tin carboxylates or platinum catalysts. The reaction could be one-step (as pictured) or two-step, wherein the first step involves hydrolysis of the hydridosilane with atmospheric water to form a silanol, and then condensation of two silanols. Both reactions could occur simultaneously; Nitzche and Wick note that insoluble rubbery material formed when a copolymer containing hydridosilane groups was mixed with dibutyl tin dilaurate (78), and a similar system based on chloroplatinic acid has been developed by Lewis (79). To the best of our knowledge, this reaction has not been previously used to form model networks.

The present investigation used the short chains from the condensation approach (300 and 2,500 g/mol, whereas longer chains would not form a gel) and the tetrakisdimethylsiloxysilane cross-linking agent used for the hydrosilylation reactions (22). Network formation reactions are very slow in this hybrid cross-linking system, and gelation can take days, with turbity development occurring in concert with gelation. This allows one to conclusively rule out the possibility that cross linking effectively traps a precipitate into a colloid-like structure. It was not suitable to use the platinum complexes as a catalyst for monolithic systems. Perhaps the preponderance of hydroxyl groups (especially with the additional contribution of atmospheric water) tends to poison the platinum complex. Addition of a large excess (~5000 ppm Pt/Si) of the platinum catalyst induced extensive cross linking in an exothermic burst that actually allowed the liberated hydrogen gas to form an open-celled foam; however, the struts of the foam were seen to be turbid. (Further analysis of this network microstructure was not conducted due to the obvious byproduct structural artifacts) Similarly, the tin(II) catalyst, stannous octoate, was found to be ineffective in gelling the system. The only reproducible method was to use dibutuyltin(IV)dilaurate catalyst in relatively large amounts (5 wt% to produce gel, 10 wt% to make a non-tacky, model network with turbity pervading the volume of the sample and reasonably small soluble fraction, ~10%, which is approximately the catalyst loading). These results are indirect proof that side reactions are not the cause of the phase separation, specifically hydrolysis and condensation of the hydridosilane in hydrosilylation cross linking, or stannasiloxane formation in the condensation cross linking (22).

One cannot rigorously say whether the phase separation necessarily occurs post-gel, or occurs in the aforementioned "Liesegang" zone due to the fact that the slow gelation could allow for reorganizations (both in bulk and in thin films). Yet even with a viscous fluid character, there was no gross segregation that made the phase separation significantly different from that seen in the hydrosilylation or condensation cases. Hence, we believe that this hybrid network-forming system shows that the phase separation mechanism is a completely general phenomenon for these network-forming systems.

The SANS of a 300 g/mol hybrid network swollen to equilibrium is shown in Figure 4. The equilibrium swelling volume fraction of this network was quite low, 0.22, which corresponds to a very low average cross-link density. Obviously in this hybrid condensation reaction the low cross-link density domain reactions predominate, which may account for the inability to make good networks with this cross-linking chemistry. Nevertheless, while the sample is seen to have the phase separation pervading the sample volume, the SANS at first glance appears like results from (assumed) homogenous gels. There is a small low q (large length scale) upturn but not as intense as would be expected for true phase separated (Porod smooth surface I~q^{-4}) structure. The scattering profile is seen to have a slight double shoulder profile, which has been seen in many differing network systems(25). In analogy with previous work on bimodal PDMS networks (20-22) and a published report on the SANS from a fluorosilicone gels (both unimodal and bimodal) (80), the lower q hump could be a signature of a nanophase domain (22,25). The unusual kinetics of formation for this network suggests that these nanophases could have "ripened" to add to already formed micron domains. As a result, the (expected) upturn in SANS might manifest below the lowest resolvable q. This is again indication that traditional scattering analyses based on assumed gross homogeneity may be in error.

The Molecular Nature of the Observed Phases

For turbidity, there must be distinct regions that have a size on the order of the wavelength of visible light (ca. microns). This size-scale requirement indicates that the phenomenon must be supramolecular, and involve at least several primary chains. It is seductive to presuppose that the observed phase separation is a secondary growth mechanism among the low molecular weight sol components that occurs in (and in fact may be mediated by) the apparently homogenous gel structure. However, it is important to note that growth within a network matrix would necessarily entail perturbing the network chains from their most favored conformation. Secondary growth should be limited to a scale on the order of the mesh (the size of a precursor chain) (38,81), and the observed phase separation must be the result of correlation and segregation of supramolecular structures.

In addition, there must be a significant difference in the refractive index of the various phases in order for them to be *seen*. Thus, some polymer blends that are known to phase separate via thermodynamic instability must be imaged with relatively exotic techniques such as phase contrast microscopy or laser scanning confocal microscopy (82-84). The simplest explanation for a difference in refractive index in this system is the chain length. For example, the refractive index of a short PDMS chain of 300 g/mole molecular weight is 1.375, while the refractive index of chains that are orders of magnitude in chain length longer (2,500 g/mole and 25,000 g/mole) is 1.4012 and 1.4035, respectively. The refractive index for all high molecular weight PDMS (>17,000 g/mol) is a relatively constant value of 1.4033 - 1.4035. Thus, the phases could be seen as regions of high and low cross-link density, having small and large chains (respectively) in each phase. This model is discussed in more detail elsewhere (22), as well as specific aspects of the refractive index difference (20). These differences could manifest themselves in the scattering response, where the scattering is due to some difference in types of contrast (refractive index for light, electron density for X-rays, spin state for neutrons, etc.)

154

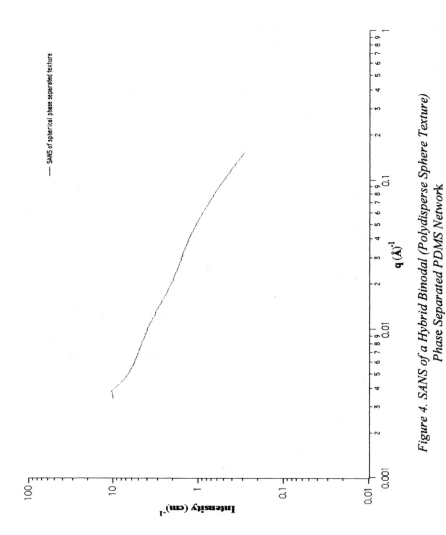

Figure 4. SANS of a Hybrid Binodal (Polydisperse Sphere Texture) Phase Separated PDMS Network

Phenomenology of the Phase Separation

The necessary and sufficient conditions for forming the phase-separated structure in these model networks appear to be (i) very short chains making up a large fraction of the composition, and (ii) high concentrations of catalyst. A logical extension of condition (i) could be use of (excess) cross-linking agent that is capable of self-condensation, which in turn would emulate the short-chain component. It is self-apparent in our model that the short chains are responsible for forming the high-cross-link density phase (and long chains, in bimodal formulations, would be relegated to the low cross-link density phase). Hence bimodal elastomers are naturally templated toward this type of phase separation. However, the phase-separated texture can develop from a unimodal, stoichiometric mixture of short chains and cross-linking agent. There is apparently some mechanism for reorganization and/or reaction of the short chains to form the phase of low cross-link density.

The most attractive possibility for generation of a low cross-link density phase is by intramolecular cyclization with another reactive site on a cross-linking molecule. This is quite likely; when one chain end becomes tethered, the shortness of the chain prevents the reactive end from sampling many potential reactive sites. The most likely reaction would be with another group on the same cross-linking molecule. Thus, this mechanism doesn't presuppose a particular cross-linking chemistry. If the cross-linking molecule is tetrafunctional, such a reaction would consume two reactive sites. If the other two sites were connected to differing chains, the net effect would be a linear, dimerized chain. After repetitive cyclizations and joining of such molecules, a rather large molecular weight chain structure would appear. The advantage of this approach is that the stoichiometric balance of functional groups is maintained. This mechanism is not directly applicable to trifunctional junctions, since an intramolecular cyclization would leave only one reactive site, which if reacted would act as a blocked site, preventing both linear chain polymerization and gel formation. Yet small-ring formation between two adjacent junctions would still allow many reactive sites. This ring formation by nature should change the course of the gelation; i.e. the gel point would be delayed since many potential branching reactions would be "wasted." In rigorous terms, ring formation would push the gel formation mechanism away from a mean-field "Flory-Stockmayer" type model and closer to a percolation model (85). In fact, the gel-growth mechanism of end-linked PDMS elastomers has been addressed by Lapp et al. (86) who have found good agreement with a percolation model.

Another possibility is chain extension. Small chains would link together to form a high molecular weight end-functionalized polymer that could then react into the network. Obviously, there is no straightforward mechanism for chain extension in the hydrosilylation case, yet high molecular weight α,ω-divinyl functionalized PDMS can be polymerized by insertion of siloxane species (even a cylic monomer) into a divinyltetramethyldisiloxane end blocker (87). It would seem that chain disproportionation would be rather random, and not favor a distinct chain length, or chain length population. Similarly, an end-linking and/or disproportionation mechanism should also tend to form stable PDMS cyclics in some equilibrium quantity. Small cyclics would not be able to be threaded by the network chains, and thus could be extracted from the networks, which would affect the soluble fraction. Chain-extension reactions consume potential reactive sites, while disproportionation does not. The weakness of this approach is that it seems that the initial stoichiometry of the reactants must be violated, and as a result the soluble fraction of un-reacted, or incompletely reacted, components should be fairly large (22).

We do not seriously consider the possibility that there is a reaction with the catalyst, and possible insertion of the catalyst or any foreign species into the network structure. Three differing cross-linking mechanisms, which have two differing catalyst systems (tin carboxylates for condensation. and platinum complexes for hydrosilylation) have shown this effect. In addition, the sizes of the phases do not appear to change much among the various systems, even though the amount of catalyst used to induce the phase separation could vary by large amounts.

Another unlikely event would be the low-cross-link density region containing a preponderance of dangling-chain species. A dangling chain does tend to lower the effective junction functionality, and would tend to lower the cross-link density. However, a dangling chain would necessarily violate the balanced stoichiometry. These dangling chains would be expected to have a marked influence on the soluble fraction and on the swelling, since the favorable polymer-solvent interactions are not counteracted by the elasticity of a chain tethered at both ends. Thus the most likely explanation is that some combination of both the chain extension and cyclization reactions causes the formation of the low-cross-link density phase (22).

The Thermodynamic Nature of the Phase Separation

There are multiple conditions for single-phase behavior in any system, including the present elastomers, in the binodal and spinodal regimes (88,89). The second law of thermodynamics implies that in order to be compatible, a mixture must have a negative free energy of mixing $\Delta G_{mix} = \Delta H_{mix} - T\Delta S_{mix}$. Since polymerization tends to lower the entropy of the system, it is not uncommon for phase separation to be induced. The assumption here is that even in the homogeneous, single-phase pre-gel stage there have been enough reactions that the system is solely branched-PDMS species. There should be no enthalpic interactions in this bulk state, making the free energy change invariably negative.

Even for negative ΔG_{mix}, one should be at the minimum of the free energy landscape, with the equilibrium phases occurring at the respective minima (88,89). However, unlike small-molecule phase separations that are very rapid, the slow diffusion in polymers allows one to infer mechanisms of the phase separation. For example, in the binodal regime, the free energy curvature is such that the small concentration fluctuations are not stable due to non-favorable interfacial contributions. A binodal "nucleation and growth" regime shows a nucleation event must precede the large-scale phase separation. These nuclei then grow by addition of material to a nucleus growth front to form the equilibrium phases (*21*). Our other published reports indicate the likelihood of a nucleation and growth mechanism (*20-22*).

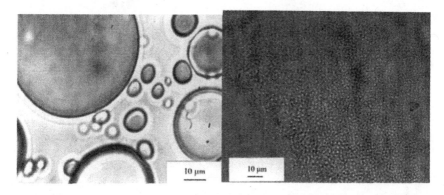

Figure 5. Binodal (Nucleation and Growth) Type Structure (300 g/mol Hybrid Network, left) and Spinodal Network (99 mol% Bimodal Network, right).

Figure 5 demonstrates that a common motif of the phase separation is micron-sized spheres of polydipserse size that appear to grow in time. This is consistent with a nucleation and growth mechanism, although not conclusively. This texture is very common for condensation and hydridosilane/silanol networks. Conversely, a small concentration fluctuation in the spinodal regime would be stable, and thus no restoring force will cause dissolution. In fact, the system will tend to establish a periodic fluctuation of some critical size throughout. The signature of spinodal decomposition would be correlation of these fluctuations giving rise to a peak in the scattering spectrum. Such a peak has, in fact, been observed (*20,22*). However, the result of the spinodal decomposition is that there should be periodic modulation in the phases, and thus they would look like randomly oriented, interpenetrating cylinders. Such a result for our system is shown in Figure 5. This micrograph is quite surprising; other studies of phase separation in polymeric systems have indicated that the interpenetrating structure akin to what is shown in Figure 5 tends to break up into droplets (thus minimizing the unfavorable interfacial contacts) in late stages of spinodal phase separation. Clearly, the narrow size distribution of phase sizes (*20-23*) suggests that there might be complicated growth mechanisms. In fact, our phase separated morphology will not reversibly return to a single-phase region nor change morphology (break up or further grow) by swelling and/or heat treatment. The fact that these structures are stable to dispersal attempts implies that this phase separation is probably not a true equilibrium event. This system appears to be a "reaction induced" phase separation that mimics an actual thermodynamic phase separation, but is trapped by the extensive cross linking.

Corresponding Results from Atomic Force Microscopy

As expected, the atomic force microscopy experiments reproduced the scale of the phase-separated networks, viz. micron-sized regions. Separate imaging modes for the spinodal textured sample show regions that are all ca. 5 microns in size, and have a characteristic spacing (22). The contrast in the phase modulated imaging would arise from highly cross-linked regions having limited viscoelastic relaxation. Conversely, the low cross-link density regions would have a more rubbery relaxation. The topographic difference (difference in height among surface regions) was somewhat surprising. These networks were grown from a quiescent solution, and we would expect that the surface should be reasonably smooth. The difference might be the result of differing surface tensions, or from the phase separation process including growth in real space! Clearly, further experiments are needed to characterize this phenomenon.

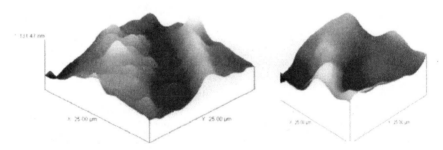

Figure 6. Topography (left) and Friction (right) AFM of Spinoda Texture Ssize Scale 25 µm x 25 µm x ca 150 nm Height)

We show a three-dimensional view of the spinodal topography in Figure 6. The relative height of the fluctuations agreed very well for both AFM instruments used, around 100 nm. This is still much larger than would be expected from random noise. The phases again are seemingly cylindrical, each about 5 microns in cross-sectional size with approximately 5 micron periodicity.

However, there is seen to be a large amount of fine scale detail in Figure 6. This particular mode of imaging had the AFM tip dragged across the surface to provide the necessary contour plot. It is possible that the tip deforms the surface as it is moved, and we would expect that the greater amount of deformation in the low cross-link density regions. Clearly, this is an area that warrants further attention. As Figure 6 illustrates, the frictional force diagram has a much smoother profile, as would be expected for a true spinodal phase separation with smooth interfaces (22).

The AFM images of the bimodal-type phase separation have also been studied, and a typical frictional force diagram is shown in 7.

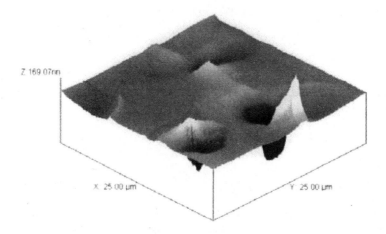

Figure 7. Frictional Force AFM Image of a Binodal Texture

The circular profiles of the nucleation and growth type phases are obvious. However, the background is generally smooth as would be expected from a quiescently cured gel. Thus, while the gross details of the morphology are reproduced, the fine scale details (as are implied in the unusual small-angle scattering scaling regimes) are not obviously related to AFM. AFM could be encouraging for further studies of this type on these interesting and important materials, as well as for understanding the rubberlike elasticity of silicone based elastomers (90-93)

Conclusions

Comparison of three differing cross-linking chemistries have shown that PDMS elastomers can generate a large-scale phase separation of clusters of high cross-link density. The necessary and sufficient conditions to cause phase separation appear to be (i) large concentrations of short PDMS chains, and (ii) large amounts of the catalyst used in the cross-linking reactions. A critical review of the PDMS end-linking literature indicates that system specific artifacts could not be the cause of this separation, and thus this phenomenon is completely general for such network-forming systems. The phase separated structures appear to emulate the structures predicted from thermodynamic instability; as a result, this system is a reaction-induced phase separation which actually traps the phase separation by extensive cross linking. The failure to observe these structures to date may arise from instrumental limitations or limited contrast as may be expected from an initially amorphous single-component system. Variants on model end-linked network themes, most generally toughened bimodal networks containing short chains can demonstrate phase separated microstructures. Further characterization work from small-angle scattering and atomic force microscopy is needed.

Acknowledgments

It is a pleasure to acknowledge the financial support provided JEM by the National Science Foundation through Grants DMR-0075198 and DMR0314760 (Polymers Program, Division of Materials Research). We also want to thank Dr. Mark Buese of PCR Chemicals for his generous donation of a sample of trimeric hydroxyl-encapped PDMS. We are also grateful to Professors Wim van Ooij and Howard Jackson of the University of Cincinnati for the use of their atomic force microscopes and to Mr. Guru Prasad Sundarajan and Ms. Susan Lindsay for their help with these AFM measurements. We would like to thank Dr. Sathish Sukumaran, Dr. Shrish Rane, Dr. Volker Urban, Dr. Kenneth Littrell, Mr. Dennis Wozniak, and Dr. Pappannan Thiyagarajan at the Argonne National Laboratories for their assistance with the SANS experiments. (The facilities of Argonne National Laboratories are supported by the Division of Materials Science, Office of Basic Energy Sciences, U.S. Department of Energy, under Contract W-31-109-ENG-38 to the University of Chicago). Finally, we wish to thank Dr. Michael Satkowski and Mr. Jeff Grothaus of Procter and Gamble for conducting the SAXS experiments.

References

1. Macosko, C. W.; Benjamin, G. S. *Pure Appl. Chem.* **1981**, *53*, 1505.
2. Sharaf, M. A.; Mark, J. E.; Al-Shamsi, A. S. *Polym. J.* **1996**, *28*, 375.
3. Llorente, M. A.; Andrady, A. L.; Mark, J. E. *J. Polym. Sci., Polym. Phys. Ed.* **1980**, *18*, 2263.

161

4. Patel, S. K.; Malone, S.; Cohen, C.; Gillmor, J. R.; Colby, R. H. *Macromolecules* **1992**, *25*, 5241.
5. Clarson, S. J.; Wang, Z.; Mark, J. E. *Eur. Polym. J.* **1990**, *26*, 621.
6. Mark, J. E. *Makromol. Chemie, Suppl.* **1979**, *2*, 87.
7. Erman, B.; Mark, J. E. *Structures and Properties of Rubberlike Networks*; Oxford University Press: New York, NY, 1997.
8. Mark, J. E. In *Applied Polymer Science - 21st Century*; Craver, C. D., Carraher, C. E., Jr., Eds.; American Chemical Society: Washington, DC, 2000; p. 209.
9. Mark, J. E.; Erman, B. In *Performance of Plastics*; Brostow, W., Ed.; Hanser: Cincinnati, OH, 2001; p. 401.
10. Mark, J. E. *J. Phys. Chem., Part B* **2003**, *107*, 903.
11. Mark, J. E. *Prog. Polym. Sci.* **2003**, *28*, 1205.
12. Mark, J. E. In *Physical Properties of Polymers. Third Edition*; Third Edition, Mark, J.E.; Ngai, K.L.; Graessley, W.W.; Mandelkern, L.; Samulski, E.T.; Koenig, J.L.; Wignall, G.D.; Eds.; Cambridge University Press: Cambridge, 2004; p. 3.
13. Kuo, C. M. Ph.D. Thesis in Materials Science, The University of Cincinnati, 1991.
14. Simon, G.; Birnsteil, A.; Schimmel, K.-H. *Polym. Bull.* **1989**, *21*, 235.
15. He, X. W.; Widmaier, J. M.; Herz, J.; Meyer, G. C. *Eur. Polym. J.* **1988**, *24*, 1145.
16. He, X. W.; Lapp, A.; Herz, J. *Makromol. Chemie* **1988**, *189*, 1061.
17. Severnyi, V. V.; Minas'yan, R. M.; Makarenko, I. A.; Bizyukova, N. M. *Polym. Sci., U.S.S.R.* **1976**, *18*, 1464.
18. van der Weij, F. W. *Makromol. Chemie* **1980**, *181*, 2541.
19. Thomas, D. R. In *Siloxane Polymers*; Clarson, S. J.; Semlyen, J. A.; Eds.; Prentice Hall: Englewood Cliffs, NJ, 1993; p. 567.
20. Viers, B. D.; Mark, J. E. *J. Inorg. Organomet. Polym.* **2006**, *16*, 00.
21. Kulkarni, A; Beaucage, G; *Polymer*, **2005**, *46*, 4454.
22. Viers, B. D.; Ph. D. Thesis in Chemistry, The University of Cincinnati, 1998.
23. Bueche, F. *J. Coll. Interfac. Sci.* **1970**, *33*, 61.
24. Stein, R. S.; *J. Polym. Sci., Polym. Lett.* **1969**, *7*, 657 .
25. Bastide, J.; Candau, S. J. in *Physical Properties of Polymeric Gels*; Cohen Addad, J.P.; Ed.; Wiley: New York, NY, 1996; p. 143.
26. Urayama, K.; Kawamura, T.; Hirata, Y.; Kohjiya, S. *Polymer* **1998**, *39*, 3827.
27. Soni, V. K.; Stein, R. S.; *Macromolecules* **1990**, *23*, 5257.
28. Hecht, A.M.; Horkay, F.; Geissler, E. *J. Phys Chem. B.* **2001**, *105*, 5637.
29. Mendes, E.; Girard, B.; Picot, C.; Buzier, M.; Boue, F.; Bastide, J. *Macromolecules* **1993**, *26*, 6873.
30. Ramzi, A.; Mendes, E.; Zielinski, F.; Rouf, C.; Hakiki, A.; Herz,J.; Oeser, R.; Boue, F.; Bastide, J. *J. de Phys. IV* **1993**, *3*, 91.

162

31. Guinier, A.; Fournet, G. *Small Angle Scattering of X-rays* John Wiley & Sons, New York, NY, 1955.

32. Glatter; O.; Kratky, O. *Small-angle X-ray Scattering* Academic Press: London, 1982.

33. (Roe, R.-J. *Methods of X-Ray and Neutron Scattering in Polymer Science* Oxford University Press: New York, NY; 2000.

34. Wignall, G. D. in *Physical Properties of Polymers*, Third Edition, Mark, J.E.; Ngai, K.L.; Graessley, W.W.; Mandelkern, L.; Samulski, E.T.; Koenig, J.L.; Wignall, G.D.; Eds.; Cambridge University Press: Cambridge, 2004; p. 424.

35. Edwards, S. F in *Polymer Networks: Structural and Mechanical Properties*; A. Chompff, A.J.; Newman, S.; Eds.; Plenum: New York, NY; 1971.

36. Falcao, A. N.; Pedersen, J. S.; Mortensen, K.; Boue, F. *Macromolecules* **1996**, *29*, 809

37. Falcao, A. N.; Pedersen, J. S.; Mortensen, K. *Macromolecules* **1993**, *26*, 5350

38. de Gennes, P. G. *Scaling Concepts in Polymer Physics*; Cornell University Press: Ithaca, NY, 1979

39. Yilgor, I.; McGrath, J. E. *Adv. Polym. Sci.* **1988**, *86*, 1.

40. Owen, M. J. In *Physical Properties of Polymers Handbook*; 2nd ed.; Mark, J. E., Ed.; Springer-Verlag: New York, NY, 1996; p. 669.

41. Vasilets, V. N.; Nakamura, K.; Uyama, Y.; Ogata, S.; Ikada, Y. *Polymer* **1998**, *39*, 2875.

42. Leite, C. A.; Soares, R. F.; Goncalves, M. D.; Galembeck, F. *Polymer* **1994**, *35*, 3173.

43. van Landingham, M. R.; McKnight, S. H.; Palmese, G. R.; Eduljee, R. F.; Gillespie, J. W.; McCulough, R. L. *J. Mater. Sci. Lett.* **1997**, *16*, 117.

44. Van Landingham, M. R.; McKnight, S. H.; Palmese, G. R.; Elings, J. R.; Huang, X.; Bogetti, T. A.; Eduljee, R. F.; Gillespie, J. W. *J. Adhes.* **1997**, *64*, 31.

45. Whangbo, M. H.; Bar, G.; Brandsch, R. *Surf. Sci.* **1998**, *411*, L794.

46. Llorente, M. A.; Mark, J. E. *Macromolecules* **1980**, *13*, 681.

47. Valles, E. M.; Macosko, C. W. *Macromolecules* **1979**, *12*, 673.

48. Meyers, K. O.; Bye, M. L.; Merrill, E. W. *Macromolecules* **1980**, *13*, 1045.

49. Meyers, K. O.; Merrill, E. W. In *Elastomers and Rubber Elasticity*; Mark, J. E., Lal, J., Eds.; American Chemical Society: Washington, DC, 1982; Vol. 193; p. 329.

50. Llorente, M. A.; Andrady, A. L.; Mark, J. E. *Coll. Polym. Sci.* **1981**, *259*, 1056.

51. Falender, J. R.; Yeh, G. S. Y.; Mark, J. E. *J. Chem. Phys.* **1979**, *70*, 5324.

52. Falender, J. R.; Yeh, G. S. Y.; Mark, J. E. *Macromolecules* **1979**, *12*, 1207.

53. Falender, J. R.; Yeh, G. S. Y.; Mark, J. E. *J. Am. Chem. Soc.* **1979**, *101*, 7353.

163

54. Speier, J. L.; Webster, J. A.; Barnes, G. H. *J. Amer. Chem. Soc.* **1957**, *79*, 974.
55. Benkeser, R. A.; Kang, J. *J. Organomet. Chem.* **1980**, *C9*, 185.
56. Valles, E. M.; Macosko, C. W. in *Chemistry and Properties of Crosslinked Polymers*; S. S., Labana, Ed.,New York, 1977; p. 401.
57. Lewis, L. N.; Lewis, N. *J. Am. Chem. Soc.* **1986**, *108*, 7228.
58. Lewis, L. N. *J. Am. Chem. Soc.* **1990**, *112*, 5998.
59. Ojima, I. In *The Chemistry of Organic Silicon Compounds*; Rappoport, Z; Patai, S., Eds.; Wiley: New York, NY, 1989; Vol. 1.
60. Gorshov, A. V.; Kopylov, Y. M.; Dontsov, A. A.; Khazen, L. Z. *Int. Polym. Sci. Tech.* **1986**, T/26.
61. Gorshov, A. V.; Kopylov, Y. M.; Dontsov, A. A.; Khazen, L. Z. *Int. Polym. Sci. Tech.* **1987**, *14*, T/42.
62. Gorshov, A. V.; Khazen, L. Z.; Kopylov, Y. M.; Dontsov, A. A. *Int. Polym. Sci. Tech.* **1987**, *14*, T/37.
63. Gorshov, A. V.; Dontsov, A. A.; Khazen, L. Z.; Ermilova, N. V. *Int. Polym. Sci. Tech.* **1987**, *14*, T/32.
64. Quan, X. Polym. Eng. Sci. **1989**, *29*, 1419.
65. Meyers, K. O.; Merrill, E. W. In *Elastomers and Rubber Elasticity*; Lal, J., Mark, J.E., Eds.; American Chemical Society: Washington, DC, 1982.
66. Macosko, C. W.; Saam, J. C. *Polym. Bull.* **1987**, *18*, 463.
67. Gustavson, W. A.; Epstein, P. S.; Curtis, M. *J. Organomet. Chem.* **1982**, *238*, 87.
68. Viers, B.D.; Sukuamaran, S.; Beaucage, G.; Mark, J.E.; *Polymer Preprints (A.C.S. Div. Poly. Chem.)* **1997**, *38(2)*, 333
69. Sukumaran, S. K.; Beaucage, G.; Mark, J. E.; Viers, B. D. *Eur. Phys. J.* **2005**, *18*, 29.
70. Mark, J. E.; Llorente, M. A. *J. Am. Chem. Soc.* **1980**, *102*, 632.
71. Mark, J. E. *Rubber Chem. Technol.* **1981**, *54*, 809.
72. Mark, J. E. *Adv. Polym. Sci.* **1982**, *44*, 1.
73. Mark, J. E. *Acc. Chem. Res.* **1985**, *18*, 202.
74. Mark, J. E.; Rahalkar, R. R.; Sullivan, J. L. *J. Chem. Phys.* **1979**, *70*, 1794.
75. Borisov, S. N.; Voronkov, M. G.; Lukevits, E. Y. *Organanosilicon Heteropolymers and Heterocompounds*; Plenum Press: New York, NY, 1970.
76. Brinker, C. J.; Scherer, G. W. *Sol-Gel Science*; Academic Press: New York, NY, 1990.
77. Mallam, S.; Hecht, A. M.; Geissler, E.; Pruvost, P. *J. Chem. Phys.* **1989**, *10*, 6447.
78. Nitzche, S.; Wick, M. : U.S. Patent 3,032,529, 1960.
79. Lewis, F. M.: U.S. Patent 3,451,965, 1969.
80. Hecht, A.M.; Geissler, E.; Horkay, F.; *J. Phys. Chem. B* **2001**, *105*, 5637

164

81. de Gennes, P. G. *J. Phys.* **1979**, *40*, 69.
82. Ribbe, A. E.; Hashimoto, T.; Jinnai, H. *J. Mater. Sci.* **1996**, *31*, 5837.
83. Ribbe, A. E.; Hayashi, M.; Weber, M.; Hashimoto, T. *Polymer* **1998**, *39*, 7149.
84. Ribbe, A. E.; Hashimoto, T. *Macromolecules* **1997**, *30*, 3999.
85. Adam, M.; Lairez, D. in *Physical Properties of Polymeric Gels*; Cohen Addad, J.P. Ed.; Wiley: New York, NY, 1996.
86. Lapp, A.; Leibler, L.; Schosseler, F.; Strazielle, C. *Macromolecules* **1989**, *22*, 2871.
87. Clarson, S. J.; Semlyen, J. A. *Siloxane Polymers*; Prentice Hall: Englewood Cliffs, N.J., 1993.
88. Doi, M. *Introduction to Polymer Physics*; Clarendon Press: Oxford, 1996.
89. Strobl, G. *The Physics of Polymers*; 2nd ed.; Springer: Berlin, 1996.
90. Erman B.; J. E. Mark *Structures and Properties of Rubberlike Networks;* Oxford University Press: New York, NY, 1997.
91. Mark, J. E. *Macromol. Symp., St. Petersburg issue* **2003**, *121*, 191.
92. Mark, J. E. *J. Phys. Chem., Part B* **2003**, *903*, 107.
93. Mark, J. E. *Acct. Chem. Res.* **2004**, *946*, 37.

Chapter 11

A Comparison of Poly(methylphenylsiloxane) and Poly(dimethylsiloxane) Membranes for the Removal of Low Molecular Weight Organic Compounds from Water by Pervaporation

Xiao Kang Zhang, Yadagiri Poojari, and Stephen J. Clarson*

Department of Chemical and Materials Engineering and the Membrane Applied Science and Technology Center, University of Cincinnati, Cincinnati, OH 45221–0012
*Corresponding author: Stephen.Clarson@UC.edu

Pervaporation experiments were conducted to remove low concentration (<10wt.%) organic liquids: pyridine, iso-propanol (IPA) and methylethylketone (MEK) respectively from binary aqueous mixtures using lab-made poly-(methylphenylsiloxane) (PMPS) and poly(dimethylsiloxane) (PDMS) membranes. The performance of these membranes (selectivity and flux) were studied and compared with commercial PDMS membranes. The PMPS membranes showed greater selectivity towards pyridine while the PDMS membranes showed higher permeation flux and selectivity towards IPA and MEK.

Introduction

Membrane separations have several advantages over competing technologies such as distillation; one example is that membrane separations are far more efficient in their use of energy. Based on the number of publications and patents over the last fifteen years, pervaporation shows great potential for commercial applications. One application for pervaporation is in surface-water or ground-water treatment or purification (1). Elastomers, and particularly silicone elastomers, seem to be the most suitable membrane materials for the removal of low molecular weight organic compounds from water (2).

When considering the mechanisms of transport in polymer-based systems the case of glassy polymers and the case of rubbery polymers are the ones most often studied (1, 12). The commercial linear silicones or polysiloxanes are fluids at room temperature and their viscosity is governed by their molar masses. The linear silicones are readily crosslinked into network structures that are highly elastomeric at room temperature and thus fall into the category of rubbery polymers. The PDMS and PMPS elastomers that we have investigated as membranes herein are well above their glass transition temperatures under ambient conditions (3, 4), thus their polymer backbones and side chains exhibit considerable mobility under the conditions investigated.

Membranes made of PDMS elastomers have been used for pervaporation of dilute organic compounds by various authors and some of the findings that are pertinent to this investigation are described here. Watson and Payne (5) have investigated a commercial dimethyl silicone rubber membrane (supplied by ESCO Ltd., Middlesex, UK) for the separation of a wide variety of organic compounds from binary aqueous organic solutions and the pervaporate analyses were performed using mass spectrometry. They also reported solvent sorption studies for the said elastomer and the value for water was given as 0.0026 (g / g) at 25°C. The corresponding sorption values for a series of alcohols was: methanol 0.0160 (g / g); ethanol 0.0190 (g / g) ; propanol 0.1200 (g / g) and butanol 0.1300 (g / g) at 25°C. From pervaporation data on silicone membranes which were crosslinked using room temperature vulcanization RTV from a commercial precursor formulation (GE RTV 615 A + B), Nijhuis, Mulder and Smolders have reported an intrinsic water permeability of 33×10^{-14} m^2 / s (5). Water fluxes for 100 μm-thick PDMS membranes were calculated from permeability data at 25°C and a value of 12 g / m^2 – h was obtained. Solvents investigated by pervaparation were trichloroethylene and toluene, respectively, in binary aqueous mixtures. They reported an excellent permeability for organic components that are comparable in value with those for completely nonpolar hydrophobic elastomers. It was concluded that the high selectivity of PDMS for the organic component over water was not caused by an increase in permeability of the organic component but rather by the very low water permeability. The separation factors from dilute binary aqueous solutions at 25°C for various organic solvents by pervaporation across PDMS membranes were given as: methanol / 9 ; phenol / 97 ; pyridine / 220 and toluene / 44,000 (6). Feng and Huang have described the separation of low concentrations of IPA (<6wt.%) from water, single-pass enrichments of 4.1-11.8 were obtained using homogenous PDMS (25 μm thickness) and silicone-coated ultrathin silicone-polycarbonate copolymer membranes by pervaporation (7). Driolo, Zhang and Basile reported that pervaporation of pyridine-water mixtures using PDMS and poly(vinyl alcohol) (PVA) composite membranes and a dense cation-exchange membrane showed different selectivity, PDMS was found to be selective to IPA while the other two membranes indicated water selectivity (8). The performance of commercially available PDMS membranes in a shell-and-tube heat-exchanger type of hollow fiber module configuration in a pervaporative mode have been investigated by Hwang and coworkers for the treatment of wastewater containing toluene, trichloroethane and methylene chloride (9, 10). Hong and Hong (11) have reported an investigation of a PDMS / ceramic composite membrane in a tubular type configuration and the results compared with a dense homogeneous membrane with an approximate thickness of 50 μm. The ceramic support material was γ-alumina and the PDMS was a commercial RTV formulation supplied by LG-Dow. Good selectivity for the separation of isopropyl alcohol (IPA) from water was seen for both the PDMS / ceramic composite membrane and the dense PDMS membrane particularly below 0.2 weight fraction IPA in the feed at 50°C but the selectivity for IPA decreased with increased IPA in the feed.

To our knowledge, no research work has been reported on the recovery of organic compounds from dilute binary aqueous solutions using poly(methylphenylsiloxane) membranes by pervaporation. In the present work we have investigated the pervaporation behaviour of dilute solutions (<10wt.%) of pyridine in water, IPA in water and MEK in water across an unfilled PMPS membrane, an unfilled end-linked PDMS membrane and a commercial PDMS Silastic® brand membrane. The permeability and permselectivity were determined for each binary system and the results are compared and discussed below.

Experimental

PMPS Membrane Synthesis

The monomer, cyclic methylphenylsiloxane tetramer, was kindly provided by the Dow Corning Corporation, Midland, Michigan. This was placed under vacuum in order to remove any trace amounts of solvent or moisture prior to polymerization. The potassium hydroxide initiator was ground to a fine powder under dry nitrogen prior to adding to flame dried reactor. The anionic ring-opening polymerization was maintained at 150°C. After the polymerization, the polymer gum was diluted using ethyl acetate and several milliliters of glacial acetic acid were added so that the potassium silanolate end-groups would be converted to terminal silanol groups. The polymer solution was washed several times with ultra filtered deionized water. After the final water layer was separated, the solution was dried using anhydrous magnesium sulphate and the ethyl acetate was then removed by distillation. Analysis by Gel Permeation Chromatography (GPC) showed the PMPS to contain some oligomeric cyclic species due to the equilibrium nature of the polymerization. Several washings with heptane were used to remove the cyclic species before the α,ω-hydroxyl terminated PMPS was placed in a vacuum oven overnight in order to remove any residual solvent. Several such samples were made and were characterized in toluene using a Waters Gel Permeation Chromatograph that was calibrated using polystyrene and PMPS standards. The number-average molar masses M_n for the linear PMPS samples ranged from 24,800 g mol^{-1} to 247,200 g mol^{-1}. The PMPS membranes investigated here were made from a sample having a number-average molar mass of 59,900 g mol^{-1}.

Room temperature vulcanization was used to prepare the PMPS membranes, with tetraethylorthosilicate (TEOS) as the cross linking agent and stannous octoate as the catalyst. The membrane precursor mixture was first degassed under vacuum and then poured onto a flat Teflon plate, before being uniformly spread using a doctor blade. After curing, the membrane was placed into a tray filled with ethanol and the membrane was peeled from the Teflon under ethanol. The PMPS membrane was then placed in a tray of toluene and allowed to swell overnight. The final PMPS membrane was deswollen using methanol, prior to drying in a vacuum oven.

PDMS Membrane Synthesis

The PDMS was cross linked using tetraethylorthosilicate (TEOS) as the cross linking agent and stannous octoate as the catalyst, as described above. The commercial elastomer membrane investigated was a Dow Corning 500-1 Silastic® brand medical-grade silicone rubber sheeting. Dow Corning also kindly provided us with α,ω-hydroxyl terminated poly(dimethyl siloxane) (PDMS) having a molar mass of 18,000 g mol^{-1}.

Compositional Analysis of the Feed and Permeate

Gas Chromatographic analysis of the feed and the permeate was performed using a Hewlett Packard 5890 Series-II GC system with a thermal conductivity

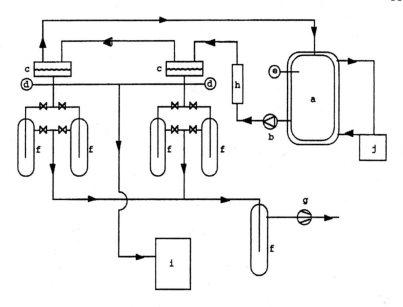

Figure 1. The pervaporation apparatus: (a) feed reservoir (5-L); (b) feed circulating pump (Baxter; 1/16 HP); (c) pervaporation cell; (d) pirani gauge (range of 10-3 to 200 Torr); (e) thermometer; (f) cold trap; (g) vacuum pump; (h) flow meter (Gilmont Accucal; maximum flow rate of 4.7 + 2% L / min; maximum pressure of 200 psi); (i) HP 5890 Series II gas chromatograph; (j) thermostated bath.

detector and a Hewlett Packard 3396 Series II integrator. A 6' x 1/8" stainless steel column packed with 60/80 Teuax TA (a porous polymer based on 2,6-diphenyl-p-phenylene oxide) was used. Nitrogen was employed as the carrier gas. Weight percentage calculations were based upon calibration plots determined for a series of known weight percentages of pyridine in water, IPA in water and MEK in water respectively.

Pervaporation

A schematic representation of the pervaporation apparatus is shown in Figure 1. The membrane was installed in a stainless steel pervaporation cell. To prevent the film being damaged during the experiments by the support screen of the membrane cell, we used a regenerated-cellulose with a pore size 3.0μm backing between the membrane and the support screen. The effective membrane area in each cell was 39.7 cm^2. The feed aqueous solution was circulated through the cell at a flow rate of 2.7 L/min. from a feed reservoir at room temperature

(26 ~ 28 °C). The pressure at the downstream side of the membrane was kept below 1 mmHg. Liquid nitrogen was used as a cooling agent for the cold traps. At -196°C the partial pressures of permeates were low enough to prevent loss of product through sublimation. The test procedure was as followed: after installation of the membrane in the cell, the feed solution was circulated through the cell for the first 12 hours and the permeate was collected in cold trap for four hours so that steady state conditions could be reached. Subsequently permeate was collected again for another 4 hours. Finally the collected permeate was analyzed using a Gas Chromatograph.

Results and Discussion

For each of the membrane and organic liquid in water systems, the flux (g / m^2 h) was determined from the weight W (g) of permeate collected over a given time t (hours) per unit area of the membrane from Equation (1), where A is the effective membrane surface area (m^2).

$$J = W/(A \times t) \tag{1}$$

The flux values were normalized to a membrane thickness of 100 μm in each case by Equation (2), where l is the membrane thickness (μm).

$$J = (W/(A \times t))(l/100) \tag{2}$$

The permselectivity ($\alpha_{A/B}$) was determined by Equation (3) from the known composition (weight %) of components A and B in the liquid feed mixture (X) and the composition of the collected pervaporate (Y) as determined by Gas Chromatography analysis.

$$\alpha_{A/B} = (Y_A/Y_B) / (X_A/X_B) \tag{3}$$

In the present study three types of membranes were used to separate three different dilute organic liquids (<10wt.%) of, (i) pyridine (ii) IPA and (iii) MEK from aqueous mixtures. The membranes used to separate these mixtures were: (a) PMPS (b) lab-made PDMS and (c) commercial PDMS Silastic® (medical grade) membranes. Pervaporation results of these membranes and their flux and selectivity are presented in Table 1, 2 and 3.

The permeation flux and the corresponding permselectivity for pyridine/water through the PMPS, lab-made PDMS and commercial PDMS Silastic® membranes are shown in Table 1. It can be seen that lab-made PDMS membrane and the

Table 1. Pervaporation measurements of the polysiloxane membranes (feed: 1 wt% of pyridine in water)

Membrane	Thickness (μm)	Total Flux (g m⁻² h⁻¹)	Normalized Total Flux (g m⁻² h⁻¹)	Wt % pyridine in permeate
Lab-made PMPS	185	10.4	19.3	31.7
Lab-made PDMS	65	34.6	29.4	52.7
Silastic® (DC 500-1)	160	16.1	25.6	53.5

Table 2. Pervaporation measurements of the polysiloxane membranes (feed: 1 wt% of IPA in water)

Membrane	Thickness (μm)	Total Flux (g m⁻² h⁻¹)	Normalized Total Flux (g m⁻² h⁻¹)	Wt % IPA in permeate
Lab-made PMPS	185	8.9	16.4	3.7
Lab-made PDMS	115	12.1	13.9	14.6
Silastic® (DC 500-1)	160	9.0	14.4	12.7

commercial PDMS Silastic® membrane had better permselectivity to pyridine than the PMPS membrane at low concentration (1 wt%).

However, the selectivity of pyridine-water liquid mixtures by PMPS membranes compared to that of commercial PDMS Silastic® membranes was superior and increased with the pyridine concentration in the feed mixture. The results are shown in Figure 2. The permeation flux of pyridine-water liquid mixtures by PMPS membranes was plotted in Figure 3. It can be seen that the flux increased with the pyridine concentration in the feed mixtures.

The flux for the IPA-water system was a little higher for the PMPS membrane than for either the commercial PDMS Silastic® membrane or the lab-made PDMS membrane. However, the permselectivity for the PMPS membrane was found to be lower than the other membranes (see Table 2). The results obtained here for the commercial PDMS Silastic® membrane are very similar to those reported by Feng and Huang (7) from pervaporation data for a 1 wt% aqueous IPA-water solutions at 25°C through a 25 μm commercial PDMS membrane (Membrane Products Co., Troy, NY).

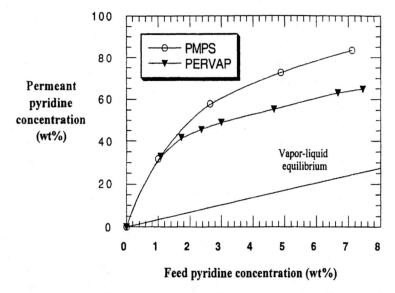

Figure 2. Effect of feed composition on the pervaporation performance for pyridine in water mixtures through (a) a PMPS membrane at 26°C and (b) a PERVAP 1160 commercial PDMS composite membrane (reference 8). The vapor-liquid line is included for purpose of comparison with the pervaporation data.

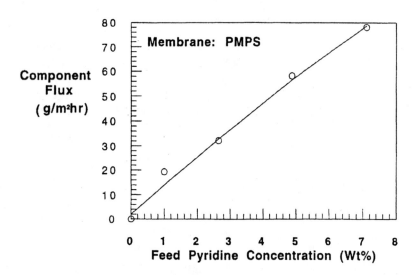

Figure 3. Permeation flux of pyridine-water mixtures through the PMPS membrane.

Table 3. Pervaporation measurements of the polysiloxane membranes
(feed: 1 wt% of MEK in water)

Membrane	Thickness (μm)	Total Flux ($g\ m^{-2}\ h^{-1}$)	Normalized Flux ($g\ m^{-2}\ h^{-1}$)	Wt % MEK in permeate
Lab-made PMPS	185	20.6	38.0	18.0
Silastic® (DC 500-1)	160	57.4	91.9	42.8

In separation of the MEK from aqueous mixtures, the performance of PMPS membranes were compared with that of commercial PDMS Silastic® membranes and the results are presented in Table 3. It can be seen that both the flux and the selectivity of the PMPS membranes were poor compared to that of commercial PDMS Silastic® membranes.

Table 4. Solubility parameters δ of the membrane polymers and the
liquid permeates

Membrane	δ $(cal/cm^3)^{1/2}$	Permeate	δ $(cal/cm^3)^{1/2}$	Boiling Point (°C)
Lab-made PMPS	9.6	Pyridine	10.0	115
Lab-made PDMS	7.5	IPA	11.5	82.4
Silastic® (DC 500-1)	7.5	MEK	9.3	79.6
		Water	23.4	100

These results can be explained from the fact that pervaporation performance is governed by three leading factors: the solubility of the feed components in the membrane; the relative diffusion rates of permeates through the membrane; the evaporation of permeates from the downstream face of the membrane (1). Though both PMPS and PDMS are considered hydrophobic membranes, the closer the solubility parameter of permeate and membrane, the higher the membrane-permeate interaction. Therefore the interactions of the pyridine, IPA and MEK, are expected to be stronger with the PMPS system than those with the PDMS system which should aid the uptake of the organic components but may hinder transport (see Table 4). Secondly, for the PMPS and PDMS membranes studied here, it should be noted that the PDMS (3) is much further above its glass transition temperature T_g than the PMPS (4) and hence it should have more free volume due to the high mobility of the siloxane backbone and the methyl substituents. Thus one may anticipate higher diffusion rates across the PDMS membranes relative to the PMPS membranes. However, at high concentrations of pyridine (>1 wt%) in the feed mixtures an increased selectivity for pyridine by the PMPS membrane was seen when compared to the PDMS membranes, as show in Figure 2. This is due to the fact that polymer-permeant interactions led to an increased preferential sorption of pyridine over water with increased concentration of pyridine in the feed mixture. This was less pronounced in case of the PDMS membranes due to the presence of a larger free volume when compared to the PMPS membranes at the temperature of the experiments carried out here.

Conclusions

Low concentrated organic liquids (<10wt.%) of: pyridine, IPA and MEK from aqueous mixtures were successfully separated using lab–made PMPS and PDMS membranes. PMPS membranes showed greater selectivity towards pyridine than lab–made PDMS and commercial PDMS Silastic® membranes, and increased with composition of pyridine in water (over 1 wt% of pyridine). On the other hand PDMS membranes showed higher permeation flux and selectivity towards IPA and MEK than PMPS membranes.

Acknowledgements

We thank Dr. James White of the Dow Corning Corporation for the kind donation of the Dow Corning 500-1 Silastic® brand silicone samples.

References

1. Huang, R. Y. M. *Pervaporation Membrane Separation Processes*, Elsevier: Amsterdam, 1991.
2. Vane, L. M. *Membrane Quarterly* **2005**, *20(4)*, 6-10.
3. Clarson, S. J.; Dodgson, K.; Semlyen, J. A. *Polymer* **1985**, *26*, 930-934.
4. Clarson, S. J.; Dodgson, K.; Semlyen, J. A. *Polymer* **1991**, *32*, 2823-2827.
5. Watson, J. M.; Payne, P. A. *J. Membrane Sci.* **1990**, *49*, 171-205.
6. Nijhuis, H. H.; Mulder, M. H. V.; Smolders, C. A. *J. Appl. Polym. Sci.* **1993**, *47*, 2227-2243.

7. Feng, X.; Huang, R. Y. M. *J. Membrane Sci.* **1992**, *74*, 171-181.
8. Drioli, E.; Zhang, S.; Basile, A. *J. Membrane Sci.* **1993**, *80*, 309-318.
9. Ji, W.; Sikdar, S. K.; Hwang, S.-T. *J. Membrane Sci.* **1994**, *93*, 1-19.
10. Ji, W.; Hilaly, A.; Sikdar, S. K.; Hwang, S.-T. *J. Membrane Sci.* **1994**, *97*, 109-125.
11. Hong, Y. K.; Hong, W. H. *J. Membrane Sci.* **1999**, *159*, 29-39.
12. Vieth, W. F. *Diffusion In and Through Polymers*, Hanser: New York, 1992.

Chapter 12

Synthesis of Carbosilane Dendrimer-Based Networks

Aleksandra V. Bystrova[1,2], Elene A. Tatarinova[1],
and Aziz M. Muzafarov[1,*]

[1]Enikolopov Institute of Synthetic Polymer Materials, Russian Academy
of Sciences, ul. Profsoyuznaya 70, Moscow 117393, Russia
[2]Physics Department, Moscow State University, Moscow 119992, Russia
*Corresponding author: Email: aziz@ispm.ru

Two different approaches of dendrimer based network
preparation are described and problems of the dendrimers
functionality's conversion control are the focus of discussion.
Dendrimers of small size (G2), used as crosslinkers to the
larger ones (G6) in the second approach, are shown to be an
ideal "difunctional" linkage. The different nature of swelling-
drying stresses in dendrimeric networks is considered and
some requirements of the initial building blocks are discussed.

Dendrimers (1) are polymer materials with tree-like, well-defined
structures. Their globular nature and spherical shape make them very
perspective objects to be employed in the construction of regular networks.

Dendrimer-based networks are of interest, first, as nanostructured objects
and, second, because investigation of networks properties allows us to learn
more about dendrimers themselves.

Carbosilane dendrimers (2) have a high potential of obtaining various
derivatives including networks on their base, owing to high reactivity and
controllable chemistry of their functional groups, which can be treated by the
use of methods of not only organic chemistry, but also the chemistry of
silicones.

There are several trends in the field of dendrimer-based networks synthesis.
In principle there are only a few examples of synthesis of macroscopic three-
dimensional truly dendrimers based networks. These are, first of all, works (3)
dealing with the synthesis the networks, based on PAMAMOS dendrimers (4),
which were widely investigated. Some efforts were based on alkoxyterminated
carbosilane dendrimers (5,6). Often dendrimers are used in combination with

linear polymers or monomers, playing a role of very high functional crosslinking agents (7), modifiers of mechanical properties of materials obtained (8) or templates for generation and regulation of nanoporosity (9). It should be noted that little attention was paid to regulation of network parameters, such as crosslinking density and ratio between intermolecular and intramolecular reactions.

Particularly, it was shown that at very low degree of crosslinking dendrimer-based networks are not stressed and even exhibit elastic properties (10). Thus dendrimers themselves are sufficiently soft spheres and the way how they are crosslinked in a network has a dramatic influence on the properties of the resulting material.

Hence the goal of this study was to compare different variants of network preparation, keeping in mind assurance of maximal control over the degree of crosslinking and conversion of functional group, moreover taking into consideration the difference between intermolecular and intramolecular reactions.

In this issue we will describe two different approaches to the networks synthesis, some properties of the obtained networks and discuss advantages and disadvantages of both methods.

Experimental

Materials. All solvents were dried over calcium hydride and distilled. Organosilanes and chlorosilanes were distilled just before use. All solvents were dried over calcium hydride and distilled.

Platinum 1,3-divinyl-1,1,3,3-tetramethyl-1,3-disiloxane complex in xylene solution (catalyst) was obtained from Aldrich and used as received.

Dendrimer Si$_{253}^{256}$(All) (or G6(All)) (upper index represents functionality of dendrimer, lower – number of silicon atoms constituted the dendrimer, type of functional groups is shown in brackets). Dendrimers were synthesized according to the routine method described in our paper (11) from the four functional branching center tetraallylsilane via the sequential iteration of Grignard reaction with the use of allylchloride and magnesium, and hydrosilylation reaction with dichloromethylsilane.

Dendrimer Si$_5^{12}$(All). The same procedure as for the preparation Si$_{253}^{256}$(All) was used except that trichlorosilane was used instead of dichloromethylsilane. ^{1}H NMR (CDCl$_3$): δ (ppm) 0.58 (t, C$\underline{H}_2$-Si), 0.68 (t, C$\underline{H}_2$-Si- CH$_2$-CH=CH$_2$), 1.37 (m, CH$_2$-C$\underline{H}_2$-CH$_2$), 1.61 (d, C$\underline{H}_2$-CH=CH$_2$), 4.90 (t, CH=C$\underline{H}_2$), 5.80 (m, C$\underline{H}$=CH$_2$)

Dendrimer Si$_{17}^{12}$(Cl). 2.5 g (3.12 mmol) of dendrimer Si$_5^{12}$(All) were dissolved in 7.85 mL of dry toluene. 16 μL of platinum catalyst and 4.52 g (47.8 mmol) of chlorodimethylsilane (28% excess relative to stoichiometry) were added to dendrimers solution under argon. The reaction mixture was stirred at room temperature in closed flask equipped with Teflon coated magnetic stirrer bar till full disappearance of allyl functionalities (controlled by ^{1}H NMR). Due to the high functionality of the product obtained it was used for the following synthetic procedures without isolation from the solution. ^{1}H NMR (CDCl$_3$): δ (ppm) 0.51 (s, Si-C$\underline{H}_3$), 0.74 (t, C$\underline{H}_2$-Si-CH$_2$), 1.00 (t, Si-CH$_2$-C$\underline{H}_2$-CH$_2$-Si-Cl), 1.42 (m, Si-CH$_2$-C$\underline{H}_2$-CH$_2$-Si), 1.58 (m, C$\underline{H}_2$-Si-Cl)

Dendrimer Si$_{17}^{12}$(H) (or G2(H)). A solution of Si$_{17}^{12}$(Cl) (6.04 g, 3.12 mmol) and chlorodimethylsilane (0.98 g, 10.36 mmol) in 7.85 mL of toluene (from previous synthesis) was added dropwise slowly into a suspension of LAH (2.05 g, 54.02 mmol, 13% excess relative to stoichiometry) in 70 mL of THF. The mixture was stirred 9 hours under reflux conditions; reaction progress was controlled via functional chlorine analysis (negative probe). After reaction completion 200 mL of toluene and 50 mL of acetic acid were added cautiously to the reaction mixture and resulting mixture was washed with H$_2$O. The organic layer was dried over anhydrous Na$_2$SO$_4$ and filtered through a Shott filter. After removal of solvents in vacuum we obtained 4 g (84%) of desired product, a clear, colorless liquid. Purity (by GPC data) 94%. ^{1}H NMR (CDCl$_3$): δ (ppm) 0.06 (d, Si-C$\underline{H}_3$), 0.61 (m, C$\underline{H}_2$-Si), 1.37 (m, Si-CH$_2$-C$\underline{H}_2$-CH$_2$-Si), 3.87 (m, $\underline{H}$-Si)

Dendrimer Si$_{509}^{512}$(OC$_2$H$_5$). 7.35 g (0.23 mmol) of dendrimer Si$_{253}^{256}$(All) were dissolved in 23.36 mL of dry toluene. 61 μL of platinum catalyst and 10.91 g (81.25 mmol) of diethoxymethylsilane (38% excess to stoichiometry) were added to the dendrimer under argon atmosphere. Reaction mixture was kept in closed flask at room temperature till full disappearance of allyl functionalities (4.7–5.0 and 5.6–5.9 ppm on the ^{1}H NMR spectrum). After evacuation of solvent 15.24 g of transparent, viscous product were obtained. ^{1}H NMR (CDCl$_3$): δ (ppm) -0.10 (s, Si-C$\underline{H}_3$) 0.08 (d, C$\underline{H}_3$-Si-O), 0.53 (m, Si-C$\underline{H}_2$-), 0.65 (m, -C$\underline{H}_2$-Si-O), 1.19 (m, Si-O-CH$_2$-C$\underline{H}_3$), 1.34 (m, Si-CH$_2$-C$\underline{H}_2$-CH$_2$-Si), 3.73 (m, Si-O-C$\underline{H}_2$-CH$_3$)

Other dendrimers with ethoxysilyl and chlorosilyl functionalities were synthesized similarly.

Networks preparation 1.
G7^m-n (m – number of functional groups, n – concentration upon synthesis). Solutions of corresponding concentrations in acetic acid were refluxed for several days. Typical loading of polymer was 1-5 g. On reaching appropriate conversion, low-molecular part was removed by evacuation. Samples were dried in vacuum oven at 100 ^{0}C. Results of elemental analysis of the networks based on dendrimer Si$_{509}^{512}$(OC$_2$H$_5$) are shown in Table I.

G7^{512}(Cl). Dendrimer was placed in Petri dish and exposed to air moisture for one day, then heated in an oven at 70 °C for 3 hours. Conversion was estimated based on IR spectra data, where one can see significant decrease of the intensity of Si-Cl band absorbance area (~484 cm^{-1}).

Table I. Elemental analysis data for dendrimer-based networks samples

Network sample	Si	C	H
$G7^{512}$-3	28.83	51.10	9.55
$G7^{512}$-10	28.99	50.86	9.72
$G7^{512}$-25	28.53	50.76	9.58
Estimated for full conversion	30.12	51.55	9.67

Networks preparation 2.
n(G2)/(G6) (n – molar ratio of G2 to G6). Calculated quantities of Si$_{253}^{256}$(All) and Si$_{17}^{12}$ (H) were mixed together, either in bulk or followed by adding toluene to obtain 30% concentration (mass). After that platinum catalyst was added and the resulting mixtures were stirred in a closed flask till gelation. The bulk reaction mixture was heated up to 100 ^{0}C for one day. Qualitative control of functional groups conversion was accomplished via comparison of IR-spectra of the initial dendrimers and the networks obtained. Anal. Calcd for 6(G2)/(G6): C, 64.60; H, 11.26; Si, 24.14. Found: C, 65.60; H, 11.28; Si, 21.70.

n(TMDS)/(G6). Calculated quantities of Si$_{253}^{256}$(All) and TMDS were mixed together, either in bulk or followed by adding toluene to obtain 30% concentration (mass). After that platinum catalyst was added and the resulting solutions were stirred in closed flask till gelation. The bulk reaction mixture was heated up to 100 ^{0}C for one day. Qualitative control of functional groups conversion was accomplished via comparison of IR-spectra of the initial compounds and the networks obtained.

Analytical techniques. The purity of the polymers was evaluated by ^{1}H NMR (Bruker WP-200 SY spectrometer (200.13 MHz)) in CDCl$_3$, internal standard – tetramethylsilane, gel permeation chromatography (SEC) in THF (detector – refractometer) and elemental analysis. FTIR was performed on a "Bruker Equinox 55/S" spectrometer.

Degree of swelling. Degree of swelling was measured in the following way (15a). A sample was placed in a tube with glass filter bottom, filled up with a solvent, and left for a day for equilibrium swelling. Then excess of solvent was removed by centrifugation (1400 g, 4 min.). The swollen sample was transferred into a weighed vial, weighed and dried in vacuum oven at 100 ^{0}C. Degree of swelling was determined as a ratio of the mass of swollen gel to the mass of dry gel. Several measurements were made for each sample, relative error didn't exceed 3%.

Results and Discussion

For the first approach (6) condensation of ethoxysilyl and chlorosilyl derivatives of carbosilane dendrimers via intermediate formation of silanol groups was used, because their high reactivity in the hydrolytic polycondensation processes is well-known. Allyl-terminated dendrimers of 6th generation Si_{253}^{256} (All) were transformed into derivatives containing ethoxysilyl functional groups through the exhaustive hydrosilylation with diethoxymethylsilane. Networks were prepared from the solutions of different concentrations (3, 10, 25 wt.%) in the acetic acid (Scheme 1).

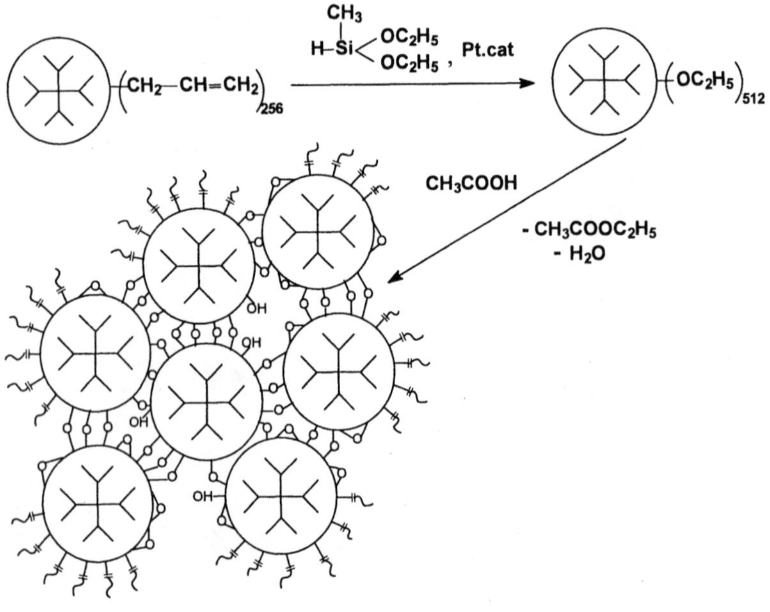

Scheme 1. Synthesis of the networks $G7^{512}$-n.

In this case, acetic acid plays the role of an active solvent; that is it serves as both solvent and co-reagent. The interaction between acetic acid and alkoxysilyl groups is worth discussing separately, but this issue is beyond the scope of this investigation. It is important, that the above process gives rise to the formation of siloxane bonds and is accompanied by evolution of low-molecular-mass products: ethyl acetate, alcohol, and water. Regarding this, the conversion of initial ethoxysilyl functional groups may be calculated from the contents of alcohol and ethyl acetate in reaction mixture. For all obtained networks, the conversion of functional groups was about 70% (Table II). It should be noted that, since the conversion was calculated from the yield of low-molecular-mass products, whose evolution under the experimental conditions was accompanied by inevitable losses, the real values of conversion can be substantially higher than those in Table II. Extraction with a Soxhlet's apparatus showed that the content of the insoluble part was above 90% for all the networks. The data obtained suggested the efficiency of the selected synthetic scheme; therefore, this scheme was further used for the synthesis of dendrimer-based networks of the first series.

The properties of the final networks are determined by parameters of the initial dendrimer structure. A decrease in the dendrimer functionality inevitably leads to a reduction in the crosslink density. In order to vary the density of the siloxane interlayer, a dendrimer of the 7th generation $Si_{509}^{256}(OC_2H_5)$ containing 256 ethoxysilyl functional groups was prepared via hydrosilylation of $Si_{253}^{256}(All)$ with ethoxydimethylsilane by analogy with the synthesis of the dendrimer $Si_{509}^{512}(OC_2H_5)$. On the basis of the seventh generation dendrimer, two crosslinked samples were prepared (at initial dendrimer concentrations in acetic acid of 25 and 50 wt.%). When a new (ethoxydimethylsilyl) derivative was used for the production of network samples, their crosslink density formally decreased by a factor of 2. Moreover, the incomplete replacement of allyl functional groups in dendrimer (All) by ethoxydimethylsilane yielded dendrimers $Si_{381}^{128}(OC_2H_5)^{128}(All)$ and $Si_{279}^{26}(OC_2H_5)^{230}(All)$, thereby making it possible to further decrease the crosslink density of the network samples in proportion to a reduction in the total functionality (in the two latter cases, the distribution of ethoxydimethylsilyl groups was random.) Owing to a noticeably lower content of ethoxysilyl groups in the aforementioned samples, their solubility in acetic acid appeared to be insufficient; therefore, the reaction was carried out in an acetic acid–toluene mixture (1:1, wt/wt). Interesting data were obtained for the synthesis of networks based on dendrimers with a reduced functionality. In 25% solutions of dendrimers $Si_{381}^{128}(OC_2H_5)^{128}(All)$ and $Si_{279}^{26}(OC_2H_5)^{230}(All)$ no macroscopic networks formed even though the number of functional groups was many times larger than the stoichiometric amount, required for their network formation. According to Flory (12), for identical functions $\alpha_{cr} = 1/(f - 1)$, given the minimum average functionality of a dendrimer is $f = 26$, we have $1/(26 - 1)$; that is, the conversion of functional groups equal to 4% is sufficient for gelation. Considering that the reaction proceeds to a high conversion values ($\geq 70\%$), we conclude that, under these conditions, the majority of functional groups are consumed for intramolecular cyclization. Indeed, it is obvious that intramolecular cyclization is the only alternative to network formation. Of greater importance is the qualitative result: the role of intramolecular cyclization may be very significant.

As was expected, an increase of the dendrimer concentration in the reaction mixture changed the ratio between intra- and intermolecular processes in favor of the latter. At a concentration of 50%, dendrimer $Si_{279}^{26}(OC_2H_5)^{230}(All)$ formed a network. However, the ^1H NMR analysis of soluble reaction products indicated that the allyl groups partly interact with each other, thus uncontrollably changing the functionality of the system as a whole. Therefore the above-mentioned sample was not studied. To investigate networks based on dendrimers containing a smaller number of functional groups, it would be reasonable to develop a different synthetic approach.

After drying in a vacuum oven all the networks were white powders, swelling in organic solvents. Some characteristics of the networks obtained are represented in Table II.

In view of the aforesaid it is obvious, that intramolecular reactions play an important role in the network formation. In this connection it was highly desirable to prepare a series of the networks, avoiding intramolecular

Table II. Synthetic conditions and characteristics of the first series of networks

Initial compounds	Sample	reaction conditions	conversion of functional groups, %	non-soluble part, %	$\alpha = m_{sw}/m_{dry}$	
					xylene	methanol
$Si^{512}_{509}(OC_2H_5)$	$G7^{512}$-3	acetic acid, 3%	71	95	1.90	1.67
$Si^{512}_{509}(OC_2H_5)$	$G7^{512}$-10	acetic acid, 10%	76	94	1.98	1.62
$Si^{512}_{509}(OC_2H_5)$	$G7^{512}$-25	acetic acid, 25%	81	95	1.70	1.59
$Si^{256}_{509}(OC_2H_5)$	$G7^{256}$-25	acetic acid, 25%	70	92	2.83	1.86
$Si^{256}_{509}(OC_2H_5)$	$G7^{256}$-50	acetic acid, 50%	>70[2]	94	1.59	1.09
$Si^{128}_{381}(OC_2H_5)^{128}(All)$	—[1]	acetic acid – toluene (1:1), 25%	>90[2]	–	–	–
$Si^{26}_{279}(OC_2H_5)^{230}(All)$	—[1]	acetic acid – toluene (1:1), 25%	>90[2]	–	–	–
$Si^{26}_{279}(OC_2H_5)^{230}(All)$	$G7^{26}$-50	acetic acid – toluene (1:1), 50%	72	87	–	–
$Si^{512}_{509}(CI)$	$G7^{512}(CI)$	bulk	>90[2]	–	1.12	1.03

[1] – no gelation occured

[2] – conversion was qualitatively estimated from spectroscopy data

cyclization, which is inevitable under homofunctional processes due to very high functionality of the dendrimers. Previously (13) we have already mentioned the idea to construct a network from two dendrimers with different functional groups capable of heterofunctional condensation. For dendrimers of high generation this suggestion appeared to be unachievable due to diffusion problems, significant when the objects have dimensions about 6 nm. That is why in this study we utilized two dendrimers of different sizes, namely 6th and 2nd generation dendrimers with allyl and hydride functional groups, respectively, and hydrosilylation reaction as heterofunctional addition process. Indeed, it is possible to crosslink 6th generation dendrimers with a simple difunctional reagent, like tetramethyldisiloxane (TMDS). However, this way does not solve the problem of intramolecular reactions, because it can utilize both functionalities for the same dendrimer. We assume that the use of the small dendrimers to connect the bigger ones instead of linear crosslinking agents allows one to eliminate cyclization of crosslinkers on one dendrimer.

It is also important, that by changing the ratio between dendrimers one can easily regulate the degree of crosslinking and correspondingly content of residual allyl groups. To the best of our knowledge such an approach involving co-condensation of two dendrimers of different sizes and functionality to produce a macroscopic three-dimensional network has not been reported.

Dendrimer G2(H) was obtained via hydrosilylation of dendrimer G1(All) with chlorodimethylsilane and following reduction of obtained product with lithium aluminum hydride (Scheme 2). Structure and purity of obtained dendrimer were approved by ^{1}H NMR and GPC.

Crosslinked samples n(G2)/(G6) (where n is the molar ratio between G2 and G6) were obtained as a result of hydrosilylation reaction between dendrimers G6(All) and G2(H) in toluene under different dendrimers ratio (G6:G2=1:1, 1:3, 1:6, 1:18) (Scheme 3).

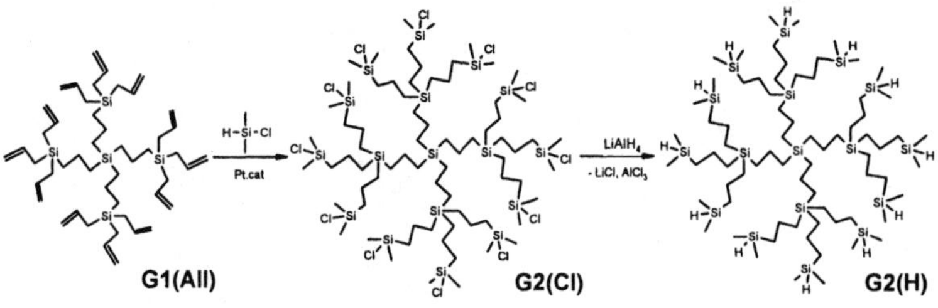

Scheme 2. Synthesis of dendrimer G2(H).

For the comparison networks from G6(All) and tetramethyldisiloxane (TMDS) were synthesized in the same way (Scheme 4). Dendrimer/TMDS ratio was 1:36, which gave the same balance between hydride and allyl functionalities as in network 6(G2)/(G6).

All the samples obtained were transparent gels, except (G2)/(G6). In the latter case we obtained fully soluble product with broad molecular weight

184

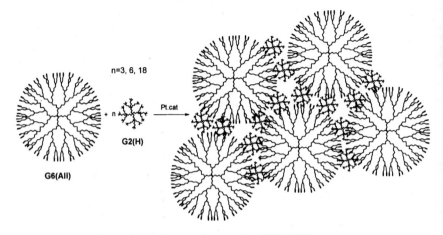

Scheme 3. Synthesis of networks n(G2)/(G6).

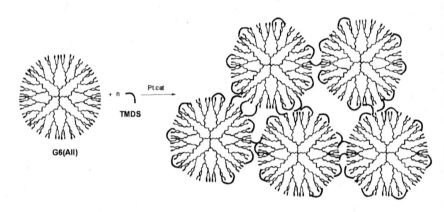

Scheme 4. Synthesis of networks n(TMDS)/(G6).

distribution. This fact supports our assumption, that the G2(H) dendrimer actually works as a difunctional crosslink relative to the dendrimers of 6th generation. Due to the steric hindrance it can react only with two dendrimers, otherwise we would have obtained network under this conditions, because number of functional groups is still high. What is also important – it could not spend all functionalities reacting with the one dendrimeric counterpart; in any case some functions are available to bind another dendrimer.

After drying procedure gels collapsed, but remained transparent unlike the first series of networks, that proves absence of heterogeneities on a hundred nanometer scale in the network structure. Samples were exposed to Soxhlet extraction procedure and the content of insoluble part in all cases exceeded 95% (Table III). Conversion of functional groups was qualitatively monitored through a comparison of the IR-spectra of the initial dendrimers and samples of networks obtained (Fig. 1). Decreasing of the dendrimers ratio did not induced any principal changes to the gelation process results; as it was mentioned above in all cases yield of crosslinked matter were quantitative. At the same time a more accurate comparison showed differences in the non reacted functional groups concentration. One can see that depending on a dendrimers ratio a lot of allyl functional groups remains in networks structure. These groups can be used for further chemical transformations (14) in order to modify network properties.

Table III. Synthetic conditions and characteristics of the second series of networks

Sample	Concentration in toluene, wt.%	non-soluble part, %	$\alpha=m_{sw.}/m_{dry}$	
			toluene	methanol
18(G2)/(G6)	30	98	1.40	1.12
6(G2)/(G6)	30	99	1.77	1.01
6(G2)/(G6)	bulk	96	1.38	–
3(G2)/(G6)	30	100	1.99	–
36(TMDS)/(G6)	30	95	3.03	1.01
36(TMDS)/(G6)	bulk	98	1.62	1.01
(G2)/(G6)	30	no gelation	–	–

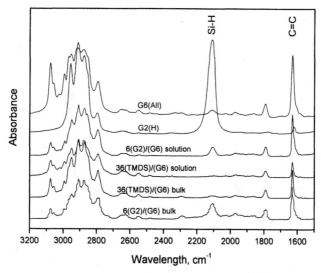

Figure 1. IR-spectra of the initial dendrimers G6 (All), G2(H) and obtained networks

Swelling.

Swelling behavior of the samples obtained was studied using the method, described in the experimental part. Analysis of network behavior over the solvents of different thermodynamic quality can give important information about its internal organization. The degree of swelling was measured in toluene or xylene and methanol (good and poor solvents for initial compounds, respectively) and the results differ significantly for the two series. For the first series swelling behavior of almost all samples doesn't change noticeably for xylene and methanol, that means totally different swelling mechanism not typical for conventional polymer networks. The closest analogue of obtained compounds in terms of swelling behavior are so called "hyper-cross-linked" polystyrene (15). Two possible explanation of such behavior is as follows. During network formation dendrimers are in the swollen state. Because of very high functionality of the dendrimers they are fixed in this conformation and can not collapse even after removing the solvent. Thus nanocavities are formed inside and between the dendrimers, which uptake solvent of any thermodynamic quality to decrease surface energy. Another version is that because of very stressed network organization, its structure could be partly destroyed with the formation of microsized channels and cages. So the reason for the swelling is a capillary effect. Most probably we have here the mixture of two mechanisms.

It is important to correlate the results for the networks composed from dendrimers of the same size, but with different number of functional groups. A peculiarity of dense dendrimer networks is the fact that the set of network cells is limited and defined by the structure of initial dendrimers. Decrease of dendrimer functionality significantly reduce the set of possible dimensions of cells and leads to the increase of bigger cells, which are reflected in swelling behavior of the networks. Comparing swelling of networks, $G7^{512}$-25 и $G7^{256}$-25, one will see that swelling degree in good solvent increased from 1.70 to 2.83 for the network with bigger cells (Table II); at the same time greatly increased the difference between swelling degree in xylene and methanol, indicating presence of more equilibrated cells capable of collapse. Figure 2 graphically illustrates difference between sets of network cells depending on functionality of initial dendrimers.

Comparing networks, obtained from dendrimers with the same number of functionalities but in different conditions, one can see that properties of the

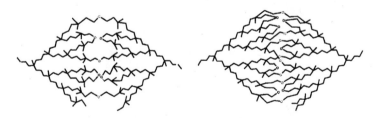

Figure 2. Simplified illustration of dependence of network cells sizes on the functionality of the dendrimers

resulting network are determined by the state of the dendrimer during vulcanization. In case of G7[256]-50 crosslinking was held upon lack of solvent, so its removal doesn't produce sufficient stress in the network structure. As a result the swelling degree in methanol is close to unity (1.09, see Table II).

To confirm influence of preliminary swelling on the formation of network structure one from the network samples was synthesized in bulk using the dendrimer Si $^{512}_{509}$(Cl) with more active chlorosilyl functionalities. Considering total lack of preliminary swelling internal structure of the dendrimers should not be enabled, because they are in a collapsed state and there should be also no voids between the dendrimers. Obtained results are in good agreement with our consideration, swelling degree of this network both in methanol and xylene is close to zero. These data are a forcible argument in favor of the assumption about determinative influence of dendrimer state before crosslinking on the structure of the resulting network.

The results obtained for the second series look more conventional, which is a consequence of their homogeneous, non-stressed structure. Swelling degree increases with decreasing of G2(H) content and dramatically increases with the change to TMDS instead of G2(H). As we assumed above G2(H) crosslinks only two dendrimers just as TMDS, at the same time as follows from the swelling behavior 36(TMDS)/(G6) network is much more sparse, than 6(G2)/(G6) in spite of the fact that the latter formally has less crosslinks, which give ground to suppose that substantial part of TMDS is consumed for intradendrimer reaction, i.e. both functions of TMDS react with one and the same dendrimer (Scheme 4). Portion of TMDS reacting in such a way is of course lower for the bulk reaction and as a consequence swelling degree goes down. The tendency of lowering of the swelling degree upon a switch to the bulk reaction is also true for a fully dendrimer network, but is not so abrupt.

As one might expect networks don't swell in methanol except for the most rigid network sample 18(G2)/(G6) that exhibited some swelling probably due to the same reasons as the networks from the first series.

Preliminary study of the specific surface of obtained networks showed, that specific surface of the networks of the first series is about 200-400 m^2/g, whereas for the second series it has zero value. These data confirm a non-stressed structure of the second series.

Conclusions

Carbosilane dendrimers can be covalently cross-linked into three-dimensional networks. Two representative series of carbosilane dendrimer-based networks with different parameters were obtained.

In the case of dense and rigid networks one could not distinguish two probable reasons of internal cavities formation. Dramatic difference between properties of preliminary swollen and non-swollen networks could be considered as a confirmation of nanocavities formation mechanism. While heterogeneous character of formed network in combination with nonspecific swelling behavior clearly shows the presence of the destructive phenomena during the swelling process. The high stress leading to the destructive processes during drying and swelling procedures in large extent could be caused by specific features of this particular dendrimers possessing low glass-transition temperatures (16) indicating chain flexibility.

The combination of the dendrimers of different size and functionality allowed us to prepare unstressed network built from dendrimeric subunits. The ratio between dendrimers with different types of functionalities is an effective tool to regulate density of the forming network and to avoid excessive stress in network sample. Thus, one could prevent heterogeneities caused by swelling stresses by elaborating the kind of equilibrated networks. However, together with stresses one loses the ability to rigidly fix dendritic units and prevent their collapse during the drying procedure. At the same time the idea of fixing dendrimer structure in the swollen state by means of network formation is still attractive and probably can be accomplished by means of the use of stiff dendrimers with a high glass-transition temperature.

References

1. Dendrimers and Other Dendritic Polymers; Frechet, J.M.J; Tomalia, D.A. Eds.; Wiley Series in Polymer Science; John Wiley & Sons, Ltd.: Chichester, UK, 2001.
2. Tatarinova, E.A.; Rebrov, E.A.; Myakouchev, V.D.; Meshkov, I.B.; Demchenko, N.V.; Bystrova, A.V.; Lebedeva, O.V.; Muzafarov A.M. Russ. Chem. Bull. 2004, 53(11), 2591-2600.
3. (a) Dvornic, P.R.; de Leuze-Jallouli, A.M.; Owen, M.J.; Perz. S.V. In Silicones and Silicone-Modified Materials; Clarson, S.J., Fitzgerald, J.J., Owen, M.J., Eds.; ASC Symp. Ser.; American Chemical Society: Washington, DC, 2000, Vol. 729, pp 241-269. (b) Dvornic, P.R.; Li, J.; de Leuze-Jallouli, A.M.; Reeves, S.D.; Owen, M.J. Macromolecules 2002, 35, 9323-9333.
4. Dvornic, P.R.; de Leuze-Jallouli, A.M.; Owen, M.J.; Perz, S.V. Macromolecules 2000, 33, 5366-5378.
5. (a) Boury, B.; Corriu, R.J.P.; Nunez, R. Chem. Mater. 1998, 10, 1795-1804. (b) Kriesel, J.W.; Tilley, T.D. Chem. Mater. 1999, 11, 1190-1193. (c) Kriesel, J.W.; Tilley T.D. Adv. Mater. 2001, 13, 1645-1648. (d) Kriesel, J.W.; Tilley, T.D. Chem. Mater. 2000, 12, 1171-1179.
6. Bystrova, A.V.; Tatarinova, E.A.; Buzin, M.I.; Muzafarov A.M. Polymer Sci. A 2005, 47(8), 820-827.
7. Zhao, M.; Liu, Y.; Crooks, R.M.; Bergbreiter, D.E. J. Am. Chem. Soc. 1999, 121(5), 923-930.
8. Jahromi, S.; Litvinov, V.; Coussens, B. Macromolecules 2001, 34(4), 1013-1017.
9. (a)Larsen, G.; Lotero, E. J. Phys. Chem. B 2000, 104(20), 4840-4843. (b) Larsen, G.; Lotero, E.; Marquez, M. Chem. Mater. 2000, 12(6), 1513-1515. (c) Jahromi, S.; Mostert, B. Macromolecules 2004, 37(6), 2159-2162. (d) Doneanu, A.; Chirica, G.S.; Remcho,V.T. J. Sep. Sci. 2002, 25, 1252-1256.

189

10. Ignat'eva, G.M.; Rebrov, E.A.; Myakushev, V.D.; Muzafarov, A.M.; Il'ina, M.N.; Dubovik, I.I.; Papkov, V.S. Polymer Sci. A 1997, 39(8), 874-881.
11. Ponomarenko, S.A.; Rebrov, E.A.; Boiko, N.I.; Muzafarov, A.M.; Shibaev, V.P. Polymer Sci. A, 1998, 40(8), 763-774.
12. Hiemenz, P.C. Polymer Chemistry; Marcel Dekker: New York, 1984, p 318.
13. Muzafarov, A.M.; Gorbatsevich, O.B.; Rebrov, E.A.; Ignat'eva, G.M.; Chenskaya, T.B.; Myakushev, V.D.; Bulkin, A.F.; Papkov, V.S. Polymer Sci. 1993, 35(11), 1575-1580.
14. (a) Vasilenko, N.G.; Rebrov, E.A.; Muzafarov, A.M.; Eßwein, B.; Striegel, B.; Möller, M. Macromol. Chem. Phys. 1998, 199(5), 889-895. (b) Getmanova, E.V.; Rebrov, E.A.; Myakushev, V.D.; Chenskaya, T.B.; Krupers, M.J.; Muzafarov, A.M. Polymer Sci. A 2000, 42(6), 610-619.
15. Tsyurupa, M.P. Ph.D. Thesis, Institute of Organoelement Compounds RAS, Moscow, Russia, 1974. (b) Shantarovich, V.P.; Suzuki, T.; He, C.; Davankov, V.A.; Pastukhov, A.V.; Tsyurupa, M.P.; Kondo, K.; Ito, Y. Macromolecules 2002, 35(26), 9723-9729.
16. Smirnova, N.N.; Stepanova, O.V.; Bykova, T.A.; Markin, A.V.; Muzafarov, A.M.; Tatarinova, E.A.; Myakushev, V.D. Thermochim. Acta 2006, 440, 188-194.

Chapter 13

The Gel Point and Network Formation Including Poly(dimethylsiloxane) Polymerizations: Theory and Experiment

J. I. Cail and R. F. T. Stepto[*]

Polymer Science and Technology Group, School of Materials,
The University of Manchester, Grosvenor Street, Manchester M1 7HS,
United Kingdom
[*]Corresponding author: robert.stepto@manchester.ac.uk

Gel points, accounting for intramolecular reaction, are predicted using Ahmed-Rolfes-Stepto (ARS) theory. They are compared with experimental gel points for polyester (PES)-forming, polyurethane (PU)-forming and poly(dimethyl siloxane) (PDMS) polymerisations. The PES and PU polymerisations were from stoichiometric reaction mixtures at different initial dilutions and the PDMS ones were from critical-ratio experiments at different fixed dilutions of one reactant. The ARS predictions use realistic chain statistics to define intramolecular reaction probabilities and employ no arbitrary parameters. Universal plots of excess reaction at gelation *versus* ring-forming parameter are devised to enable the experimental data and theoretical predictions to be compared critically. Significant deviations between experiment and theory are found for the PU systems and the PDMS systems with higher molar-mass reactants. However, ARS theory fits well the results for the PES systems and the PDMS systems having lower molar-mass reactants. Possible reasons for these differences in behaviour are discussed in terms of the PU-forming reaction mechanism and the effects of entanglements in PDMS. Whilst ARS theory provides a good basis for gel-point predictions and can be applied to many types of polymerisation, more experimental systems at different initial dilutions and ratios of reactants still need to be studied and the various methods used for detecting gel points still need to be compared to enable a definitive assessment of its performance to be made.

Introduction

The present paper summarises and combines the results of previous papers (1-6) on gel-point prediction directly from formation conditions and reactant structures accounting for intramolecular reaction. Gelation results from a wide range of reaction systems, forming polyesters (PESs), polyurethanes (PUs) and poly(dimethyl siloxane)s (PDMSs), are interpreted using a universal approach based on Ahmed-Rolfes-Stepto (ARS) theory (7). For a given system, the comparison of experimental and theoretical gel points at various dilutions of preparation enables the effective values of the ring-forming parameter, P_{ab}, to be evaluated. These effective values are compared with those expected from reactant structures and deviations discussed in terms of the ability of ARS theory to provide *ab initio* gel-point predictions and also possible special effects features of the poymerisations studied.

It should be mentioned that the experimental and theoretical gel points are interpreted in terms of deviations from Flory-Stockmayer (F-S) random-reaction statistics due to intramolecular reaction. The theory does not purport to account for the detailed molecular growth in the critical percolation regime, where all reaction groups are not equally accessible over the timescale of a polymerisation [8,9]. However, it will be seen that any effects of percolation on the observed gel points cannot explain the wide differences in agreement between the predictions of ARS theory and the experimental results for the various polymerisation systems.

Interpretation and Prediction of Gel Points – Ahmed-Rolfes-Stepto (ARS) Theory

It has been shown (10) that the accurate, direct Monte-Carlo modelling of gel points is not possible for finite populations of reactive groups. The best approach is to use a statistical theory that accounts reasonably completely for the ring structures that form during a polymerisation and can be applied to a wide range of polymerisations. To date, the statistical theory that accounts most completely for intramolecular reaction is ARS theory (7).

In its most general form, ARS theory treats polymerisations of mixtures of reactants bearing A groups and B groups, *i.e.* $RA_{faw} + R'B_{fbw}$ polymerisations, where f_{aw} and f_{bw} are the mass(weight)-average functionalities, defined originally by Stockmayer (11). For such polymerisations, the theory evaluates the probability of growth between statistically equivalent points B^1 and B^2, in the molecular structure shown in Figure 1. Ring structures of all sizes are accounted for.

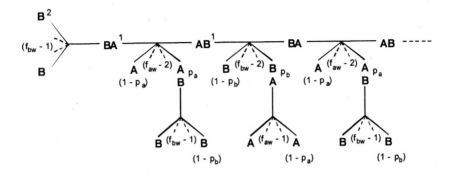

Figure 1: Sequence of structural units used to define gelation in an RA_{faw} + RB_{fbw} polymerisation in ARS theory (7). Side chains up to two units long are considered. p_a and p_b are extents of reaction.

The resulting expression for the gel point, defining unit probability of growth from B^1 to B^2, is

$$r_a p_{ac}^2 (f_{aw} - 1)(f_{bw} - 1) =$$

$$[1 + (f_{aw} - 2)\Phi(1, \tfrac{3}{2})\lambda_{a0} + (f_{bw} - 2)(f_{aw} - 1)r_a \Phi(1, \tfrac{3}{2})\lambda_{a0} p_{ac}] \cdot$$

$$[1 + (f_{bw} - 2)r_a \Phi(1, \tfrac{3}{2})\lambda_{a0} + (f_{aw} - 2)(f_{bw} - 1)r_a \Phi(1, \tfrac{3}{2})\lambda_{a0} p_{ac}], \quad (1)$$

where
$$r_a = \frac{c_{a0}}{c_{b0}} = \frac{p_b}{p_a} , \quad (2)$$

$$\lambda_{a0} = \frac{P_{ab}}{c_{a0}} \quad (3)$$

and
$$\Phi(1, \tfrac{3}{2}) = \left(\sum_1^\infty 1^i i^{-3/2} \right) = \sum_1^\infty i^{-3/2} = 2.612 . \quad (4)$$

p_{ac} and p_{bc} are the extents of reaction of A and B groups at the gel point and r_a is the initial molar ratio of A and B groups. λ_{a0} is a ring-forming parameter, with c_{a0} the initial concentration of A groups and P_{ab} represents the mutual concentration of A and B groups at the ends of the shortest sub-chain that can react intramolecularly. The structure of this sub-chain, consisting of ν skeletal bonds, and of root-mean-square end-to-end distance $\langle r^2 \rangle^{1/2}$, is illustrated in Figure 2. In detail (1,4,12),

$$P_{ab} = \frac{P(r = 0)}{N_{Av}} , \quad (5)$$

where $P(\underline{r} = \underline{0})$ is the probability-density of an end-to-end vector of magnitude

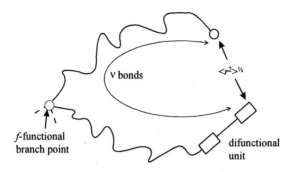

Figure 2: Sub-chain forming a smallest loop structure illustrated with respect to an $RA_2+R'B_f$ polymerisation. The diagram shows the two arms of a star reactant, one arm having reacted with a difunctional monomer; the root-mean-square end-to-end distance of the chain of v bonds between the reactive terminal groups is $<r^2>^{1/2}$.

equal to zero between reactive groups. If it is assumed that the end-to-end distance distribution can be represented by a Gaussian function, P_{ab} is given by

$$P_{ab} = \frac{1}{N_{Av}} \left\{ \frac{3}{2\pi < r^2 >} \right\}^{3/2} \tag{6}$$

and

$$<r^2> = vb^2 , \tag{7}$$

with b the effective bond length, a measure of chain stiffness. P_{ab} can be described as the mutual *internal concentration*, c_{int}, of a pair of reactive groups on the same molecule that can react intramolecularly to form the smallest loop. Thus, the parameter λ_{a0} of eq 3 is a useful measure of the propensity of a system at a given ratio of reactants for intramolecular reaction. It captures the combined effects of reactant structure and reactive-group concentration on intramolecular reaction. A decrease in chain length or chain stiffness (i.e., a decrease in $<r^2>$) results in an increase in P_{ab} and, hence, in the probability of intramolecular reaction and the formation of loop structures. Similarly, decreasing the concentration of reactive groups enhances the probability of intramolecular reaction.

Returning to eq 1, $\Phi(1,3/2)$ sums over one opportunity for forming each size of ring structure. The actual numbers of opportunities are accounted for by the various factors in f_{aw} and f_{bw} in eq 1. Finally, when $\lambda_{a0} = 0$, the r.h.s. of eq 1 equals 1 and the F-S expression for the gel point is obtained.

Eq 1 is simple to apply. It is quadratic in both p_{ac} and λ_{a0} and may be arranged to solve analytically for either of these quantities (7). To *interpret* experimental gel points, eq 1 may be solved for λ_{a0} using the measured values of

p_{ac}. The values of λ_{a0} so obtained may then plotted *versus* c_{a0}, or another measure of dilution of the polymerisation, and P_{ab} for the reactants evaluated from the slope of the linear plot, or the initial slope of the slightly curved plot so obtained. The value of P_{ab} can then used with eq 6 to evaluate $<r^2>$, characteristic of the chain forming the smallest loop for the given reactants, and the value can be compared with that expected from chain conformational calculations.

The principal aim of the present paper is the analysis of gel-point results in a *predictive* rather than *interpretative* manner (*cf.* (*13*)). Universal plots of extents of reaction at gelation *versus* the ring-forming parameter λ_{a0} will be used. Theoretically, according to ARS theory, for a given value of r_a and given functionalities of reactants, all extents of reaction at gelation should lie on a single curve. Experimentally, knowing the molar masses and the functionalities of the reactants, the values of λ_{a0} required may be derived from the values of c_{a0} of the reactions, together with the values of $<r^2>$ of the finite sub-chains forming the smallest loop structure (see eqs 3, 6 and 7). In the present work, the values of $<r^2>$ have been calculated rigorously using rotational-isomeric-state (RIS) statistics and the RIS Metropolis Monte-Carlo (RMMC) software of Molecular Simulations Incorporated (now Accelrys) (*14*). By considering gel points from eighteen PES, PU and PDMS systems covering two values of branch-point functionality and wide ranges of reactant molar masses, chain structures, dilutions and ratios, it is possible to look critically at the predictive capability of ARS theory and also to comment on the consistency of experimentally determined gel points across the systems.

Experimental Systems and Gel Points

The pairs of reactants used, together with their values of ν, deduced from their molar masses, and the calculated values of $<r^2>$ and effective bond length b are given in Table I. As stated, the calculations were made using the RMMC software of MSI. The values of $<r^2>$ for the more complex PES and PU chain structures, comprising two polyoxypropylene polyol arms and a difunctional residue, were based on detailed RIS modelling of the atactic polyoxypropylene chain (*15,16*), linked with RMMC analysis of the *complete*, branched structures of the polyols (*17*).

The PES experiments used stoichiometric mixtures reacting at 60 °C in bulk and various dilutions in diglyme (*1,2,4*). The PU experiments used stoichiometric mixtures reacting at 80 °C in bulk and various dilutions in nitrobenzene (*1-5*).The PDMS experiments (*6*) used reactions at 25 °C in bulk and various dilutions in inert, linear PDMS. 3-functional and 4-functional DMS endlinkers (R'B_f) with reactive H groups were reacted with linear PDMS fractions with vinyl end groups (RA$_2$). At a given dilution of the RA$_2$ reactant, the reactive-group ratio, r_a, was adjusted systematically, by increasing the initial concentration of the minority R'B_f endlinker until gelation was observed at complete reaction. The critical ratio, r_{ac}, for zero gel fraction at complete reaction was determined.

The experimental results for the stoichiometric PES and PU polymerisations are shown in Figure 3, where α_{rc} is plotted *versus* the average initial dilution of reactive groups. α_{rc} is the excess of the product of the extents of reaction at gelation over the value predicted by F-S theory. That is,

$$\alpha_c = p_{ac} \cdot p_{bc} , \tag{8}$$

$$\alpha_c^0 = \frac{1}{f-1} \tag{9}$$

and
$$\alpha_{rc} = \alpha_c - \alpha_c^0 . \tag{10}$$

The results show the expected trends. There are significant delays beyond the F-S gel point. For each group of systems (PES $f = 3$, PU $f = 3$ and PU $f = 4$), intramolecular reaction increases with dilution and decreases as ν increases. Also more intramolecular reaction occurs in the PU polymerisations for $f = 4$ than for $f = 3$. A higher functionality means more opportunities for

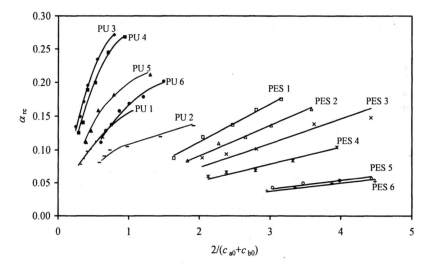

Figure 3: Experimental gel points for stoichiometric PES f = 3, PU f = 3 (PU1,2) and PU f = 4 (PU3-6) polymerisations. α_{rc}, the excess of the product of the extents of reaction at gelation as defined in eq 10, versus 2/(c$_{a0}$ + c$_{b0}$), the average initial dilution of reactive groups.

Table I. Experimental systems and their values of v, $<r^2>$ and b characterising the chains forming the smallest loops.

System	Reactants[a]	v	$<r^2>$/nm^2	b/nm
Polyesters ($f = 3$)				
PES 1	Adipoyl chloride + LHT240	37	3.68	0.315
PES 2	Sebacoyl chloride + LHT240	41	4.39	0.327
PES 3	Adipoyl chloride + LHT112	66	7.00	0.326
PES 4	Sebacoyl chloride + LHT112	70	7.52	0.328
PES 5	Adipoyl chloride + LG56	129	13.67	0.325
PES 6	Sebacoyl chloride + LG56	133	14.21	0.327
Polyurethanes ($f = 3,4$)				
PU 1 ($f = 3$)	HDI + LHT240	33	3.72	0.337
PU 2 ($f = 3$)	HDI + LHT112	61	6.88	0.336
PU 3 ($f = 4$)	HDI + OPPE1	29	2.75	0.308
PU 4 ($f = 4$)	HDI + OPPE2	33	3.63	0.332
PU 5 ($f = 4$)	HDI + OPPE3	44	4.61	0.324
PU 6 ($f = 4$)	HDI + OPPE4	66	6.58	0.316
Poly(dimethyl siloxane)s ($f = 3,4$)				
PDMS 1 ($f = 3$)	5k + B3	141	22.52	0.400
PDMS 2 ($f = 4$)	5k + B4	141	22.52	0.400
PDMS 3 ($f = 4$)	5k + B4 (short)	141	22.52	0.400
PDMS 4 ($f = 4$)	7k + B4	192	30.66	0.400
PDMS 5 ($f = 4$)	11k + B4	315	50.31	0.400
PDMS 6 ($f = 4$)	13k + B4	354	56.53	0.400

[a] LHT112, LHT240 and LG 56 are trifunctional polyoxypropylene (POP) triols; OPPE 1-4 are tetrafunctional POP tetrols; HDI is hexamethylene diisocyanate; 5k to 13k are vinyl terminated linear PDMS chains of molar masses 5 to 13 kg mol^{-1}, approximately; B3 and B4 are trifunctional and tetrafunctional (R'B$_f$) DMS H-functional endlinkers; "short" denotes the use of a linear PDMS diluent of low molar mass.

intramolecular reaction. The PU systems undergo more intramolecular reaction than the PES systems. This difference is seen more clearly in the next section on the basis of the universal ARS plots. The reason for it is not understood, especially as Table 1 shows that for similar lengths of chain for $f = 3$ systems, the PES chains have smaller values of b than the PU chains and should, therefore, have larger values of P_{ab} and undergo more intramolecular reaction at a given dilution.

The experimental results for the PDMS systems are shown in the next section on the basis of universal ARS plots.

Universal Plots - ARS Modelling

To convert the experimental gel-point results into universal plots of α_{rc} versus λ_{a0} for given reactant functionalities and ratios of reactive groups, the values of λ_{a0} need to be calculated. The values of $<r^2>$ given in Table I were used to calculate, using eq 6, values of P_{ab} for each of the reaction systems. For the PES and PU reactions at $r_a = 1$, the values of P_{ab} and the initial dilutions of reactive groups were used in eq 3 to define values of λ_{a0} for the individual polymerisations studied. The experimental values of α_{rc} were then plotted versus λ_{a0}. The results for the trifunctional systems are shown in Figure 4(a), together with the theoretically predicted universal curve, obtained by solving eq 1 for p_{ac} in terms of λ_{a0} with $f_{aw} = 2$ and $f_{bw} = 3$ and using eqs 2, 8 and 10 to evaluate p_{bc}, α_c and, finally, α_{rc}. Significantly, the use of λ_{a0} as a variable reduces the distinct plots in Figure 3 for the individual PES and PU systems to essentially single curves. This reduction means that the combined effects of dilution, functionality (f), molar mass (v) and chain structure (b) are, in relative terms, correctly accounted for in λ_{a0}. For the PES systems, good agreement with the predicted ARS universal behaviour is found, especially at the lower values of λ_{a0}. For the PU systems, the measured values of α_{rc} are higher than predicted by ARS theory. This is a real difference between the results of the PES and PU systems that cannot be resolved without investigations into the possible effects of the detailed reaction mechanisms on the probabilities of intramolecular reaction. Other possible factors should have similar influences on both types of system. For example, nominally the same triols were used in the PU systems as in four of the PES systems (LHT240 and LHT112) so that any slight reduction in the functionality below the value of 3 should affect both types of system similarly. In addition, preliminary calculations (17) evaluating $P(\underline{r} = \underline{0})$ directly (eq 5), avoiding the Gaussian approximation of eq 6, show that the values of P_{ab} for the PU and PES chains are affected approximately equally.

The experimental and theoretical results for the trifunctional PDMS system are compared in Figure 4(b). For the non-stoichiometric polymerisations used

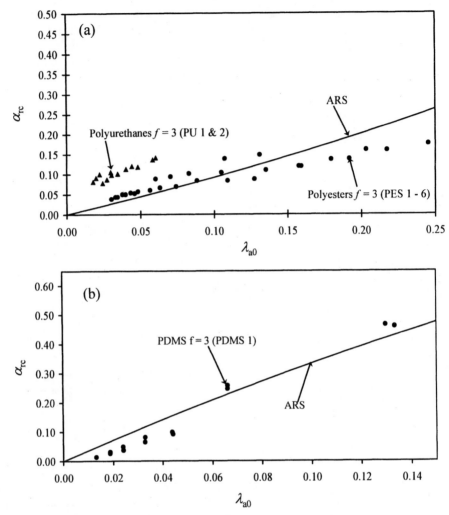

Figure 4: Comparison of ARS theory and experiment for f = 3 systems. Universal representations of α_{rc} versus λ_{a0} for (a) PES and PU, $r_a = 1$ polymerisations and (b) PDMS, $r_a > 1$ critical-ratio polymerisations.

with gelation at complete reaction of the minority B group on the $R'B_f$ reactant, the experimental values of α_{rc} were evaluated using the relationships

$$p_{bc} = 1 \; ; \; p_{ac} = r_{ac}^{-1} \; , \tag{11}$$

$$\alpha_c = p_{ac} \cdot p_{bc} = r_{ac}^{-1} \tag{12}$$

and
$$\alpha_{rc} = r_{ac}^{-1} - \alpha_c^0 \quad , \tag{13}$$

with α_c^0 given by eq 9. The reactions were carried out at fixed values of c_{a0} of the RA_2 reactant. Hence, λ_{a0} is still the natural ring-forming parameter to use for these non-stoichiometric polymerisations. To derive the predicted values of α_c ($= r_{ac}^{-1}$), eq 1 was used with $r_a \cdot p_{ac}^2 = r_{ac}^{-1}$, and $p_{ac} = r_{ac}^{-1}$. Figure 4(b) shows that there is good agreement between the universal ARS plot and experiment for the trifunctional PDMS system.

The results for the tetrafunctional PU and PDMS systems are shown in Figures 5(a) and 5(b). Significantly, the results for the PU systems from Figure 3 again normalise to essentially a single curve and again it lies above the universal curve predicted by ARS theory. The results for the PDMS systems split into two, with the values of α_{rc} for the linear PDMS chains of higher molar mass (11k and 13k) lying above the curve predicted by ARS theory and those for the chains of lower molar mass (5k and 7k) generally lying below, but in reasonable agreement with the theoretical curve. Again, the points for the two groups of systems (PDMS 2-4 and PDMS 5,6) reduce to single curves. No effect of the molar mass of the PDMS diluent is apparent (PDMS 3). The splitting of the PDMS results into two branches seems to indicate a reduced accessibility of the reactive groups of the higher molar mass PDMS chains for intermolecular reaction.

Discussion and Conclusions

The gel point is neither an easy quantity to measure nor to model and, by analysing experimental and theoretical gel points for several systems in terms of universal plots, it has been possible to highlight significant variations in the closeness of agreement between experiment and theory. ARS theory attempts to provide an absolute prediction of the gel point, with its ring-forming parameter established (independent of the polymerisation) from the conformational behaviour of reactant sub-chains. It performs well for the PES systems (Figure 4(a)), the $f = 3$ PDMS system and the $f = 4$ PDMS systems of lower molar mass (PDMS 2-4) (Figures 4(b) and 5(b)). Also, as mentioned previously, the reduction of the results for the various systems in Figure 3 to single curves for the PES and PU systems in Figures 4(a) and 5(a), and the reduction of the results for systems PDMS 2-4 and PDMS 5,6 to single curves in Figure 5(b) show that the use of ARS theory and the single parameter λ_{a0} accounts well in relative terms for the combined effects of dilution (c_{a0}), reactant functionalities (f), molar masses (v) and chain structures (b).

The relatively larger amounts of intramolecular reaction for the PU systems and the PDMS systems of higher molar mass would seem to indicate special

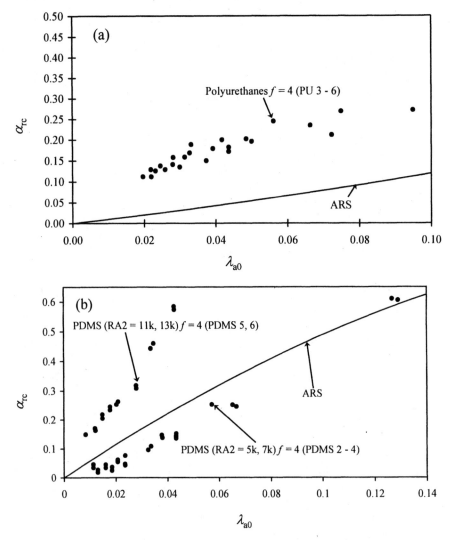

Figure 5: Comparison of ARS theory and experiment for f = 4 systems.
Universal representations of α_{rc} versus λ_{a0} for (a) PU, $r_a = 1$ polymerisations
and (b) PDMS, $r_a > 1$ critical-ratio polymerisations.

effects; such as groups being inaccessible for intermolecular reaction in the PDMS systems or, for the PU systems, side reactions or a bias towards intramolecular reaction due to the complex reaction mechanism. In this respect, it could be significant that the molar masses of the linear (RA_2) reactants of the higher molar-mass systems, PDMS 5,6 ($\approx$11,000 and 13,000 g mol^{-1}) are approximately equal to the entanglement molar mass of 12,000 g mol^{-1} *(18)*.

The unexpectedly high extents of reaction at gelation shown by the PU systems may be summarised in terms of the larger values of P_{ab} needed to fit the experimental results as plotted in Figure 3 on the basis of the ARS expression (eq 1) compared with the values predicted from the independently-evaluated values of $<r^2>$ in Table I. The comparison between the predicted and ARS-gel-point values of P_{ab} is made in Table II, where it can be seen that the values of P_{ab} consistent with the experimental data are four to seven times larger than the predicted values. The corresponding comparison between fitted and predicted values of P_{ab} for the results for the PDMS systems of higher molar mass (PDMS 5,6) in Figure 5(b) shows similar relative values of P_{ab} from experiment and prediction.

It does appear that experimentally determined gel points are extremely sensitive to what may be termed "non-random" effects that lead to excess intramolecular reaction. In the PU-forming polymerisations, the reaction mechanism probably causes a bias that favours intramolecular reaction. In the PDMS polymerisations using the higher molar-mass linear reactants, entanglements probably reduce the mutual accessibility of pairs of groups for intermolecular reaction. The PES-forming polymerisationa and the PDMS polymerisations using reactants below the entanglement threshold appear to show negligible "non-random" effects. ARS theory performs well for these systems and any effects of percolation phenomena on the extents of reaction at gelation are not seen. To improve the agreement between theory and experiment for such systems, the development of ARS theory, taking account of a more complex structure than that shown in Figure 1 and improvements to the assumption of Gaussian statistics for the ring-forming sub-chains would seem to be the best way forward.

Regarding the experimental gel points considered in this paper, it should be recalled that these were obtained using two different methods; the onset of the Weissenberg effect in a stirred reaction flask (PES and PU systems) and the measurement of gel fraction (PDMS systems). Although the differences between the predictions of ARS theory and experiment do not depend on which method

Table II. Predicted and fitted values of P_{ab} characterising ring formation in the PU-forming polymerisations of Table I. The predicted values were derived using eq 6 and the values of $<r^2>$ given in Table I. The fitted values are averages for each system, derived using the experimental extents of reaction at gelation in the ARS expression (eq 1) to evaluate λ_{a0} and , hence, P_{ab} from eq 3 and the known value of c_{a0}.

System	Reactants	ν	P_{ab}/ mol l^{-1} predicted from $<r^2>$	P_{ab}/ mol l^{-1} gel-point fitting
PU 1 ($f=3$)	HDI + LHT240	33	0.076	0.267
PU 2 ($f=3$)	HDI + LHT112	61	0.030	0.150
PU 3 ($f=4$)	HDI + OPPE1	29	0.120	0.409
PU 4 ($f=4$)	HDI + OPPE2	33	0.079	0.337
PU 5 ($f=4$)	HDI + OPPE3	44	0.033	0.228
PU 6 ($f=4$)	HDI + OPPE4	66	0.032	0.153

202

was used, the experimental methods do need to be examined critically and compared with the modulus self-similarity criterion of Winter and Chambon (see (*19*)). The present comparison of theory and experiment has highlighted the fact that the measurement of gel points is not fully understood. The methods used need to be reviewed critically and the gel points of even more systems as functions of dilution and ratios of reactants need to be studied.

Acknowledgement

The provision of their Polymer Software by MSI is gratefully acknowledged.

References

1. Stepto, R. F. T. In *Comprehensive Polymer Science, First Supplement;* Aggarwal, S.; Russo, S., Eds.; Pergamon Press: Oxford, 1992; Chap. 10.
2. Stepto, R. F. T. *Progress in Rubber & Plastics Technology* **1994**, 10, 130.
3. Dutton, S.; Rolfes, H.; Stepto, R. F. T. *Polymer* **1994**, *35,* 4521.
4. Stepto, R. F. T. In *Polymer Networks – Principles of their Formation Structure and Properties;* Stepto, R. F. T., Ed.; Blackie Academic & Professional: London, 1998; Chap. 2.
5. Stepto, R. F. T. In *The Wiley Polymer Networks Review Series, Vol. 1;* te Nijenhuis, K.; Mijs, W. J., Eds.; John Wiley & Sons: Chichester, 1998; Chap. 14.
6. Stepto, R. F. T.; Taylor, D. J. R.; Partchuk, T.; Gottlieb, M. In *ACS Symposium Series 729, Silicones and Silicone-Modified Materials;* Clarson, S. J.; Fitzgerald, J. J.; Owen, M. J.; Smith, M. D., Eds.; Amer. Chem. Soc.: Washington DC, 2000; Chap. 12.
7. Rolfes, H.; Stepto, R. F. T. *Makromol. Chem., Makromol. Symp.* **1993**, 76, 1.
8. Stauffer, D.; Coniglio, A.; Adams, M. *Adv. Polymer Sci.* **1982**, 44, 103.
9. Stauffer, D. *Introduction to Percolation Theory;* Taylor & Francis: Philadelphia, PA, 1985.
10. Cail, J. I.; Stepto, R. F. T.; Taylor, D. J. R. *Macromol. Symp.* **2001**, *171,* 19.
11. Stockmayer, W. H. *J. Polymer Sci.* **1952**, 9, 69.
12. Stepto, R. F. T.; Taylor, D. J. R. In *Cyclisation and the Formation, Structure and Properties of Polymer Networks in Cyclic Polymers, 2nd edition;* Semlyen, J. A Ed.; Kluwer Academic Publishers: Dordrecht, 2000; Chap. 15.
13. Stepto, R. F. T.; Taylor, D. J. R. *Polymer Gels and Networks* **1996**, 4, 405.
14. *Molecular Simulations Incorporated,* 9685 Scranton Road, San Diego, CA 92121, USA
15. Stepto, R. F. T.; Taylor, D. J. R. *Computat. and Theor. Polymer Science* **1996**, 6, 49.
16. Stepto, R. F. T.; Taylor, D. J. R. *Coll. Czech. Chem. Comm.* **1995**, 60, 1589.
17. Stepto, R. F. T.; Taylor, D. J. R. *unpublished work.*
18. Fetters, L. J.; Lhose, D. J.; Milner, S. T.; Graessley, W. W. *Macromolecules* **1999**, 32, 6847.
19. Ilavsky, M. In *Polymer Networks – Principles of their Formation Structure and Properties;* Stepto, R. F. T., Ed.; Blackie Academic & Professional: London, 1998; Chap. 8.

Chapter 14

Use of Solubility Parameters for Predicting the Separation Characteristics of Poly(dimethylsiloxane) and Siloxane-Containing Membranes

Alexander R. Anim-Mensah[1], James E. Mark[2], and William B. Krantz[3]

Departments of [1]Chemical and Materials Engineering and [2]Chemistry, University of Cincinnati, Cincinnati, OH 45221
[3]Department of Chemical and Biomolecular Engineering, National University of Singapore, Republic of Singapore 117576

The separation characteristics of poly(dimethylsiloxane) (PDMS) and siloxane-containing membranes can be predicted by the use of the solubility parameter (δ) differences between the membranes and solvents and/or the solutes. The membranes considered were PDMS and siloxane derivatives such as polysiloxaneimide (PSI) and poly[1-(trimethylsilyl)-1-propyne] (PTMSP), respectively. PSI is a copolymer made of PDMS and polyimide (in this case using 3,3,4,4-benzophenonetetracarboxylic dianhydride (BTDA) while silicone is a substituent in PTMSP. The system considered was the separation of PDMS and PSI using different aqueous alcohol feed solutions through pervaporation. The separation characteristics of PSI membranes having varying PDMS content were also investigated using an aqueous ethanol feed solution. PDMS was investigated for the different alcohol separations from water while PSI was investigated for the recovery of ethanol or water from different aqueous ethanol solutions. It was observed that an increase in the δ-difference between the membrane and solvent resulted in a general decrease in the separation factor (SF). In all the cases investigated, the closeness in the δ-difference between the membrane and solvent compared to that of the membrane and solute indicated good membrane performance. However, an increasing concentration of ethanol in an aqueous ethanol feed resulted in membrane swelling that caused an increased flux and decreased SF of ethanol relative to water. It could be concluded from the systems considered here; that a small difference between the δ of the membrane and solvent compared with that of the membrane and solute corresponds to an improved SF. However if this difference is too small, it might cause excessive membrane swelling. Thus, this approach provides a quick way of selecting membranes including PDMS and siloxane-containing polymers for specific membrane applications.

Introduction

This chapter will provide an overview of how the solubility parameter (δ) can be used as a guide for selecting membranes, especially PDMS and siloxane-containing membranes for a specific separation task. Emphasis is placed on membranes where the separation mechanism involves solution-diffusion. Hence, the focus is on improving nanofiltration (NF)[1], reverse osmosis (RO) and pervaporation membrane processes. It is well known that PDMS is the most studied siloxane[2] and has been used in addition to other siloxane-containing polymers for gas separations [2-6]. The siloxanes have also been used for the recovery of organics from aqueous systems[7-10]. In membrane separations involving liquids, three components are encountered: the solvent, the solute, and the membrane. In solution-diffusion membrane processes, the solute or solvent undergoes four sequential steps: (1) adsorption on the membrane surface at the high pressure side; (2) solubilization in the membrane; (3) diffusion through the membrane under the driving force; and (4) desorption at the low-pressure side of the membrane. In most cases, the extent of solubilization of the solute in the membrane is dictated by the partition or distribution coefficient of the solute between the solvent and membrane at a particular temperature [11].

The solubility parameter has been extensively used for identifying suitable solvents, swelling agents, plasticizers and non-solvents for polymers. It provides useful semi-quantitative information that can be used as a rule-of-thumb when experiments are unavailable for predicting miscibility of polymer solutions. The foundation for using δ is deeply rooted in macroscopic thermodynamics[12]. The theory behind δ is based on the principles of solubility or miscibility, that is, "like dissolves like". For two or more components to mix at constant temperature and pressure, the Gibb's free-energy-of-mixing (ΔG_m) should be negative. The change in ΔG_m is defined as the difference between the Gibb's free energy (ΔG) of the fully mixed solution and the weighted sum of the individual Gs of the pure components as given by Equation (1) [12]:

$$\Delta G_m = \Delta G - \sum_i x_i G_i \tag{1}$$

The change in the Gibb's free energy of-mixing (ΔG_m) is related to the enthalpy- and entropy-of mixing as given by Equation (2) [12]:

$$\Delta G_m = \Delta H_m - T\Delta S_m \tag{2}$$

The change in the entropy-of-mixing ΔS_m is always positive since mixing implies increased disorder; hence, for ΔG_m to be negative requires that the enthalpy-of-mixing ΔH_m, which is usually positive owing to intermolecular attraction, not be too large. ΔH_m is related to the internal-energy-of-mixing ΔU_m and volume-of-mixing ΔV_m at constant pressure by Equation (3).

$$\Delta H_m = \Delta U_m - \Delta P V_m = \Delta U_m - P\Delta V_m \tag{3}$$

Although, ΔV_m is generally negative for polymers solutions, its effect on ΔH_m is generally less than that of ΔU_m and ignored in the δ-approach to assess the miscibility of polymer solutions; that is, the following assumption is made in Equation (2):

$$\Delta G_m \approx \Delta U_m \tag{4}$$

Hildebrand in his work related ΔU_m to the change in the internal-energy-of-vaporization Δu_i^v, the partial molar volumes v_i, and the volume fractions ϕ_i of the individual components in the mixture, as well as the total volume of the mixture V by using a 6-n intermolecular potential energy function that can be shown to result in Equation (5) below [13]:

$$\Delta U_m = V\left[\left(\frac{\Delta u_i^v}{v_i}\right)^{1/2} + \left(\frac{\Delta u_j^v}{v_j}\right)^{1/2}\right]^2 \phi_i \phi_j \tag{5}$$

The solubility parameter is related to the energy-of-vaporization (Δu_i^v) and the partial molar volumes (v_i) as shown in Equation (6) [14]:

$$\delta_i = \left(\frac{\Delta u_i^v}{v_i}\right)^{1/2} \tag{6}$$

In the absence of experimental data, group contribution methods can be used to estimate δ for polymers in which case it is given by the following [15]:

$$\delta_i = \sum_j \frac{F_j}{v_i} = \sum_j \frac{\rho_i F_j}{M_r} \tag{7}$$

where the group contribution F_j can be obtained from the standard references[15], ρ_i is the density of the polymer and M_r is the molecular weight of the repeat group in the polymer.

The solubility parameter can be calculated from knowledge of the polar and hydrogen-bonding (h) contributions. The polar contribution is subdivided into dispersive (d) and permanent dipole-dipole (p) components. The overall δ is related to the polar and hydrogen-bond contributions as follows [15]:

$$\delta^2 = \delta_d^{\,2} + \delta_p^{\,2} + \delta_h^{\,2} \tag{8}$$

The Hanson radius (R_{ij}) defined by Equation (9) can be used to predict the miscibility or compatibility between two substances [15], particularly that of a polymer and a liquid component.

$$R_{ij} = \left[4\left(\delta_i - \delta_j\right)_d^{\,2} + \left(\delta_i - \delta_j\right)_p^{\,2} + \left(\delta_i - \delta_j\right)_h^{\,2} \right]^{1/2} \tag{9}$$

The $(\delta_i\text{-}\delta_j)^2$ terms represent only the difference in the interactions between the molecules of different polarities. Smaller R_{ij} values for a polymer and possible solvent suggest that stronger interaction that can result in swelling or miscibility. Smaller R_{ij} values for a polymer and a solute again suggest stronger interactions that can be manifested by fouling of a polymer membrane by the solute. Smaller R_{ij} values for any two (2) components generally correspond to an increased flux of either the solute or solvent a polymeric membrane whose transport is governed by a solution-diffusion mechanism. Conversely, larger R_{ij} values generally imply less interaction and corresponding diminished swelling or immiscibility, reduced fouling, and lower flux.

A quick approach of estimating δ from knowledge of only the overall δ is to use the difference between the δs of the two substances, i and j, as indicated in the following:

$$\Delta\delta_{i-j} = \left|\delta_i - \delta_j\right| \tag{10}$$

Stronger interaction between the two (2) components corresponds to smaller values of $\Delta\delta_{ij}$. Although Equation (10) fails for some solute-solvent systems for which $\Delta\delta$ is small (<1) it has been shown to work well for many polymer systems. Hence, it will be used here for predicting the membrane performance.

In dealing with filled or blended polymers, the additive properties of the two phases can be used to predict the resulting δ of the filled materials. It is also known that the transport properties of a solute or solvent through a polymeric membrane containing a filler or blended polymer can be described by the additive properties of the two phases involved. The δ can be calculated for membranes made from polymer blends based on the following equation [16, 17]:

$$\delta_m = \sum_i^n \delta_i \phi_i \tag{11}$$

where δ_m, and δ_i are the solubility parameters of the resulting polymeric membrane and individual polymer components, respectively. ϕ_i is the volume fraction of component i.

Experimental Procedure

Membrane Materials: Poly(dimethylsiloxane) (PDMS), poly[1-(trimethylsilyl)-1-propyne (PTMSP) and polysiloxaneimide (PSI) are the membrane materials considered in this investigation.

The structure of PDMS is shown in Figure 1.

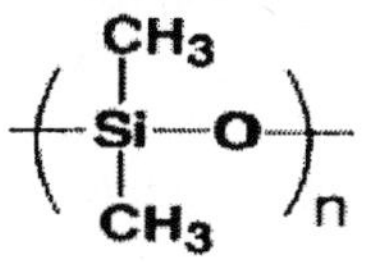

Figure 1. Structure of Poly(dimethylsiloxane) [18]

PTMSP membranes are made through solvent casting from solutions using cyclohexane, toluene, and tetrahydrofuran[3]. The structure of PTMSP is shown in Figure 2.

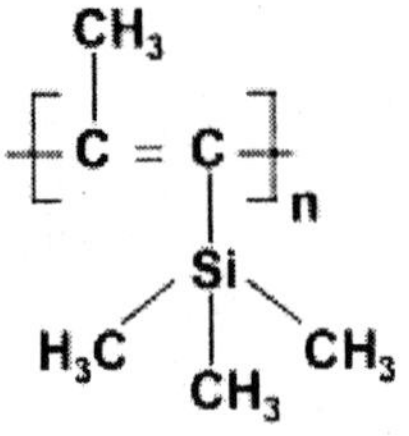

Figure 2. Structure of Poly(1-trimethylsilyl-1-propyne) (PTMSP) [5]

The incorporation of a siloxane group into the polyimide backbone yields processable polysiloxaneimide (PSI) [19]. For this investigation the PSI studied was synthesized from amine-terminated PDMS and BTDA[20]. The typical structure of PSI is shown in Figure 3 [21]:

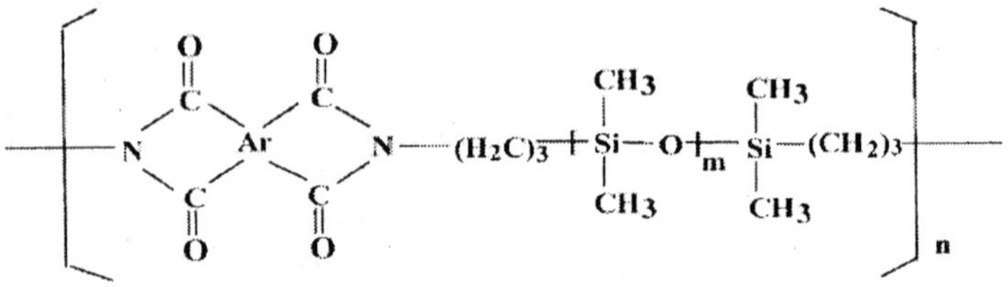

Figure 3. Structure of Polysiloxaneimide (PSI)[21].

Liquid feed: Various aqueous alcohol solutions made from methanol, ethanol, propanol, butanol, pentanol, hexanol, heptanol, octanol, nonanol and decanol.
Procedure: The δs of some of the membrane polymers were calculated using Equation (7) while those of the polymer blends were calculated using both Equations (7) and (12).

The difference in the solubility parameters was calculated using Equation (11) and used to predict the separation characteristics of the membrane. In most membrane separations the solvent is often permeated while the solutes are retained. In this situation, it is required that the δ-difference between the membrane polymer and solvent be smaller than that between the membrane polymer and solute. For high separation factors, the δ-difference between the membrane polymer and the solute must be large compared to that of the membrane polymer and solvent. A situation where the membrane polymer and the solute have a smaller δ-difference than that of membrane polymer and the solvent could result in the solute fouling the membrane. This will impair the membrane performance. Moreover, a smaller δ-difference between the membrane polymer and solvent is more likely to lead to swelling. Excessive swelling can lead to poorer membrane performance. Hence, a good membrane process design requires a careful balance between the opposing effects of a small δ-difference between the membrane polymer and component being separated. The solubility parameters for the PDMS (δ_{PD}) and PTMSP (δ_{PT}) membranes used here were found to be 15.0 and 18.2 MPa$^{1/2}$, respectively [15].

Results and Discussions

1.0 Separation of Alcohol from Aqueous Alcohols Solution using a PDMS Membrane

Here the separation of alcohols from aqueous alcohol solutions using a PDMS membrane by means of pervaporation is predicted. The aqueous alcohol solutions of methanol, ethanol, propanol, butanol, pentanol, hexanol, heptanol, octanol, nonanol and decanol are considered here. For the purpose of this comparison the alcohols are considered as the solvents and water as the solute.

Table 1 shows the δ values of the alcohols and separation factors (SF's) of the alcohols with respect to water, whereas Table 2 shows the δ-differences between the various alcohols and the membrane calculated using Equation (11) as well as the SF's. Figure 4 shows the variation of SF's of the various alcohols with respect to water along with the associated δ-differences between the alcohol and the membrane.

Table 2 indicates that smaller δ-differences between the alcohol and the PDMS membrane are associated with higher SF's. This follows from the fact that a small δ-difference implies a greater affinity between the two (2) components, which in turn leads to an enhanced selectivity.

2.0 Separation Characteristics of Butanol from Aqueous Butanol Solutions using PDMS and PTMSP Membranes

Here we report on the use of the solvent δ to assess the separation of butanol from aqueous butanol solutions using PDMS and PTMSP through pervaporation at different temperatures. From the data reported in Table 2 and Figure 4 we expect butanol to permeate through the membrane relative to water.

Figure 5 shows a plot of permeation (left ordinate) and SF (right ordinate) as function of temperature for the separation of butanol relative to water from an aqueous solution using PDMS and PTMSP membranes. Table 3 was generated from Figure 5 and shows the δ-difference between butanol and the PDMS and PTMSP membranes. It also show the SF comparison of butanol relative to water

Table 1. Various Alcohol Solubility Parameters ($MPa^{1/2}$) and Separation Factors with Respect to Water [11]

Materials	Feed Conc.	γ	δ_d	δ_p	δ_h	δ_s	SF alcohol/ water
Methanol	1	2	15.1	12.3	22.3	29.7	9
Ethanol	1	5	15.8	8.8	19.4	26	17
Propanol	1	15	16.0	6.8	17.4	24.3	67
n-Butanol	1	50	16.0	5.7	15.8	23.3	74
n-Pentanol	1	200	16.0	4.5	13.9	21.7	265
n-Hexanol	0.5	1000				21.9	1050
n-Heptanol	0.1	3000				21.7	1600
n-Octanol	0.05	14000	17.0	3.3	11.9	21.1	3100
n-Nonanol	0.01						4000
n-Decanol	0.005					20.5	5000
H_2O						47.9	
PDMS						15.0	

γ- activity ; alcohol/water- alcohol relative to water
SOURCE: Reproduced with permission from reference 11. Copyright 1990 Elsevier.

Table 2. Various Alcohol Solubility Parameters Differences ($MPa^{1/2}$) and Separation Factors [11]

Aqueous Solution of Alcohols (s)	δ_S	PDMS Membrane δ_{PD}	$\Delta\delta_{PD-S}$	SF alcohol/water
Methanol	29.7	15.0	14.7	9
Ethanol	26.0	15.0	11.0	17
Propanol	24.3	15.0	9.3	67
Butanol	23.3	15.0	8.3	74
Pentanol	21.7	15.0	6.7	265
Hexanol	21.9	15.0	6.9	1050
Heptanol	21.7	15.0	6.7	1600
Octanol	21.1	15.0	6.1	3100
Nonanol		15.0		4000
Decanol	20.5	15.0	5.5	5000
H_2O	47.9	15.0	32.9	

$\Delta\delta_{PD-S} = \delta_{PD} - \delta_S$; δ_{PD} and δ_S are solubility parameter of PDMS membrane and alcohols
SOURCE: Reproduced with permission from reference 11. Copyright 1990 Elsevier.

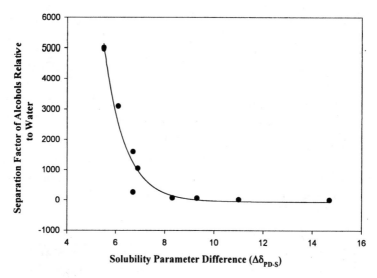

Figure 4. Separation Factor Variation with Solubility Parameter Difference from Decanol (highest) through to Methanol (lowest)

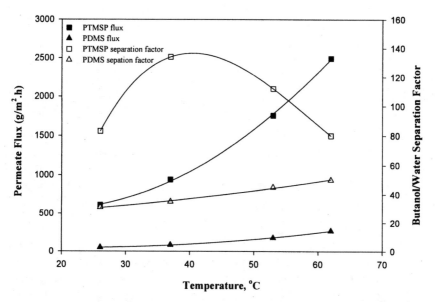

Figure 5. Permeation Flux and Separation Factor for the Separation of Butanol from Aqueous Solution using PTMSP and PDMS membranes[22] (Reproduced with permission from reference 22. Copyright 2000 Elsevier.)

using PDMS and PTMSP membranes for two (2) temperatures 26.0 and 62.0°C, respectively.

From Figure 5 and Table 3 it can be seen that both the flux and the SF of butanol relative to water were higher for PTMSP than for PDMS. These observations can be explained by the fact that the δ-difference between butanol and PTMSP is smaller than that of butanol and PDMS. Hence, the greater affinity of butanol for PTMSP relative to PDMS manifests itself in the higher SF's of PTMSP relative to PDMS membranes.

Table 3. Solubility Parameter Differences (MPa$^{1/2}$) for Butanol relative to PDMS and PTMSP and the Corresponding Separation Factors

Aqueous Butanol Solution	δ_s	$\Delta\delta_{PD-S}$	$\Delta\delta_{PT-S}$	SF Butanol/Water PDMS/PTMSP @ 26.0°C	SF Butanol/Water PDMS/PTMSP @ 62.0°C
Butanol	23.3	8.30	5.10	40/88	52/80
H$_2$O	47.9	32.90	29.70	-	-

$\Delta\delta_{PT-S} = \delta_{PT} - \delta_S$; δ_{PT} is the solubility parameter of PTMSP membrane

3.0 Effect of PDMS Composition in Polysiloxaneimide (PSI) Membranes on the Separation Characteristics for Ethanol or Water from Aqueous Ethanol Solutions

This section deals with the effect of PDMS content in the PSI membrane on the following: (1) Ethanol separation from a 10.0 wt % aqueous ethanol feed solution using pervaporation at 80.0°C; (2) Water separation from a 90.0 wt % aqueous ethanol feed solution using pervaporation at 20°C; and, (3) Ethanol separation from an aqueous ethanol feed comprised of varying concentrations of ethanol using pervaporation at 68.0°C.

Figure 6 shows the variation of the SF's of ethanol relative to water for PSI membranes with different PDMS content for a 10.0 wt % ethanol feed. Table 4 is generated from Figure 6 and shows the calculated δ's for the various PSI membranes. Table 4 also shows the various SF's for ethanol relative to water corresponding to the δ-difference between ethanol and the resulting PSI membranes. Figure 7 shows the variation of the SF as a function of the δ-difference between the PSI membranes and ethanol.

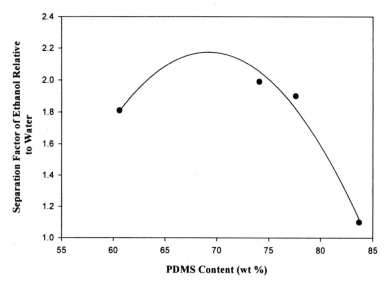

Figure 6. Effect of PDMS Content in PSI Membranes on the Separation Factor
for Ethanol using a 10 wt % Aqueous Ethanol Solution at 80°C [20]
(Reproduced with permission from reference 20. Copyright 1994 Elsevier.)

Table 4. PDMS Copolymer Compositions in the Polysiloxaneimide (PSI)
Membranes and the Separation Factors for Aqueous Ethanol Solutions

PDMS Weight (g)	BTDA Weight (g)	PSI Membrane δ_m (MPa$^{1/2}$)	Ethanol δ_e (MPa$^{1/2}$)	$\Delta\delta_{m-e}$ (MPa$^{1/2}$)	SF EtOH/Water
60.6	39.4	17.87	26.00	8.13	1.80
74.1	25.9	16.87	26.00	9.13	2.00
77.6	22.4	16.59	26.00	9.41	1.90
83.7	16.3	16.16	26.00	9.84	1.10

$\Delta\delta_{m-e} = \delta_m - \delta_e$; δ_m and δ_e are solubility parameters of PSI membrane and ethanol

Table 4 and Figures 6 and 7 indicate that increasing the PDMS content in the PSI membrane resulted in an increment in the δ-difference. These translate to a slight but non-systematic decrease in the SF.

Table 5 summarizes data for the study in which the permeation rate of aqueous ethanol and SF of water relative to ethanol were measured for PSI

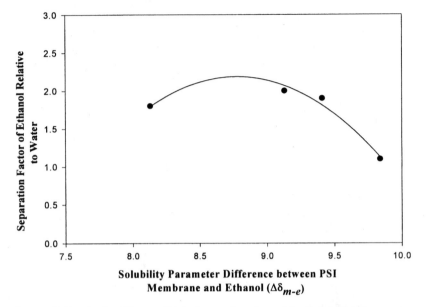

Figure 7. Separation Factor Variation as function of the Solubility Parameter Difference between Ethanol and a PSI Membrane having progressively Increasing PDMS Content.

membranes with four PDMS compositions. These same data are plotted in Figure 8. Table 6 is generated from Table 5 and shows the calculated δ's of the resulting PSI membranes from the different PDMS compositions and, the δ-differences between water and the resulting PSI membranes. Figure 9 shows the variation of the SF of water relative to ethanol and the δ-difference between the resulting PSI membranes and water.

From Table 5 and Table 6 it can be seen that there is a general increase in the δ-difference between the resulting PSI membrane and water as the PDMS content in the PSI membrane increases. Also the SF of water relative to ethanol decreases as the PDMS in the PSI membrane increases. This is because increasing the PDMS content in the PSI membrane results in a decreased affinity between the membrane and water. Hence the membrane becomes more hydrophobic.

Figure 10 shows the variation of the permeation rate and SF of ethanol relative to water for the separation of aqueous ethanol solutions using a PSI membrane (with 83.7 wt % PDMS) as a function of ethanol concentration. Table 7 shows calculated values of δ for the resulting aqueous ethanol feed with different ethanol concentrations, the δ-difference between the PSI membrane (83.7 wt % PDMS) and the aqueous ethanol feed with varying compositions of ethanol, and the δ-difference between PSI and ethanol.

Table 5. Permeation Rate and SF for Polysiloxaneimide (PSI) Membranes having different PDMS Copolymer Composition at 20°C [20]

PDMS Content (wt %)	SF Water/EtOH	Permeation Rate (g/m²h)
60.6	7.5	207
74.1	5.2	517
77.6	5.2	437
83.7	5.6	421

Feed concentration: 90 wt % EtOH
SOURCE: Reproduced with permission from reference 20. Copyright 1994 Elsevier.

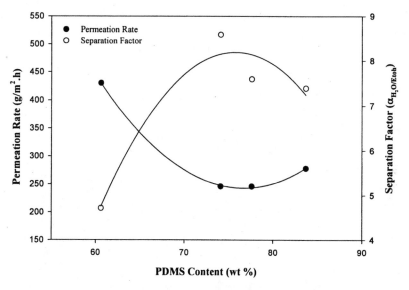

Figure 8. Permeation Rate (•) and Separation Factor (○) as function of the PDMS content in PSI membrane [20]
(Reproduced with permission from reference 20. Copyright 1994 Elsevier.)

Table 6. PDMS and BTDA Compositions in the Polysiloxaneimide (PSI) Membranes and the Separation Factor Characteristics.

PDMS Weight (g)	BTDA Weight (g)	PSI Membrane δ_m	Water δ_w	$\Delta\delta_{m\text{-}w}$	SF Water/EtOH
60.6	39.4	17.87	47.9	30.03	7.5
74.1	25.9	16.87	47.9	31.03	5.2
77.6	22.4	16.59	47.9	31.31	5.2
83.7	16.3	16.16	47.9	31.74	5.6

$\Delta\delta_{m\text{-}w} = \delta_m - \delta_w$; δ_w solubility parameter of water

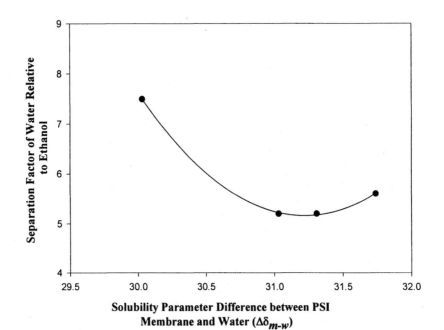

Solubility Parameter Difference between PSI Membrane and Water ($\Delta\delta_{m\text{-}w}$)

Figure 9. Separation Factor for Water relative to Ethanol as function of the δ- Difference between Water and PSI Membranes having Increasing PDMS Content.

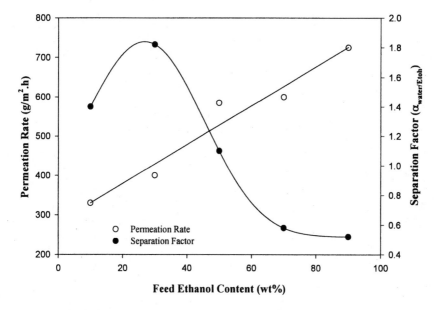

Figure 10. Effect of the Ethanol Feed Concentration on Flux (○) and Separation Factor (●) using PSI membrane made up of 83.7 wt % PDMS at 68.0 °C [20]. (Reproduced with permission from reference 20. Copyright 1994 Elsevier.)

Table 7. PDMS Copolymer Compositions in the Polysiloxaneimide (PSI) Membranes and the Separation Factor Characteristics

Weight of Ethanol in Aqueous Solution (g)	δ of Aqueous Ethanol Solution δs	$\Delta\delta_{m\text{-}s}$ (MPa$^{1/2}$)	$\Delta\delta_{m\text{-}e}$ (MPa$^{1/2}$)	SF EtOH / water
10	45.20	29.04	9.11	1.4
30	40.19	24.03	9.11	1.8
50	35.66	19.50	9.11	1.0
70	31.53	15.37	9.11	0.6
90	27.77	11.61	9.11	0.5

δ_m of PSI membrane is 16.16 MPa$^{1/2}$; δ_e is δ of ethanol

From Figure 10, it is observed that the flux increased while the SF increased initially and then decreased. It is also seen from Table 7 that as the ethanol content in the feed solution increased the δ-difference between the resulting solution and the membrane decreased. The increasing ethanol content in the feed made the solution hydrophobic enough to swell the PSI membrane, thereby increasing the flux and decreasing the SF for ethanol relative to water.

Conclusions

We have shown here that the selectivity of a liquid mixture through a polymer membrane can be related to the difference in solubility parameters between the solute, solvent, or solution of interest and the membrane. In addition, more hydrophobic membranes have a higher capacity to repel water. The hydrophobicity of the PSI membranes increases with increasing PDMS content of the copolymer, thus making them useful in water-organic separations.

Acknowledgements

The authors recognize Arseni Radomyselski and Scott Powell (Procter and Gamble, Cincinnati, Ohio, USA) for their contributions to the science and engineering of siloxane membranes. The first author also expresses his thanks to Dr. Stephen Clarson (University of Cincinnati, Ohio, USA) for his inspiration and assistance with this chapter.

References

1. Aerts S.; Vanhulse, A.; Buekenhoudt, A.; Weyten, H.; Kuypers, S.; Chena, H.; Bryjak, M.; Gevers, L. E .M.; Vankelecom, I. F. J.; Jacobs, P. A. *J. Membr. Sci.* **2005**, *275*, 212.
2. Mark, J. E. *Acc. Chem. Res.* **2004**, *37*, 946.
3. Bi, J.; Simon, G. P.; Yamasaki, A.; Wang, C. L.; Kobayashi, Y.; Griesser, H. J. *Rad. Phy. and Chem.* **2000**, *58*, 563.
4. Morlière, N.; Vallières, C.; Perrin, L.; Roizard, D. *J. Membr. Sci.*, **2006**, *270*, 123.
5. Nagai, K.; Kanehashi, S.; Tabei, S.; Nakagawa, T. *J. Membr. Sci.* **2005**, *251*, 101.
6. Gomes, D.; Nunes, S. P. ; Peinemann, K. *J. Membr. Sci.* **2005**, *246*, 13.
7. Drioli, E.; Zhang S.; Basile, A. *J. Membr. Sci.* **1993**, *80*, 306.
8. Ulutan, S.; Nakagawa, T. *J. Membr. Sci.* **1998**,*143*, 275.
9. Chang, C. L.; Chang, M. S. *J. Membr. Sci.* **2004**, *238*, 117.

10. Wu, P.; Brisdon, B. J.; England, R.; Field, R. W. *J. Membr. Sci.* **2002**, *206*, 265.

11. Watson, J. M.; Payne, P. A. *J. Membr. Sci.* **1990**, *49*, 171.

12. Hildebrand, J. H.; Scott, R. L.; *The Solubility of Nonelectrolyte*, Dover Publishers Inc, NY, 1964.

13. LaPack, M. A.; Tou, J. C.; McGuffinb, V. L.; Enkeb, C.G. *J. Membr. Sci.* **1994**, *86*, 263.

14. Pazuki, G. R.; Nikookar, M. *Fuel*, **2006**, *85*, 1083.

15. Brandrup, J.; Inmergut, E.H.; Grulke E. A., *Polymer Handbook, 4th Edition*, Wiley, New York ,1999.

16. Stannett, V.; Simple gases, in Crank, J.; Park; G. S. (Eds.); *Diffusion in Polymers*, Academic Press, New York, NY, 1968.

17. Mark, H. F.; Bales, N. M.; Overberger, C. G.; Menges, G.; Kroschwitz, J. I. (Eds.); *Encyclopedia of Polymer Science and Engineering*, Vol. 3, Wiley-Interscience, New York, NY, 1985.

18. Mishima, S.; Kaneoka, H.; Nakagawa, T. J. Appl. Polym. Sci., **1998**, *71*, 273.

19. Chang, T. C.; Wu, K. H.; Liao, C. L.; Lin, S. T.; Wang, G. P. *Polym. Degrad. and Stab.* **1998**, *62*, 299.

20. Lai, J. Y.; Li, S. H.; Lee, K. R. *J. Membr. Sci.* **1994**, *93*, 273.

21. Chang, Y. H.; Kim, J. H.; Lee, S. B.; Rhee, H. W. *J. Appl. Polym. Sci.* **2000**, *77*, 2691.

22. Fadeev, A. G.; Meagher, M. M.; Kelley, S. S.; Volkov, V. V. *J. Membr. Sci.* **2000**, *173*, 133.

Surfaces and Interfaces

Chapter 15

Manipulating Siloxane Surfaces: Obtaining the Desired Surface Function via Engineering Design

Julie A. Crowe, Kirill Efimenko, and Jan Genzer[*]

Department of Chemical and Biomolecular Engineering,
North Carolina State University, Raleigh, NC 27695
[*]Corresponding author: Jan_Genzer@ncsu.edu

We present a synopsis of recent accomplishment in our group in the area of surface-functionalized silicone elastomer networks. Specifically, we show that by combining mechanical manipulation of poly(dimethylsiloxane) (PDMS) networks with activation via ultraviolet/ozone (UVO) treatment and subsequent chemical modification of the pre-activated surfaces, one can generate so-called mechanically assembled monolayers (MAMs). This technology can be successfully applied to create a variety of surfaces, including dense polymer brushes, long-lived superhydrophobic surfaces, molecular gradients comprising tunable length scales and two-dimensional chemical gradients. In addition, the UVO modification of mechanically strained PDMS sheets provides a convenient route for creating "buckled" elastomeric sheets. These surfaces have a multitude of applications; ranging from anti-fouling surfaces to directed particle assembly. Finally we demonstrate control of surface wettability and responsiveness using poly(vinylmethylsiloxane) (PVMS) networks. We have shown that PVMS surfaces can be chemically tailored by reacting with thiols. The degree of response (including response rate) of such surfaces to hydrophobic-hydrophilic interactions can be adjusted by varying the chemical and structural properties of the side-group modifiers.

Introduction

Commercial interest in siloxane elastomers has lasted for over 50 years. The most-widely used material in this category is poly(dimethylsiloxane) (PDMS) (R = -CH$_3$ in Figure 1). The high flexibility of the Si-O bond, the greater Si-O-Si bond angle and length relative to the C-C-C bond, and low energy barriers to rotation, contribute to very low glass transition temperature of PDMS (T$_g$ ≈150 K). This low T$_g$ value gives rise to a very high chain flexibility even at a room temperature. In addition, the presence of two stable methyl groups on the silicon atom produces a chemically inert PDMS. These two characteristics make siloxanes a popular choice for thermal or electrical insulators and as barrier sealants or coatings.

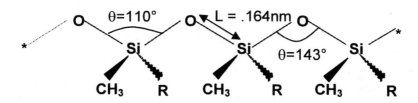

Figure 1: The siloxane backbone has a high degree of flexibility unparalleled in organic polymers due to its large Si-O-Si bond angle and length.

We have been able to harvest the underlying properties of the extreme flexibility of siloxane elastomers for growth in understanding and manipulating surface design. In order to command a desired response from a surface, altering its chemical make-up is necessary. The inherent inertness of PDMS, however, makes this task very difficult. Surface modifications involving 1) physical treatments such as plasma/corona irradiation and ultraviolet (UV) and UV/ozone exposure, 2) chemical oxidation or 3) a combination of both physical and chemical treatments have been utilized in the past to render the original PDMS either more hydrophobic or hydrophilic. One challenge associated with chemical oxidation of siloxane polymers is that the required high basic or acidic conditions promote backbone chain scission or degradation. Yet another challenge is the surface stability of a converted hydrophilic PDMS substrate due to the so-called "hydrophobic-recovery". This Chapter discusses the methods of conversion for both hydrophilic and hydrophobic silicone elastomers and our subsequent routes for modifying their surfaces. Some of the established technologies for PDMS surface modification gave us the capability to explore basic design manipulations leading to surface stability, controlled topography and surface reconstruction and reversibility.

Surface Modification of PDMS

Obtaining Hydrophilicity

There are many instances, where the low surface energy PDMS surface is not always desired. For instance, in blood-contacting devices and adhesive coatings it is paramount to alter the PDMS surface properties.[1-3] In addition, surface transformation from hydrophobic (e.g., methyl) to hydrophilic (e.g., hydroxyl-, carboxy-) groups enables the use of coupling techniques with organic modifiers of varying functionality. In applications where minimal surface reconstruction is required, decreasing polymer bulk and surface mobility is also necessary. One way to achieve this goal is to use a polymer with a glass transition temperature (T_g) much higher than its application temperature. This was elegantly demonstrated by Rouse et al.'s work on "frustrated" glassy surfaces where they manipulated the polymer interface composition by controlling the system temperature.[4] After oxidizing the surface of a model glassy polymer, poly(4-methylstyrene), to obtain hydroxy- and carboxy- surface groups, they observed that this surface was "locked" and remained hydrophilic when exposed to air below its bulk T_g. However, for temperatures above T_g, the surface was no longer "frozen" as the chains were allowed to resume their preferred random state. For polymers, such as PDMS, where the T_g is much lower than most (if not all) of its applications, hydrophilic stability is difficult to achieve without some change in crystallinity or surface orientation order.

Traditional transformation of the hydrophobic PDMS to a stable hydrophilic state has been achieved via techniques such as corona and plasma treatments.[5, 6] Conventional plasma treatments comprise exposing the substrate to a glow discharge between two electrodes or from a radio or microwave frequency generator at low pressures in various gases.[5] X-ray photoelectron spectroscopy (XPS) analysis on oxygen plasma treated samples confirmed a rapid substitution of carbon atoms by oxygen atoms, which led to the formation of hydrophilic surfaces.[7] This treatment will propagate several hundred nanometers below the surface with irreversible chemical changes at the near-surface region.[7-9] Simultaneously, a PDMS surface "skin" is transformed into a thin and brittle silica-like layer. The creation of a silica-like layer leads to crack formation causing changes in the PDMS mechanical properties. The cracks also allow the diffusion of low molecular weight PDMS species to migrate to the surface reducing the initial hydrophilic state. While not thoroughly discussed in this chapter, it is recognized that the outcome of corona treatments are very similar to plasma-treated surfaces.

A detailed stability study on plasma-treated silicon rubber was performed by Williams and co-workers.[10] The low-powered plasma treatment occurred in the presence of either argon, oxygen, nitrogen, or ammonia gas. A battery of analytical techniques confirmed that the modified surfaces were hydrophilic. For O_2- and Ar-treated samples, there was an increase in wettability as determined by contact angle, formation of a brittle silica-like layer, and a recovery to a hydrophobic state ("hydrophobic recovery") after aging for 1 month in air. When aged in phosphate-buffered saline (PBS) the hydrophobic recovery of O_2- and Ar-treated samples was reduced. Compared to their O_2- and Ar-treated counterparts, the N_2- and NH_3-treated samples showed minimal roughness as determined by atomic force microscopy (AFM), an increase in hydrophilic properties (increase in nitrogen species content), and a similar hydrophobic recovery. Interestingly, there was a difference seen in their biocompatibility despite their hydrophobic recovery patterns, which may be attributed to the degree of roughness and functionality. Compared to bare PDMS, the O_2- and Ar-treated samples exhibited decreased haemocompatibility; the time it takes for a foreign blood-contacting device to induce thrombosis. In contrast, the N_2- and NH_3-modified samples achieved longer blood-contacting times without platelet activation and coagulation than untreated PDMS. This illustrates that surface modification is a multi-faceted design problem involving chemical composition, topography, and morphology considerations for a given application. Since control of the modification process is difficult with plasma or corona technology, there is a need for other techniques.

We chose to modify PDMS surfaces using ultraviolet/ozone treatment (UVO). The UVO method involves a photo-sensitized oxidation process in which the molecules of the treated material are excited and/or dissociated by the absorption of short-wavelength ultraviolet (UV) radiation ($\lambda_1 = 184.9$ nm and $\lambda_2 = 253.7$ nm). The organic products of this excitation react with atomic oxygen to form simpler, volatile molecules, which desorb from the surface. Therefore, while atomic oxygen is continuously generated, ozone is continually formed and destroyed by this process. [11] Compared to oxygen plasma, UVO treatment is recognized as a milder system for physical modification to polymer surfaces with similar surface changes but with an order of magnitude reduction in processing time.[12, 13] This slower rate allows for better control of the surface conversion as different degrees of hydrophilicity will be obtained for different treatment times. The PDMS elastomer network system that we widely use in our research involves both model PDMS as well as industrial-grade Sylgard-184 manufactured by Dow Corning. The latter material comes in a two-part kit that makes Sylgard-184 readily applicable for commercial use. Figure 2 depicts the difference in the contact angle of water (a), the contact angle hysteresis (b), and the surface energy (c) for Sylgard-184 surfaces treated with UVO with two different lamps; one that transmits about 65% (UVO60) and the other transmits about 90 % (UVO90) of the shorter wavelength radiation.

In addition to contact angle measurements we also used XPS and near-edge x-ray absorption fine structure (NEXAFS) spectroscopy to characterize the

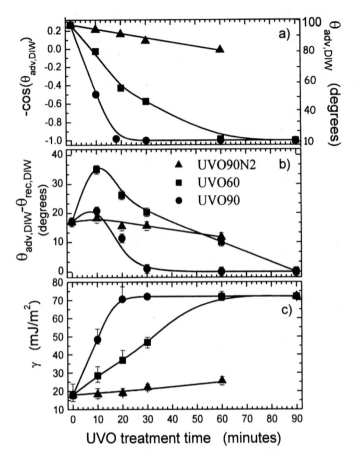

Figure 2: Dependence of a) the advancing contact angles of deionized water ($\theta_{adv,DIW}$), b) the contact angle hysteresis of deionized water ($\theta_{adv,DIW}$- $\theta_{rec,DIW}$), and c) the surface energy (γ) on the UVO treatment time for Sylgard-184 exposed to UVO60 (squares), UVO90 (circles), and UV90N2 (up-triangles). In part a) each point represents an average over 5 measurements on various areas of the same sample (the error associated with these measurements is smaller than ±1.5°). The lines are guides to the eye. Copyright © 2002 Elsevier Inc. All rights reserved. [11]

formation of the silica-like layer. Our results showed that when exposed to UV, Sylgard-184 underwent chain scission of both the main backbone and the methyl side-groups. The radicals formed during this process recombined forming a network whose wetting properties were close to those of untreated Sylgard-184. In contrast to the UV radiation, the UVO treatment created a large number of hydrophilic groups (*e.g.*, -OH) to the surface and sub-surface structure of

Sylgard-184. The material density distribution within the first ≈5 nm of Sylgard-184 was most affected by UVO treatment. The electron density near-surface region reached about 50 % of that of silica as measured by the x-ray reflectivity technique.[11]

An important observation from our work was that the silica fillers in Sylgard-184 did not alter its surface properties during and after UVO treatment. Comparison of our results with previous studies on silica filler containing PDMS materials concluded that the silica fillers were rarely seen in the sub-surface region of oxygen plasma PDMS films.[14-19] As Sylgard-184 is a convenient commercial material to use, we combined its fabrication with a mechanical technique to create highly-hydrophobic surfaces that resist reconstruction.

Surface Stability

We mentioned earlier that hydrophobic recovery from the physically-modified surfaces created with UVO or oxygen plasma treatment prevented surface stability as the surfaces lost their original hydrophilic properties. Ironically, this phenomenon is advantageous for outdoor insulator applications. PDMS elastomers used in areas with high air-born pollution are subject to contamination deposits, which cause electrolyte film formation on the insulator resulting in current leakage and dry-band arcing. The corona discharge causes an initial wettability increase but recovers to its original hydrophobic state within 100 minutes, as tracked using AFM.[20] In addition, the adhesive force was found to increase immediately after the electrical exposure but eventually returned to it original value of the unexposed surface.[20] While this type of "oscillating behavior" with PDMS elastomers allows the long-term vitality of the application even with the intermittent surface instability, there is still a great need for a surface with intrinsic stability.

The attraction of naturally occurring superhydrophobic surfaces (*e.g.*, lotus leaf effect) has inspired researchers to develop clever surface designs. For example, thermoresponsive polymers and manipulation of surface roughness has resulted in surfaces with contact angles exceeding 140°.[21, 22] Recently, Jin and co-workers used a laser-etching method to produce super-hydrophobic self-cleaning surfaces with PDMS.[22] They created PDMS with micro-, submicro-, and nano-composite surface structures by etching micropillars into PDMS and also spin-coating PDMS on polycarbonate nanopillar surfaces. These etched surfaces showed extremely high water contact angles of 160°C and minimal wettability hysteresis (*cf.* Figure 3).

Attaching various organic modifiers to a surface is another way to alter surface properties of materials. One common technique is the deposition of self-assembled monolayers (SAMs) producing high quality and structurally-defined surfaces of specific chemical functionality.[23, 24] Traditionally, SAMs have been formed with thiol or disulfide molecules on gold substrates or chlorosilanes molecules on silica substrates (molecules are of various ω-terminated functionality). The deposition of

228

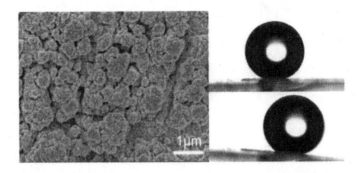

Figure 3 : Etched PDMS surface containing micro-, submicro-, and nano-composite structures shows a self-cleaning effect with water contact angle as high as 162° and sliding angle lower than 5°. Copyright © 2005 WILEY-VCH Verlag GmbH & Co. KGaA, Weinheim. [22]

SAMs onto polymer surfaces allows for specific tailoring of surface properties, such as wetting and lubrication. Recognizing the fact that wetting properties of SAMs and their subsequent stability depend on a delicate interplay between the chemical functionality and the degree of order (including packing) within the SAM, our laboratory capitalized on the inherent siloxane elasticity to manipulate monolayer packing and surface modification of PDMS to control the grafting density of organosilane SAMs.[25]   Specifically, we developed a method leading to the formation of mechanically assembled monolayers (MAMs). This technique (cf. Figure 4) is based on mechanically stretching PDMS sheets, and "activating" the surface via UVO treatment. Surface-bound hydroxyl groups on the UVO-modified PDMS were subsequently used as attachment points for semifluorinated organosilanes (SFO). After releasing the initial strain, the sheets returned to their original size thus "squeezing" the SFOs into densely populated arrays.

In Figure 5 we plot the results for the monolayers formed with $F(CF_2)_8(CH_2)_2SiCl_3$ (F8H2) on three substrates: PDMS, PDMS with an elongation (Δx) of 70%, and PDMS ($\Delta x=70\%$) after aging for six months in a humid environment. Not only did the water contact angle increase to values above 120°, but it also lost very little hydrophobicity after prolonged water exposure for the MAMs-PDMS fluorinated sample. While some hydrophobicity is lost for the aged sample, it is minimal and still far more hydrophobic than the SAMs-PDMS fluorinated sample (no stretching involved). Experiments using NEXAFS established that the high hydrophobicity and stability of the surfaces resulted from close packing of the semifluorinated groups on the surface that hindered the motion of the surface-grafted molecules.

MAMs of hydrocarbon-based silanes, $H(CH_2)_nSiCl_3$ (Hn-MAM), were also evaluated for hydrophobic performance.[26] It is known that as the length n of

A)

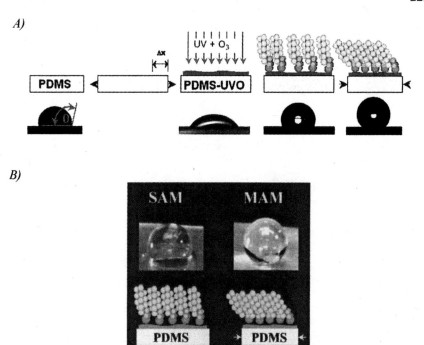

B)

Figure 4: A) Schematic illustrating the mechanism of increased packing density of a self- assembled monolayer after mechanically stretching the substrate. Copyright © 2000 American Association for the Advancement of Science. All Rights Reserved. B) Representation showing the higher water contact angle for MAMs on PDMS versus SAMs on PDMS.[25]

the hydrocarbon increases, the structure of a self-assembled monolayer transitions from a liquid-like to a solid-like state.[27-29] An increase in crystallinity should result in reduced mobility of the polymer surface. We evaluated this concept in hydrocarbon MAMs by probing the boundary between the liquid- and solid-like transition with "short" (n=8) and "long" (n=16) alkane molecules grafted onto PDMS surfaces. Grafting densities of the alkane silanes were varied by stretching the substrate to different elongations (Δx) prior to the UVO treatment. The higher the elongation, the greater the assumed packing density of the alkane chain. We hypothesized that if the alkane chains were densely packed on the PDMS surface, neighboring chains should restrict their mobility even if the alkane chain lengths were less than the commonly accepted crystalline transition state of n = 12. To test this theory, we studied the behavior of MAMs prepared from octyltrichlorosilane (H8-MAM). The data in Figure 6 depict the dependence of the contact angles of deionized (DI) water on H8-

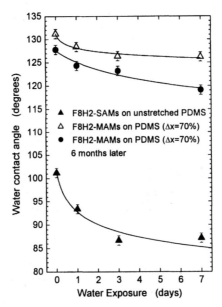

Figure 5: Water contact angles for mechanically-assembled monolayers made of semifluorinated alkanes on siloxane elastomers. Copyright © 2000 American Association for the Advancement of Science. All rights reserved.[25]

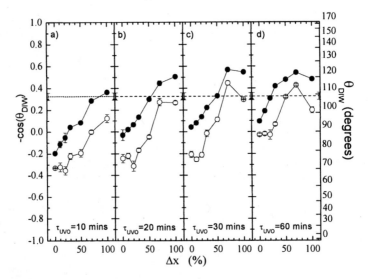

Figure 6. Advancing (●) and receding (○) contact angles of DI water, θ_{DIW}, on H8-MAMs prepared by vapor deposition of OTS as a function of the stretching of the PDMS substrate, Δx. The dashed line denote the value of θ_{DIW} for a crystalline array of $-CH_3$ groups.[24] Copyright © 2002 Materials Research Society Symposium Proceedings. [26]

MAM substrates as a function of the deposition time (τ) of the n-octyltrichlorosilane (OTS) and the substrate stretch, Δx. Both advancing (solid circles) and receding (open circles) contact angles are shown for each OTS treatment time. As with previous results for the H16-MAMs, where crystalline behavior is expected due to the van der Waals forces between neighboring methylene groups, we see an increase in the contact angle with increasing τ and Δx for the H8-MAMs. We also detect the same decrease in contact angle hysteresis until a minimum is reached; after this point the hysteresis starts to increase (Δx > 70%), which is contributed to increased roughening of the surface due to overcrowding of the alkane chain moieties.

In addition, NEXAFS experiments on the H8-MAM samples reveal very interesting information. While no chain orientation was detected at Δx = 0% and Δx > 30% for all deposition times, several H8-MAM samples prepared on PDMS-UVO pre-stretched to 0% < Δx < 30% and exposed to OTS molecules for τ < 30 minutes exhibited a non-negligible orientation order within the H8-MAM. Detailed analysis of the NEXAFS data revealed that the chains were tilted on average ≈35-45° away from the sample normal. This observation provides important evidence that the liquid-to-solid transition in substrate-anchored alkanes can be fine-tuned by tailoring the molecules packing density.

In our previous work we have established that when exposed to water, self-assembled monolayers that contain structural defects usually lose their structural order.[25] In contrast, we have shown that MAMs possess excellent barrier properties since after their exposure to polar environments, such as water, contact angles do not deteriorate dramatically with time. We also demonstrated that the high stability of the MAMs was a result of close chain packing. Thus we performed similar stability experiments on the Hn-MAMs. In Figure 7, we present the results of the contact angle measurements from "dry" samples and samples exposed to water for 3 and 5 days. Data for both H8-MAM and H16-MAM are shown. For comparison we also include the results obtained from the F8H2-MAM specimens.[25] It is apparent that the surface properties of Hn-MAMs remain very stable even after prolonged exposures to water with values similar to those as obtained by mechanically assembling semifluorinated moieties. In addition, the stability of the MAMs is much higher than those of SAMs. This work demonstrated that hydrophobic surfaces with long-lasting barrier properties can be prepared by fabricating hydrocarbon-based MAMs. The packing density of Hn-MAMs can be controlled by adjusting the strain on the elastic substrate allowing for a high grafting density. Finally, we have shown that a liquid-to-solid transition in surface-anchored alkanes can be induced by increasing Δx providing excellent surface stability. Considering that hydro-carbon-based moieties are much cheaper than fluorocarbons, the hydrocarbon-based MAMs may provide convenient routes for generating stable, water resistant surfaces in various industrial applications.

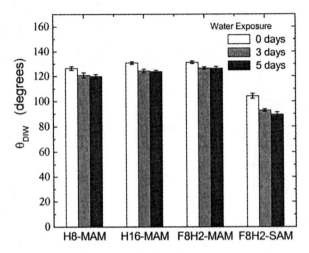

Figure 7:   Water stability data for H8-MAM, H16-MAM, F8H2-MAM,
structures, all stretched to Δx=70 %.   Also included are data for F8H2-SAM.
Copyright © 2002 Materials Research Society Symposium Proceedings.[26]

Chemically Heterogeneous Surfaces with MAMS

Surfaces with heterogeneous chemical compositions represent valuable platforms for chemical separations, selective adsorption, and probing cell/substrate interactions. Accessibility to heterogeneous surfaces was limited thus we implemented our technique of MAMs in several facets of our research to create molecular gradients. These included controlled/"living" radical polymerization initiated from silicone substrates, forming composition gradients with chemical vapor deposition on UVO-activated PDMS, and in the production of two-dimensional (2D) molecular gradients. First, a "grafting from" approach was carried out by surface-polymerization of acrylamide monomer on the PDMS substrate using atom transfer radical polymerization (ATRP).[30-32] We deposited an ATRP-chlorosilane initiator on UVO-activated PDMS to subsequently form well-defined "living" polyacrylamide (PAAm) polymer chains (cf. Figure 8).[33-35]   The advantage of combining MAMs with surface-initiated polymerization is to form polymer "brushes" where the chains are stretched perpendicular to the surface versus maintaining their preferred coiled-like conformations. It is well-known that this transition to the brush regime on a substrate occurs at sufficient concentrations of end-anchored polymers. While the "grafting from" approach eliminates the entropic penalties that would occur if polymer chains were "grafted to" the substrate, we find we can further adjust the surface density by maximizing the concentration of deposited initiators via MAMs. We call this technique "mechanically assisted polymerization assembly" (MAPA).

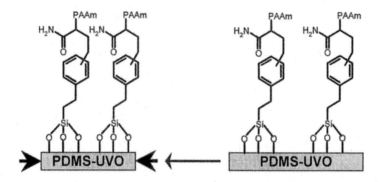

Figure 8: Polyacrylamide on PDMS substrates. Surface polymerization occurred on the stretched siloxane substrate.

The same principles of MAMs for surface coverage of the semifluorinated chlorosilane molecules on PDMS hold true for the deposition of the ATRP initiator on PDMS. After the initiator deposition, the polymerization of PAAm takes place while the PDMS is in the stretched state. Upon release of the strain, the substrate returns to its unstretched size and the grafted polymer chains are forced into a densely organized brush formation. Evidence of brush formation was obtained by Fourier transform infrared spectrometry in the attenuated total reflection mode (ATR-FTIR). Using the band assignments for the symmetric amide C=O stretching mode, C-N stretching/N-H deformation modes, and asymmetric and symmetric N-H amide stretching modes [36], close inspection of data in Figure 9 reveals the appearance of these peaks for both PAAm-MAPA samples as compared to their absence in the PDMS-UVO sample with just initiator deposited to its surface. In addition, increasing the sample elongation from 0 to 20%, resulted in an increase in intensities for both the amide and carbonyl signals thus confirming successful production of densely packed polymer chains. By controlling the degree of stretching on the elastomeric substrate, we can tailor the density distribution of the polymer chains on the substrate.

Our second application of the MAMs technique enabled us to create tunable planar molecular gradients on flexible supports.[37] After stretching the PDMS network to the desired elongation, we deposited gradients of ω-$(CH_2)_n$-$SiCl_3$ by following a chemical vapor deposition method outlined by Chaudhury and Whitesides.[38] We have shown that the width of the molecular gradient can be tailored by varying the strain on the PDMS sample before gradient formation and the concentration of the chlorosilane in the diffusing source. The samples' wettability profiles are depicted in Figure 10. The abscissa denotes the distance along the PDMS substrate with 0 mm being the closest point to the evaporating dish. Inspection of the data shows a steeper concentration profile at $\approx$5 mm and plateaus at constant contact angle at $\approx$12 mm. In essence, the majority of

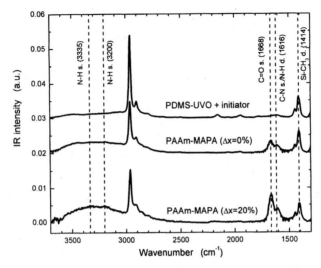

Figure 9: ATR-FTIR spectra for the mechanically assisted polymerization assembly.
Copyright © 2001 American Chemical Society. All rights reserved. [32]

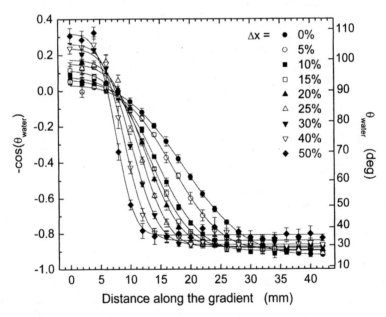

Figure 10:Contact angles of deionized water along gradient substrates
prepared on PDMS network films that were previously extended by Δx for 0% to
50% and treated with UVO for 30 minutes. The gradients were deposited from a
vapor source exposed to the substrates for 5 minutes. Copyright © 2001
WILEY-VCH Verlag GmbH, D-69469 Weinheim. [37]

hydrophobic octyltrichlorosilane molecules were packed in the 0-15 mm region in contrast to a more gradual distribution of hydrophobic moieties in the unstretched sample.

While the surface characteristics of MAMs can be tailored by simply adjusting the applied strain, we were able to take the 1-D planar molecular gradients to a second dimension by applying in-plane asymmetric stretching in combination with UVO treatment of PDMS.[39] By varying the shape of the PDMS substrate, we created different strain profiles throughout the stretched sample. Figure 11 illustrates the sample shape and stretching apparatus. The PDMS "dog-bone" sheet was inserted in the apparatus in two different ways. The first orientation (S1) is depicted in the top photos and schematic of Figure 11 where the strain will be concentrated primarily in the middle of the sample. The second orientation is shown in the bottom photos and a schematic where the strain will be concentrated predominately at the ends of the specimen. After UVO activation of the elongated PDMS sample, we uniformly deposited F8H2 organosilane. Due to the different spatial distributions of strains between the stretched samples, the degree of packing of the F8H2 molecules in the relaxed sample state was higher in the center for S1 as compared to S2. Combinatorial NEXAFS spectroscopy [40] was used to determine the in-plane F8H2 concentration on the PDMS substrates. The peak intensities for $1s \rightarrow \sigma^*$ C-F and C-F' excitations were collected in both the vertical and horizontal directions on the substrate. After normalizing the data, we obtained the C-F peak intensity map for each substrate. The darker regions in Figure 12A denote higher concentrations of fluorine. As it appeared that the higher areas of strain in each

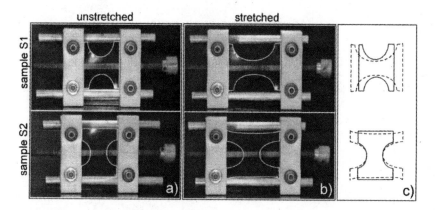

Figure 11: Photographs of the PDMS sheets clamped in the stretching apparatus before stretching (a) and after imposing the 40% uniaxial strain (b). The contours of the unstretched (solid lines) and stretched (dashed lines) samples are reproduced in (c) of the figure. Two sample geometries were used: Sample S1(top row) and S2 (bottom row). Copyright © 2003 WILEY-VCH Verlag GmbH, D-69469 Weinheim.[39]

236

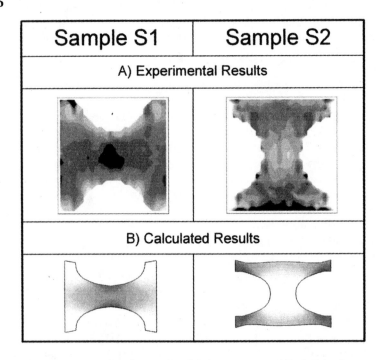

Figure 12: A) Normalized partial yield electron NEXAFS C-F peak intensity data for fluorinated-PDMS samples subjected to different variations of strain. The darker regions indicated higher concentrations of fluoring. B) Calculated stress distributions of the two samples. Copyright © 2003 WILEY-VCH Verlag GmbH, D-69469 Weinheim.[39]

sample did indeed contain higher degrees of packing of F8H2 molecules, we calculated the strain profiles in each sample. The calculated strain components in the stretching direction are illustrated in Figure 12B. Again, the darker regions in sample S1 agree with the expectations that the strain would be highest in the center and the converse is true for sample S2. This in-situ methodology for generating molecular gradients successfully created gradual in-plane variations of the F8H2 molecules via simple mechanical deformations.

In summary, MAMs has proven to be a convenient tool to generate a wide array of research applications thus providing us access to the design of tailorable chemically heterogonous surfaces. During the course of our work, we did notice one critical factor: if elongated PDMS substrates were over-exposed to UVO radiation, the samples would "buckle" upon release of the strain. Namely, when uniaxially stretched silicone elastomer specimens were exposed to UVO for prolonged time periods (tens of minutes to hours), the surfaces of the samples appeared slightly hazy after releasing the strain. An inspection of the sample with optical microscopy revealed that the reason for the haziness was roughness on the sample surfaces causing light to reflect and scatter from the surface rather than transmit through otherwise transparent silicon elastomer bulk. Hence we set off to examine the phenomena associated with the formation of these surface topologies more closely.[41]

Buckled Siloxane Surfaces

We had previously established that long exposure of PDMS to UVO led to conversion of the top-most part of PDMS to a silica-like material.[11] This created a bilayer comprising a silica-like film resting on top of a flexible PDMS substrate. When the bilayer formation is combined with mechanical deformation, interfacial instabilities give rise to the generation of wrinkles/buckles on the substrate's surface.[42, 43] We investigated the buckle formation and realized the wrinkling behavior could be manipulated to create ordered patterns with PDMS for applications such as particle sorting, anti-fouling release and selective interfacial ordering.

Optical microscopy and atomic force microscopy (AFM) experiments confirmed that the surfaces were originally flat in the presence of strain. After the UVO treatment, the strain was removed from the specimen and the skin buckled perpendicularly to the direction of the strain. A detailed analysis of the buckled surface with AFM and profilometry uncovered that the buckling patterns were hierarchical. Representative images depicting the various buckle generations are presented in Figure 13. Buckles with smaller wavelengths (and amplitude) rested parallel to and within larger buckles, forming a nested structure. Figure 13e summarizes the AFM and profilometry results of the buckle periods. The data in Figure 13e reveal that at least 5 distinct buckle generations (G) were present: the wavelengths of the generations (λ) are G1: $\approx$50 nm, G2: $\approx$1 μm, G3: $\approx$5 μm, G4: $\approx$50 μm, and G5: $\approx$0.4 mm. The mechanism of formation of the first generation of buckles, at the smallest scale, i.e. $\approx$50 nm, is as follows: The UVO treatment densifies the upper surface of the PDMS skin by proving additional cross-links and leads to an equilibrium (strain-free) configuration of the skin (S) that resides on top of the flexible substrate (B), which is still under tensile strain. When this strain is relieved from the specimen, the substrate attempts to contract back to its strain-free configuration. However, the mismatch between the equilibrium strains of the stiff skin and the soft substrate prevent this from happening uniformly throughout the depth of the material. The competition between the bending-dominated deformations of the skin and the stretching/shearing-dominated deformations of the substrate cause the skin to wrinkle in response to the relaxation of the applied strain. The characteristic wavelength of the wrinkles (λ) thus formed can be estimated from:

$$\lambda = 2\pi h \left[\frac{\left(1 - v_B^2\right) E_S}{3\left(1 - v_S^2\right) E_B} \right]^{1/3} . \tag{1}$$

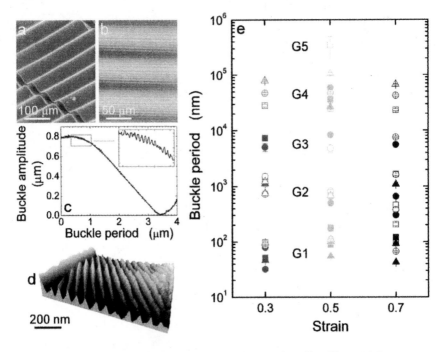

Figure 13: Characterization of the nested hierarchy of buckles. (a) Scanning electron microscopy image of a buckle on PDMS substrate revealing the G4 generation of buckles. (b) Optical microscopy image in the transmission mode of G3 and G4 generations of buckles. (c) Topography profile collected with profilometry on G2 (inset) and G3 (main figure) generations of buckles. (d) Scanning force microscopy image revealing the structure of G1 buckles. (e) Buckle period as a function of the strain imposed on the samples before the UVO treatment lasting for 30 (squares), 60 (circles), and 90 (up-triangles) minutes as measured by scanning force microscopy (filled symbols) and profilometry (open symbols). Copyright © 2005 Nature Publishing Group.[41]

In Equation (1) h is the thickness of the top skin, E_S and E_B are the elastic moduli of the skin and the elastic base, respectively, and v_S and v_B are the Poisson's ratios of the skin and the base, respectively. The ratio of Young's moduli E_S/E_B measured experimentally for UVO-treated unstretched PDMS films using AFM indentation was ≈15 and ≈87 for PDMS specimens treated with UVO for 30 and 60 minutes, respectively, which using Equation (1) gives λ≈12 and 22 nm, respectively, for the two treatment times. From the data in Figure 13, the estimated λ corresponds to the experimentally measured periods of the first generation of buckles (G1).

The basic driving force behind wrinkling is the mismatch in the equilibrium states of the skin and the substrate. Even if the skin/substrate system is subject to a simple compressive load, the skin will wrinkle in the same way due to the competition between the effects of bending the skin, which penalizes short wavelength buckles, and stretching the unmodified substrate (foundation), which penalizes long wavelengths.[43] This sets the stage for the amplitude of the primary wrinkles to grow as the applied strain is further relieved. Eventually the amplitude saturates owing to nonlinear effects in stretching and shearing the substrate. The composite of the wrinkled skin and the stretched substrate leads to the formation of an "effective skin" that is now thicker and much stiffer than the original skin. Further release of the applied strain leads to additional effective compression; as a result, the composite skin buckles on a much larger length scale, creating a hierarchical buckled pattern (*cf.* Figure 14). The formation of higher generation buckles continues until the strain is removed from the substrate. In an infinite system there is clearly no limit to this hierarchical patterning. Even in this finite system, up to five generations of these hierarchical buckles are arranged in a nested manner; each buckle generation is a scaled-up version of the primary buckle (*cf.* Figure 14). The smallest buckles are a few nanometers in wavelength while the largest ones are almost a millimeter in size, thus spanning nearly five orders of magnitude in dimension. Cerda and Mahadevan recently established that the amplitude of a wrinkle (ζ) scales as $\zeta = \lambda \varepsilon^{1/2}$, where ε is the implied strain.[44] In Figure 15 we plot the scaled experimental buckle amplitude $\zeta/\varepsilon^{1/2}$ as a function of λ; all data collapse roughly on a master curve.

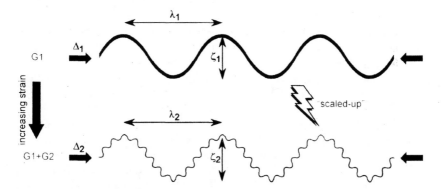

Figure 14: Schematic illustrating the formation of multigeneration-wrinkled topology. The generation 1 (G1) wrinkles form because of the competition between the bending-dominated deformations of the top layer (=skin) and the stretching/shearing-dominated deformations of the substrate cause the skin to wrinkle in response to the relaxation of the applied strain.

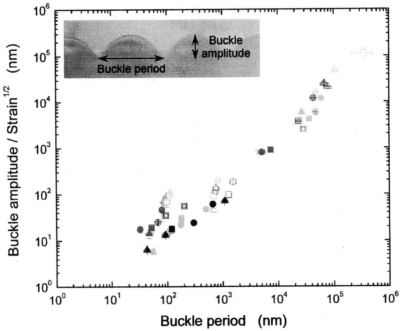

Figure 15: Ratio of the buckle amplitude to the square root of the strain plotted as a function of the buckle period on a log-log plot. We see that the data collapse onto a straight line consistent with the prediction of equation (1). Copyright © 2005 Nature Publishing Group.[41]

In the original publication, we have demonstrated that the hierarchically-wrinkled substrates may provide a convenient platform for separating mixture of particles based on their size.[41] Additional applications of wrinkled surfaces can be found in a recent review.[42] Recently, our group has starting exploring the application of the PDMS buckled sheets as antifouling release coatings. We have performed initial biofouling tests using coatings that contained the hierarchical buckles decorated with a fluorinated SAM. The experiments were done at the marine fouling testing site, owned by Microphase Coatings, in Wrightsville Beach, near Wilmington, NC. Figure 16 shows the photographs collected from the four coatings after immersion in sea water for one month. The samples are i) Sylgard-184 (Figure 16a and 16e), ii) Sylgard-184 treated with UVO for 30 minutes and covered with a F8H2 SAM (Figure 16b and 16f), iii) buckled Sylgard-184 covered with F8H2 MAM (Figure 16c and 16g), and iv) buckled Sylgard-184 treated with F8H2 MAM before the buckles were formed (Figure 16d and 16h). Pictures of samples taken immediately after removing the specimens from the ocean indicate the presence of a considerable amount of adsorbed biomass on flat samples (Figures 16a and 16b). Significantly smaller fouling with no presence of barnacles was seen on buckled specimens (Figure 16c and 16d). The difference between the two sample categories becomes even more apparent when we wash the samples with water ($\approx$60 psi pressure). While the flat Sylgard-184 samples remain fouled rather considerably, the surface of the samples with buckled geometries were cleaned almost completely. We attribute this observation to the ability of hierarchical buckles to minimize the attachment of biomolecules.

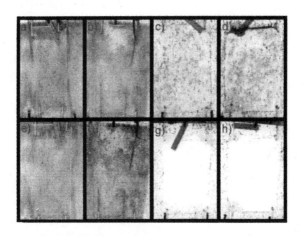

Figure 16: *Images of samples immersed in sea water (Wilmington, NC, April-May 2005) for 1 month before (a-d) and after (e-h) washing with water (pressure less than 60 psi): a)+e) Sylgard-184, b)+f) Sylgard-184 treated with UVO for 30 minutes and covered with a F8H2 SAM, c)+g) buckled Sylgard-184 treated with F8H2 SAM, and d)+h) buckled Sylgard-184 treated with F8H2 SAM (where the F8H2 molecules were incorporated before the buckles were formed).*

In parallel to the studies involving PDMS, we were also interested in finding a siloxane surface capable of retaining its elasticity during modification. As the UVO/exposure treatment depends on the generation of radicals, we expected that a substituent requiring lower activation energy than the methyl groups of PDMS would be more efficient in generating radicals that result in the formation of hydrophilic groups on surfaces. Evaluating poly(vinylmethylsiloxane) (PVMS) was a logical choice; an intermediate was available for synthesizing PVMS of varying molecular weights and the vinyl moiety provides a convenient route for preparing a wide variety of functionalized silicone elastomers using simple chemical methods.[45, 46] Besides the physical surface treatments such as ultraviolet oxidation, radiation, plasma or corona exposure, direct covalent attachment of the desired functionality through addition reactions such as hydrosilation, hydrosulfidation, hydrophosphination, epoxidation, and alkyl halide addition reactions allows for a myriad of possibilities.[47-50]   The remainder of this chapter describes our work on modifying PVMS and its resultant properties. Most notable is the generation of controlling wettability and the ability to switch surface composition states by changing the external environment.

Modification of PVMS

Surface Reconstruction and Reversibility

A surface with dual-functionality can adjust its physico-chemical characteristics in response to some external stimulus, such as electrical, chemical, or mechanical. In most cases, the responsiveness of the surface is a result of the rearrangement of the various chemical functionalities present close to or directly at the surface characterized by the degree of change of the surface properties after the external trigger was applied, and the rate at which these variations occur.   The required degree of control over these characteristic changes is dependent on the end-use for the surface.

We have recently reviewed several papers pertaining to the dynamic nature of polymer surfaces.[51]   One example of dynamic switching behavior was demonstrated with segmented polyurethanes containing both hydrophilic and hydrophobic characteristics (cf. Figure 17). A fabric sample was coated with a triblock polymer comprising perfluoropolyether, poly(dimethylsiloxane), and poly(ethylene glycol) components. Contaminants from air allowed oil to deposit on the treated sample. Once the fabric was rinsed with water, however, the soil release from the fabric was at a 90-95% level (as compared to 60% soil release of an untreated sample).[52]   Our goal was to reproduce some of the characteristic behavior seen in other materials by utilizing siloxanes.

Our initial observations of reversible PVMS siloxane surfaces was demonstrated with deionized water contact angle (θ_{DIW}) measurements .[53] As the vinyl group has a slightly higher surface energy than methyl group, it hides beneath the surface when exposed to hydrophobic environments, such as air. Exposure to water leads to a rearrangement of the surface moiety make-up as the preferred state is orientation of the vinyl groups at the surface as illustrated in Figure 18. As contact angle measurements are sensitive to just the first ≈ 5 Å of the polymer surface,[54] the composition rearrangement can be observed by tracking contact angle changes over time.   Figure 19 shows the change of wettability over time of a PVMS network as compared to a model network of PDMS and the commercially available Sylgard-184 elastomer.   The data for PVMS indicate a surface rearrangement; the slope of $d\theta_{DIW}/dt$ is not zero after initial contact with water while the other two networks do not show any time-dependent variation in slope. This change in slope correlated with a $\Delta\theta$ of about 10 degrees.[53]

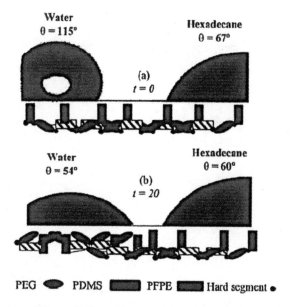

Figure 17: Drops of Water (left) and hexadecane (right) separated by a distance
of 2mm on a perfluoropolyether- modified segmented polyurethane surface at
time = 0 (a) and after 20 minutes (b). Segments besides polyurethane are PDMS
and polyethylene glycol (PEG). Copyright © 2002 Elsevier Science (USA).[51]

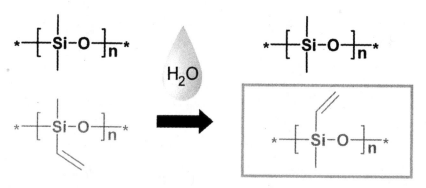

Figure 18: Illustration of the differences in surface rearrangement between
PDMS and PVMS elastomers. The vinyl moiety has a higher affinity for water
relative to the methyl group.

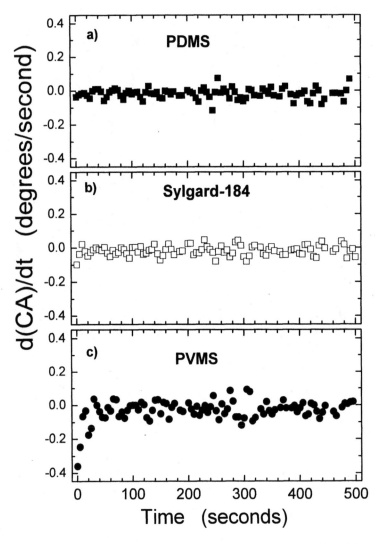

Figure 19: Change of the contact angle with increasing time for model PDMS networks (solid squares), Sylgard-184 networks (open squares), and bare PVMS networks (solid circles). Copyright © 2005 Elsevier Ltd All rights reserved.
[52]

Rapid Formation of Hydrophilic Surfaces

To further accentuate the difference in surface energy effects, we modified the vinyl moiety on the PVMS substrate by first UVO/exposure (PVMS-UVO) and then through a thiol-ene addition reaction. Similar to our work with Sylgard-184, we studied the mechanism of surface modification of PVMS-UVO via static contact angle, FTIR-ATR, NEXAFS, and AFM.[53] As mentioned earlier, modification to PVMS with UVO would be much more efficient as the strained vinyl functionality is activated more easily than the more stable Si-O and Si-C bonds.[11, 53] We confirmed this with the FTIR-ATR experiments as the signature Si-O-Si peak (1055-1090 cm^{-1}) remained unchanged after UVO exposure. In contrast, the PDMS UVO-exposed samples showed a distinct decrease in the Si-O-Si peak intensity indicative of chain scission. Comparing the hydrophilicity of the treated samples as a function of UVO exposure time (*cf.* Figure 20), showed that treatment times of 300 seconds resulted in water contact angles of 62°. In addition, comparisons of surface and bulk modulus measurements revealed that the PVMS elasticity was essentially unchanged after 2 minutes of UVO treatment.[53] As the resulting shorter UVO treatment times for PVMS (an order of magnitude lower as compared to PDMS) give rise to a considerably reduced modification to the bulk modulus, we could produce very flexible networks capable of rapid surface rearrangement under appropriate conditions. This is also demonstrated in Figure 20; the water contact angle was seen to decrease between 20-50° per given substrate. Although not completely evaluated, the variation in the rate of change between these substrates was attributed to differences in surface modulus. Hence, if we could keep the flexibility of the PVMS backbone intact during modification, then obtaining rapid reorientation of the surface to its preferred configuration is possible. By investigating the kinetics of surface restructuring utilizing the capability to functionalize PVMS, we gained an understanding of the transition between a repeatable responsive surface and a surface with an inferred stable chemistry.[55]

PVMS Modification with a Thiol-ene Reaction—Transitioning from Crystalline-like to Liquid-like

Modification of a vinyl substituent via a thiol-ene reaction had been successfully carried out by both Ferguson's and Chojnowski's groups.[46, 56] Implementing a similar addition route [51], we modified PVMS substrates with 3-mercaptopropionic acid and 11-dodecanethiol. The addition to the vinyl bond was confirmed via the elimination of the C=C peaks at 960, 1407 and 1587 cm^{-1} in ATR-FTIR spectroscopy. The initial observations of surface responsiveness, as measured by static contact angle of water, are illustrated with the data in Figure 21a. While the PVMS-UVO sample was hydrophilic, the initial water

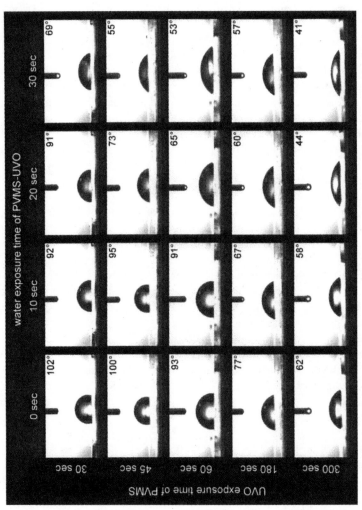

Figure 20: Images of water droplet spreading on PVMS-UVO surfaces treated for various UVO times (ordinate) collected at various time intervals after depositing the droplet (abscissa). The numbers indicate the water contact angle that was evaluated from the images of the water droplets. Copyright © 2005 Elsevier Ltd All rights reserved.[52]

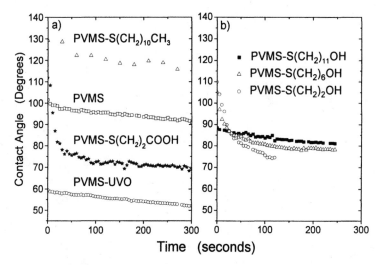

Figure 21: a) Static contact angle of deionized water on various PVMS sustrates. b) and on three mercaptoalkanol modified PVMS substrates. Copyright © 2005 American Chemical Society. All rights reserved.[54]

contact angle (≈60°) did not vary greatly from the final contact angle (≈55°). This can be attributed to the lack of flexibility in the "silica-like" layer. PVMS does reconstruct as shown by the decrease in contact angle over time (*cf.* Figure 19c), but it is minimal compared to the contact angle change of PVMS-(CH2)₂COOH. The exceedingly fast drop in contact angle (dθ/dt = 2°/second) accentuates the surface energy differences between the hydrophobic methyl substituent and the hydrophilic carboxy-terminated alkane substituent.

One notable aspect of the data from Figure 21a is that while the UVO and 3-mercaptopropionic acid surface modifications are made from the same PVMS substrate, the thiol-ene reaction with 11-dodecanethiol took place in the bulk siloxane fluid. This modified PVMS fluid was subsequently crosslinked through its hydroxyl end-groups. The high contact angle (≈130°) of the PVMS-S-(CH₂)₁₀CH₃ substrate is likely due to both the hydrophobicity of 11-dodecane thiol and the inherent roughness of crosslinking such a heavy substituted siloxane chain. This substrate, however, does lose some of its hydrophobicity after exposure to water but at a slower rate than the PVMS-S-(CH₂)₂COOH substrate. In this case, the methyl substituents have a higher surface energy compared to the 11-dodecanethiol substituents. In order to fully understand the effects of chain length on the rearrangement kinetics, PVMS sheets were modified with three different mercaptoalkanols: HS(CH₂)₂OH (-S-C₂-OH), HS(CH₂)₆OH (-S-C₆-OH), and HS(CH₂)₁₁OH (-S-C₁₁-OH). Evaluating the surface response of mercaptoalkanols versus mercaptoalkanoic acids eliminated any possible complications of acid-base interactions due to the carboxy groups.[57-59] As

248

the data in Figure 21b show, the response with PVMS-S-C_2-OH and PVMS-S-C_6-OH was fast compared to PVMS-S-C_{11}-OH. It was difficult, however, to obtain consistent results for repeated oscillations between a wet and dry state of the sample. Utilizing the Wilhelmy Plate method with a Cahn dynamic contact angle (DCA) apparatus, eliminated the need for sample handling between runs.[55] After each run, the sample was simply dried with a nitrogen purge. Repeatability testing was done on three independent samples to insure consistent results. The initial cycle of DCA measurements performed for each substrate agreed with the static contact angle results.

Figure 22 shows the DCA results illustrating the restructuring of the modified mercaptoalkanol substrates for ten oscillations between a wet and dry state.[55] We attributed the sluggish nature of the mercaptododecanol substrate to ordering of the -C_{11}- alkane chains due to van der Waals interactions. This transition to crystallinity upon water exposure prevented the rapid reconstruction observed with the other two modified substrates. PVMS-S-C_2-OH and PVMS-S-C_6-OH remained in a flexible liquid-like state, allowing continued oscillations for at least ten cycles. Our assumption that a transition to a semi-crystalline regime was supported with three independent measurements: the level of opaqueness, a change in the storage modulus (G'), and FTIR-ATR results.

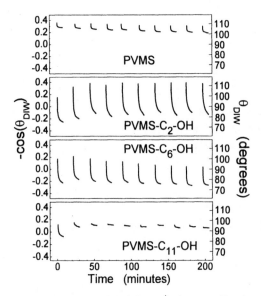

Figure 22: Dynamic contact angle of deionized water (θ_{DIW}) as a function of time reported for ten cycles on each n-alkane hydroxyl-terminated elastomer. For each n-alkane series the data represent an average over three separate specimens. The error for θ_{DIW} is ±3.8° Copyright © 2005 American Chemical Society. All rights reserved.[54]

All PVMS, PVMS-S-C$_2$-OH, and PVMS-S-C$_6$-OH samples were transparent, elastic and tacky (adhered well to a polystyrene Petri dish). In contrast, the PVMS-S-C$_{11}$-OH specimens were opaque, rigid, and non-adhering to polystyrene. Upon immersion into hot water (temperature $\geq$ 70°C), PVMS-S-C$_{11}$-OH became transparent but turned opaque again upon cooling. When the sample was quenched to room temperature in water immediately after the heated water treatment, the surface 'froze' at a water contact angle of $\approx$80°. If allowed to slowly cool, however, its contact angle of water increased above 95°.

Dynamic rheology was performed on PVMS and the three mercaptoalkanol-modified PVMS samples. It was obvious through sample handling that the PVMS-S-C$_{11}$-OH substrate was more rigid than the other modified samples. This observation was confirmed with a 10-fold increase in the storage modulus G' (*cf.* Figure 23a). Although the modulus did increase, the G' data for PVMS-S-C$_{11}$-OH is frequency-dependent, indicating an imperfect network (frequency sweep performed within the linear visco-elastic regime at 0.5% strain). In contrast, PVMS, PVMS-S-C$_2$-OH and PVMS-S-C$_6$-OH are considered perfectly elastic as a zero slope was obtained for the G' versus frequency data as shown in Figure 23a.

Upon heating semi-crystalline material, ordered polymer chains will relax into their preferred random coil configurations above a melting transition. To determine if this held true for the PVMS-S-C$_{11}$-OH substrate, more detailed

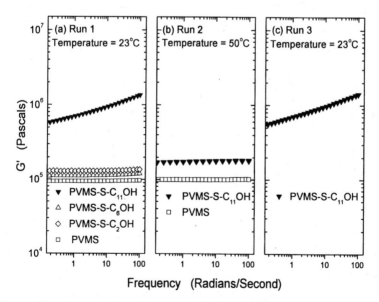

Figure 23 : Dynamic rheology on PVMS and mercaptoalkanol-modified PVMS substrates. Runs performed at sequential operational temperatures (T): a) T = 23°C, b) T = 50°C, and c) T=23°C. Average standard deviation is ± 33 kPa.

250

analysis of the rheological properties was performed by varying the run temperature for the dynamic frequency sweep tests as shown in Figures 23b-c. At an operating temperature of 50°C, the melting transition (melting point of 11-mercaptoundecanol is between 33-37° [60]) had been reached in Run 2 as this temperature-induced polymer relaxation state resulted in negligible frequency dependency. This indicated the PVMS-S-C_{11}-OH network had reached near-perfect elasticity. Figure 23c illustrates that upon cooling, there is full network recovery as the Run 3 data replicates Run 1 data. From this work it appears that the ordered alkane chains were acting as filler-like reinforcers at room temperature disrupting the elastic nature of the siloxane network.

Our last measurement, FTIR-ATR spectroscopy, provided evidence to support our claim that the 11-mercapto-1-undecanol modified PVMS surface had undergone a phase transition with the formation of semi-crystalline domains. Figure 24 depicts the decreasing frequency shift for both asymmetric and symmetric C-H stretches of the methylene group. The characteristic frequency for the C-H stretching vibrations (asymmetric) occurs at 2920 cm^{-1} for crystalline polymethylene chains. For the liquid polymethylene state, it is known that C-H stretching vibrations (asymmetric) occur at a frequency 2928 cm^{-1}.[29] A similar downward shift for the symmetrical C-H stretching vibrations is observed for liquid (2856 cm^{-1}) to crystalline (2850 cm^{-1}) transition of the polymethylene chain.[29] Our results for the modified PVMS substrates show that while the -C_6-OH substituted surface is in a liquid-like state, the -C_{11}-OH group is in a crystalline-like state. Our observed frequency shifts between these two substrates closely agree with the reported polymethylene transition. Our

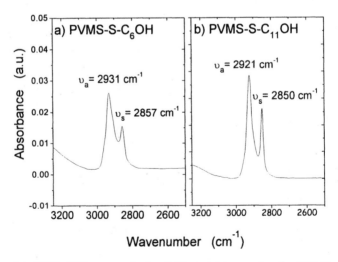

Figure 24: FTIR-ATR spectra in the C-H stretching region for PVMS modified with a) 6-mercapto-1-hexanol and b) 11-mercapto-1-undecanol. Copyright © 2005 American Chemical Society. All rights reserved.[54]

observations are also in accord with the results of Chaudhury and Owen [28] who showed that adhesion hysteresis can be tuned by varying the surface density of hexadecyltrichlorosilane (HTS) chemisorbed films on PDMS. Highly packed films of HTS were found to be crystalline ($\upsilon_a(CH_2)$=2919 cm^{-1} and $\upsilon_s(CH_2)$= 2850 cm^{-1}) with high adhesion hysteresis. The opposite was true for a less dense film of HTS. We also believe that FTIR-ATR studies for PVMS–C$_{11}$-OH performed at temperatures above the melting transition will result in C-H stretching vibrations characteristic of a liquid-like state. These studies will be the subject of future work.

Conclusions

Our work with siloxane surfaces evolved from super-hydrophobic PDMS with inherent longevity to rapidly reconstructive PVMS surfaces capable of acting in an on/off state due to environmental changes. We have utilized MAMs on PDMS for antifouling applications and particle sorting. As we continue to investigate the interactions of nature with our synthetic surfaces, our tools for manipulating the surface structure make our investigations capable of crossing broad interdisciplinary research for applications only limited by our imagination.

Acknowledgements

We gratefully acknowledge the U.S. National Science Foundation and the Office of Naval Research for financially supporting various aspects of the work. We also thank several people for their collaboration related to the research described in this chapter. Dr. Daniel A. Fischer (NIST) for collaboration on NEXAFS studies performed at the Brookhaven National Laboratory, Dr. Dwight W. Schwark (Cryovac SealedAir) for his work on UVO treatment of PVMS, and Professors Evangelos Manias (Penn State University) and L. Mahadevan (Harvard University) for working with us on the buckled elastomer surfaces. We also thank Dr. Trey Schimendinger (Microphase Coatings) for providing access to their test site in Wrightsville Beach, NC.

References

1. Buddy Ratner, A.H., Fredrick Schoen, and Jack Lemons, ed. *Biomaterials Science: An Introduction to Materials in Medicine.* 1996, Academic Press: San Diego.

2. Chen, H., M.A. Brook, and H. Sheardown, *Silicone elastomers for reduced protein adsorption.* Biomaterials, 2004. **25**(12): p. 2273-2282.

3. Gordon, G.V., et al., *Silicone release coatings: An examination of the release mechanism.* Adhesives Age, 1998. **41**(11): p. 35-+.
4. Rouse, J.H., P.L. Twaddle, and G.S. Ferguson, *Frustrated reconstruction at the surface of a glassy polymer.* Macromolecules, 1999. **32**(5): p. 1665-1671.
5. Owen, M.J., *Plasma/Corona Treatment of Silicones.* Australian Journal of Chemistry, 2005. **58**: p. 433-436.
6. Ferguson, G.S., et al., *Monolayers on Disordered Substrates - Self-Assembly of Alkyltrichlorosilanes on Surface-Modified Polyethylene and Poly(Dimethylsiloxane).* Macromolecules, 1993. **26**(22): p. 5870-5875.
7. Hillborg, H., et al., *Crosslinked polydimethylsiloxane exposed to oxygen plasma studied by neutron reflectometry and other surface specific techniques.* Polymer, 2000. **41**(18): p. 6851-6863.
8. Hillborg, H. and U.W. Gedde, *Hydrophobicity changes in silicone rubbers.* Ieee Transactions on Dielectrics and Electrical Insulation, 1999. **6**(5): p. 703-717.
9. Owen, M.J. and P.J. Smith, *Plasma Treatment of Polydimethylsiloxane.* Journal of Adhesion Science and Technology, 1994. **8**(10): p. 1063-1075.
10. Williams, R.L., D.J. Wilson, and N.P. Rhodes, *Stability of plasma-treated silicone rubber and its influence on the interfacial aspects of blood compatibility.* Biomaterials, 2004. **25**: p. 4659-4673.
11. Efimenko, K., W.E. Wallace, and J. Genzer, *Surface modification of Sylgard-184 poly(dimethyl siloxane) networks by ultraviolet and ultraviolet/ozone treatment.* Journal of Colloid and Interface Science, 2002. **254**(2): p. 306-315.
12. Huck, W.T.S., et al., *Ordering of spontaneously formed buckles on planar surfaces.* Langmuir, 2000. **16**(7): p. 3497-3501.
13. Ouyang, M., et al., *Conversion of some siloxane polymers to silicon oxide by UV/ozone photochemical processes.* Chemistry of Materials, 2000. **12**(6): p. 1591-1596.
14. *As reported by Kim and coworkers , these fillers are rarely detected within the first 5 nm below the film substrate despite their large concentration in Sylgard-184. In their study, the researchers stated that the above observation did not imply a macroscopic depletion of the filler in the surface region of PDMS. The filler particles could easily be coated by a polymer film of 5 nm thickness and even uncoated and close-packed in the surface they would still constitute only less that 1% of the volume of the top 5 nm thick layer for a flat geometry of special micrometer sized particles with PDMS filling the interstices.*
15. Kim, J., M.K. Chaudhury, and M.J. Owen, *Hydrophobicity loss and recovery of silicone HV insulation.* Ieee Transactions on Dielectrics and Electrical Insulation, 1999. **6**(5): p. 695-702.

16. Kim, J., M.K. Chaudhury, and M.J. Owen, *Hydrophobic recovery of polydimethylsiloxane elastomer exposed to partial electrical discharge.* Journal of Colloid and Interface Science, 2000. **226**(2): p. 231-236.

17. Kim, J., et al., *The mechanisms of hydrophobic recovery of polydimethylsiloxane elastomers exposed to partial electrical discharges.* Journal of Colloid and Interface Science, 2001. **244**(1): p. 200-207.

18. Kim, J., M.K. Chaudhury, and M.J. Owen, *Modeling hydrophobic recovery of electrically discharged polydimethylsiloxane elastomers.* Journal of Colloid and Interface Science, 2006. **293**(2): p. 364-375.

19. Hillborg, H. and U.W. Gedde, *Hydrophobicity recovery of polydimethylsiloxane after exposure to corona discharges.* Polymer, 1998. **39**(10): p. 1991-1998.

20. Meincken, M., T.A. Berhane, and P.E. Mallon, *Tracking the hydrophobicity recovery of PDMS compounds using the adhesive force determined by AFM force distance measurements.* Polymer, 2005. **46**(1): p. 203-208.

21. Sun, T.L., et al., *Reversible switching between superhydrophilicity and superhydrophobicity.* Angewandte Chemie-International Edition, 2004. **43**(3): p. 357-360.

22. Jin, M., et al., *Super-Hydrophobic PDMS Surface with Ultra-Low Adhesive Force.* Macromolecular Rapid Communications, 2005. **26**: p. 1805-1809.

23. Chaudhury, M.K., Materials Science & Engineering Reports, 1996. **16**: p. 97.

24. Ulman, A., *An Introduction to Ultrathin Organic Films from Langmuir-Blodgett to Self Assembly.* 1991, New York: Academic Press.

25. Genzer, J. and K. Efimenko, *Creating long-lived superhydrophobic polymer surfaces through mechanically assembled monolayers.* Science, 2000. **290**(5499): p. 2130-2133.

26. Efimenko, K. and J. Genzer, *Tuning the surface properties of elastomers using hydrocarbon-based mechanically assembled monolayers.* Materials Research Society Symposium Proceedings, 2002. **DD10.3.1**: p. 710.

27. Allara, D.L., A.N. Parikh, and E. Judge, *The existence of structure progressions and wetting transitions in intermediately disordered monolayer alkyl chain assemblies.* Journal of Chemical Physics, 1994. **100**(2): p. 1767-1764.

28. Chaudhury, M.K. and M.J. Owen, *Correlation between Adhesion Hysteresis and Phase State of Monolayer Films.* The Journal of Physical Chemistry, 1993. **97**: p. 5722-5726.

29. Snyder, R.G., H.L. Strauss, and C.A. Elliger, *C-H Stretching Modes and the Structure of n-Alkyl Chains. 1. Long, Disordered Chains.* The Journal of Physical Chemistry, 1982. **90**: p. 5623-5630.

30. Patten, T.E., et al., *Polymers with very low polydispersities from atom transfer radical polymerization.* Science, 1996. **272**(5263): p. 866-868.

31. Patten, T.E. and K. Matyjaszewski, *Atom transfer radical polymerization and the synthesis of polymeric materials.* Advanced Materials, 1998. **10**(12): p. 901-+.

32. Wu, T., K. Efimenko, and J. Genzer, *Preparing high-density polymer brushes by mechanically assisted polymer assembly.* Macromolecules, 2001. **34**(4): p. 684-686.

33. Huang, X. and M.J. Wirth, *Surface initiation of living radical polymerization for growth of tethered chains of low polydispersity.* Macromolecules, 1999. **32**(5): p. 1694-1696.

34. Huang, X.Y., L.J. Doneski, and M.J. Wirth, *Make ultrathin films using surface-confined living radical polymerization.* Chemtech, 1998. **28**(12): p. 19-25.

35. Huang, X.Y., L.J. Doneski, and M.J. Wirth, *Surface-confined living radical polymerization for coatings in capillary electrophoresis.* Analytical Chemistry, 1998. **70**(19): p. 4023-4029.

36. Gaboury, S.R. and M.W. Urban, *Quantitative Attenuated Total Reflectance Fourier-Transform Infrared-Analysis of Microwave Plasma Reacted Silicone Elastomer Surfaces.* Langmuir, 1994. **10**(7): p. 2289-2293.

37. Efimenko, K. and J. Genzer, *How to prepare tunable planar molecular chemical gradients.* Advanced Materials, 2001. **13**(20): p. 1560-+.

38. Chaudhury, M.K. and G.M. Whitesides, *How to Make Water Run Uphill.* Science, 1992. **256**(5063): p. 1539-1541.

39. Genzer, J., D.A. Fischer, and K. Efimenko, *Fabricating two-dimensional molecular gradients via asymmetric deformation of uniformly-coated elastomer sheets.* Advanced Materials, 2003. **15**(18): p. 1545-+.

40. Genzer, J., D.A. Fischer, and K. Efimenko, *Combinatorial near-edge x-ray absorption fine structure: Simultaneous determination of molecular orientation and bond concentration on chemically heterogeneous surfaces.* Applied Physics Letters, 2003. **82**(2): p. 266-268.

41. Efimenko, K., et al., *Nested self-similar wrinkling patterns in skins.* Nature Materials, 2005. **4**: p. 293-297.

42. Genzer, J. and J. Groenewold, Soft Matter, 2005, in press.

43. Allen, H.G. *Analysis and design of structural sandwich panels.* 1969. Pergamon, New York.

44. Cerda, E. and L. Mahadevan, *Geometry and physics of wrinkling.* Physical Review Letters, 2003. **90**(Art. No. 074302).

45. Holmesfarley, S.R., et al., *Reconstruction of the Interface of Oxidatively Functionalized Polyethylene and Derivatives on Heating.* Langmuir, 1987. **3**(5): p. 799-815.

46. Carey, D.H. and G.S. Ferguson, *Synthesis and Characterization of Surface-Functionalized 1,2-Polybutadiene Bearing Hydroxyl or Carboxylic-Acid Groups.* Macromolecules, 1994. **27**(25): p. 7254-7266.

47. Boutevin, B., F. Guida-Pietrsanta, and A. Ratsimihety, *Side Group Modified Polysiloxanes,* in *Silicone-Containing Polymers,* J. Chjnowski, Editor. 2000, Kluwer Academic Publishers: Dordrecht. p. 79-112.

48. Bauer, J., N. Husing, and G. Kickelbick, *Synthesis of new types of polysiloxane based surfactants.* Chemical Communications, 2001(01): p. 137-138.

49. Bauer, J., N. Husing, and G. Kickelbick, *Preparation of functional block copolymers based on a polysiloxane backbone by anionic ring-opening polymerization.* Journal of Polymer Science Part a-Polymer Chemistry, 2002. **40**(10): p. 1539-1551.

50. Cai, G.P. and W.P. Weber, *Synthesis and chemical modification of poly(divinylsiloxane).* Polymer, 2002. **43**(6): p. 1753-1759.

51. Crowe, J.A., et al., *Responsive Siloxane-Based Polymeric Surfaces,* in *Responsive Polymer Materials: Design and Applications,* S. Minko, Editor. 2006, Blackwell Publishing. p. 184-205.

52. Vaidya, A.C., MK, *Synthesis and surface properties of environmentally responsive segmented polyurethanes.* Journal of Colloid and Interface Science, 2002. **249**(1): p. 235-245.

53. Efimenko, K., et al., *Rapid formation of soft hydrophilic silicone elastomer surfaces.* Polymer, 2005. **46**(22): p. 9329-9341.

54. Bain, C.D. and G.M. Whitesides, *Depth Sensitivity of Wetting - Monolayers of Omega-Mercapto Ethers on Gold.* Journal of the American Chemical Society, 1988. **110**(17): p. 5897-5898.

55. Crowe, J.A. and J. Genzer, *Creating responsive surfaces with tailored wettability switching kinetics and reconstruction reversibility.* Journal of the American Chemical Society, 2005. **127**(50): p. 17610-17611.

56. Chojnowski, J., et al., *Controlled synthesis of vinylmethylsiloxane-dimethylsiloxane gradient, block and alternate copolymers by anionic ROP of cyclotrisiloxanes.* Polymer, 2002. **43**(7): p. 1993-2001.

57. Holmesfarley, S.R. and G.M. Whitesides, *Reactivity of Carboxylic-Acid and Ester Groups in the Functionalized Interfacial Region of Polyethylene Carboxylic-Acid (Pe-Co2h) and Its Derivatives - Differentiation of the Functional-Groups into Shallow and Deep Subsets Based on a Comparison of Contact-Angle and Atr-Ir Measurements.* Langmuir, 1987. **3**(1): p. 62-76.

58. Holmesfarley, S.R., et al., *Acid-Base Behavior of Carboxylic-Acid Groups Covalently Attached at the Surface of Polyethylene - the Usefulness of Contact-Angle in Following the Ionization of Surface Functionality.* Langmuir, 1985. **1**(6): p. 725-740.

59. Holmesfarley, S.R., C.D. Bain, and G.M. Whitesides, *Wetting of Functionalized Polyethylene Film Having Ionizable Organic-Acids and Bases at the Polymer Water Interface - Relations between Functional-Group Polarity, Extent of Ionization, and Contact-Angle with Water.* Langmuir, 1988. **4**(4): p. 921-937.

60. http://www.sigmaaldrich.com/catalog/search/ProductDetail/ALDRICH/.

Chapter 16

Proteins at Silicone Interfaces

Paul M. Zelisko[1,2], Amro M. Ragheb[1], Michael Hrynyk[1],
and Michael A. Brook[1,*]

[1]Department of Chemistry, McMaster University, 1280 Main Street West,
Hamilton, Ontario L8S 4M1, Canada
[2]Current address: Department of Chemistry, Brock University, 500
Glenridge Avenue, Saint Catharines, Ontario L2S 3A1, Canada
*Corresponding author: mabrook@mcmaster.ca

Although enzymes are normally exploited in aqueous,
biological environments, certain enzymes can be made to
operate in organic media. We report the utilization of
enzymes in active form in a yet more hydrophobic silicone
medium, in both liquid and elastomeric forms, by using
suitable non-ionic surfactants. The enzymatic activity,
depending on the protein, can be shown to be even higher than
in organic media, if the concentration of water is appropriately
controlled. α-Chymotrypsin, an enzyme readily denatured by
simple silicone oils when bulk water is present, can be
rendered highly active in both silicone oils and silicone
elastomers when stabilized with poly(ethylene oxide) based
surfactants: human serum albumin was similarly unaffected by
the liquid silicone when surfactants were present. Lipase was
shown to have a higher activity in silicone elastomers than in
hydrocarbon solvents, but only if PEO constituents were not
added. The factors necessary to stabilize protein structure and
retain enzymatic activity, such as water content and the
presence of PEO, in a silicone-rich environment are described.

Introduction

Siloxane polymers are among the most hydrophobic polymers available: only fluoropolymers have lower surface energies. Proteins in contact with exceptionally hydrophobic silicones[1] can undergo unfolding, with consequent loss of activity in the case of enzymes.[2,3] Two factors can mitigate these denaturing events: control of available water, and the presence of surface active molecules.

Klibanov, Halling, and others have shown that some proteins can operate effectively in organic solvents that have minimum water content.[4-6] At low water levels, internal hydrogen bonding interactions maintain protein tertiary structure, yet permit sufficient flexibility around the active site for reactions to be catalyzed.[7-9] An alternative approach in organic solvents stabilizes enzymes by using surfactants that control the degree to which the hydrophobe can come into contact with the protein. Thus, the addition of high molecular weight (>25000 MW) silicones, with a few hydrophilic groups, to the silicone oil/aqueous protein mixture can prevent protein denaturation, even though the protein is in close proximity to the silicone oil in an emulsion.[10]

We wished to develop silicone based emulsion and elastomeric systems in which proteins were stabilized for use as immobilized enzymes, or could be released on demand as bioactives. It was reasoned that control of both water content and the interfacial structure could be achieved using poly(ethylene oxide)(PEO) as a constituent. PEO is widely understood to be a "protein-friendly" polymer, as demonstrated by improved potency of protein drugs when "pegylated"[11] and improved biocompatibility of surfaces when modified with PEO.[12,13] Enzyme activities of two very different enzymes, α-chymotrypsin and lipase, were used to report on the interfacial stabilization provided to proteins in emulsion[14,15] and elastomeric[16] environments, respectively.

Experimental

Reagents

The surfactant DC3225C was obtained from Dow Corning; decamethylpentacyclopentasiloxane (D_5), octamethylcyclotetrasiloxane (D_4), silanol-terminated poly(dimethylsiloxane) (($HO-SiMe_2(OSiMe_2)_nOSiMe_2OH$, n $\approx$ 484), 2000 cSt, 36000 MW), dibutyltin dilaurate, tetraethyl orthosilicate (TEOS), poly(ethylene oxide) (PEO) (2000 MW), sodium hydride (60% dispersed in oil), Celite, allyl bromide, triethoxysilane, lauric acid, dimethylsulfoxide (DMSO), and octanol were purchased from Aldrich. Vinyl-terminated polydimethylsiloxane (1000 cSt and 2-3 cSt) and platinum divinyl-tetramethyldisiloxane complex (3-5% platinum concentration, Karstedt's catalyst) in vinyl-terminated polydimethylsiloxane were obtained from Gelest.

N,N-dimethylformamide (DMF), isooctane, dichloromethane, acetone, and cyclohexane were purchased from Caledon. Human serum albumin (HSA) 96-99% purity, blue dextran, α-chymotrypsin, lipase from *C. rugosa* (EC 3.1.1.3), lysozyme, protein assay kits (Lowry method) and Sephadex G-25 were purchased from Sigma. 5-(Bromomethyl)fluorescein (5-BMF) was supplied by Molecular Probes and used to label HSA.

TES-PDMS (bis(triethoxyethyl)dimethylsilicone),[17] TES-PEO (bis(triethoxy-silylpropyl) poly(ethylene oxide) including MW 2000, TES-PEO2000), and TES-mPEO 350 (triethoxysilylpropyl, methyl poly(ethylene oxide)),[18] and 5-BMF-HSA[19] were prepared using literature procedures.

Emulsion Formulation Procedure: Example using 5-BMF-Labelled HSA and TES-PDMS[17]

Emulsions were formulated using established procedures.[17] HSA (0.612 g) and 5-BMF-HSA (1.0 mL, 1.99 x 10^{-4} mg/mL) were dissolved in 1.0 M Tris-HCl buffer (30.0 mL, pH 7.8). Triethoxysilylethyl-terminated PDMS (4.14 g)) was dissolved in D_4 (16.1 g, continuous phase) in the mixing vessel. The aqueous solution was added to the silicone oil phase in a continuous, drop-wise manner over a period of 2 h under turbulent mixing conditions at 2780 rpm using a Caframo BDC6015 mixer. The emulsion was allowed to mix for an additional 2 h and stored in a sealed container at ambient temperature. Emulsions were imaged using a Zeiss LSM 510 confocal microscope.

Elastomeric Composite Preparation: Example Lipase in Silicone[18]

Lipase (100 mg) and, optionally, water were sonicated for 1-2 min in CH_2Cl_2 (~ 1 mL) then added promptly to a vial containing silanol-terminated polydimethylsiloxane (HO-PDMS, 2.85 g), and crosslinker (TEOS, 0.15 g). TES-mPEO 350 (~350 MW) was optionally included (0.3 g in CH_2Cl_2). The mixture was mixed vigorously with a spatula and then dibutyltin dilaurate catalyst (0.05 g, 0.08 mmol) was added. Mixing was continued for 4-5 min after which the mixture was poured onto small Petri dish (60 mm x 15 mm) and left uncovered for ca. 30 min, then covered and kept at room temperature to cure overnight. Cylindrical pellets of ca. 6.0 mm diameter and 1.8-2.2 mm thickness were cut from the cured film and, after washing with solvents (acetone, water, acetone, cyclohexane, then acetone) and air drying, granules of the composite were prepared by grinding the composite using a mortar. After swelling in isooctane, enzymatic activity was established using transesterification, as discussed in the text.

Results

Emulsions Human Serum Albumin and α-Chymotrypsin

Two different w/o emulsion systems were used to study the nature of protein/silicone interactions at an aqueous/oil interface: one employing the use of a silicone surfactant with terminal triethoxysilyl moieties (TES-PDMS, Figure 1A), and the second with a commercially available silicone surfactant possessing pendant poly(ethylene oxide) (PEO) groups (commercially available DC3225C). Emulsions were formulated by dissolving the silicone surfactant of interest in octamethylcyclotetrasiloxane (D$_4$), and subsequently adding drop-wise the aqueous phase, which optionally contains proteins, under dual-blade, turbulent mixing conditions. With the functional, poorer surfactant TES-PDMS, the presence of a surface active protein, human serum albumin (HSA), was required in order to form a stable w/o emulsion.[10,15]

Water-in-oil emulsions were formed with both surfactants, with protein dissolved in discrete water domains suspended in bulk silicone oil.[20,21] Confocal microscopy experiments performed on w/o emulsions containing fluorescently labelled proteins demonstrated that the protein preferentially resided at the water/silicone oil interface, indicating that contact between the biomolecule and the silicone oil was possible (Figure 1B).

The natural fluorescence of the single tryptophan residue provides a mechanism to assess the unfolding of HSA. During the unfolding processes, the fluorescence maximum is first blue and then red shifted.[22,23] It was shown that comparable rates of unfolding were observed for HSA in buffer, and in emulsions stabilized with DC3225C and TES-PDMS, respectively: all emulsions were blue shifted by day 10.

Despite the fact that the protein molecules are subjected to a great deal of mechanical stress during the emulsification process (6 h at 2780 rpm), and that the proteins are encountering silicone molecules at the emulsion interface, the proteins are protected by surfactants from silicone contact.[2,3] In some cases, emulsion stability correlates with HSA unfolding, for example, TES-PDMS emulsions were stable for 1-6 months, at which point the HSA was completely denatured. With the better surfactant DC3225C, enzyme unfolding plays no role in emulsion stability: stable emulsions are readily prepared using either native or denatured proteins, or no proteins at all.[24]

The exposure of an aqueous solution of chymotrypsin in buffer to silicone oil (octamethylcyclotetrasiloxane, (Me$_2$SiO)$_4$, D$_4$), in the absence of a surfactant, led to a rapid loss of enzymatic activity (85% loss). After exposure to D$_4$, the enzyme was unable to transesterify ethyl N-acetyl-L-phenylalanate with butanol. Under similar conditions but with addition of the surfactant DC3225C, comparable enzyme activities between the control and the emulsified proteins were observed (Figure 2A).[25] Activity remained despite the significant mechanical energy required to prepare and break the emulsion, and the possibility of contact with silicone during mixing. This is surprising given the fragility of the enzyme.[8,9]

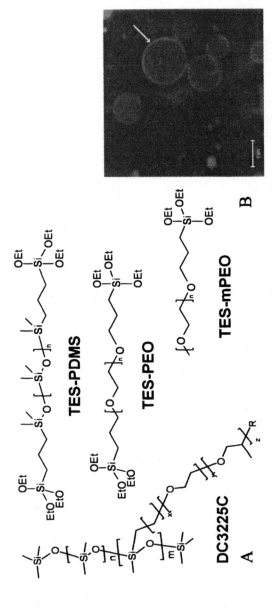

Figure 1. A: Structures of silicone surfactants; B: 5-BMF-labelled HSA shown as a corona at the interface of a w/o emulsion

These observations, based on two distinctly different proteins, suggest that a strong interaction exists between the surfactant and protein that either protects the protein from contact with unfunctionalized silicone oil during mixing, or which stabilizes protein tertiary structure.

Silicone Elastomers: *Chymotrypsin and Lipase*

Silicone-chymotrypsin elastomers were prepared using room temperature vulcanization (RTV) with the optional inclusion of functional PEO (e.g., $MeO(CH_2CH_2O)_n(CH_2)_3Si(OEt)_3$) using silanol terminated silicones $(HO(SiMe_2O)_nH, ca.$ 36000 MW) and TEOS $(Si(OEt)_4)$, catalyzed by dibutyltin dilaurate in the presence of moisture (RTV conditions[26]). In isooctane, optimum enzyme activity occurred when the enzyme was ultrasonicated in the presence of small quantities of water (no water, no sonication, activity = 0.001 mg/h per mg enzyme; sonicated with 0.2% water (a_w ~1.0), activity = 0.014 mg/h per mg of enzyme). At the same low water concentrations, the activity of chymotrypsin in D_4 was approximately the same as in isooctane, a result quite different from that noted above in excess water. The activity of the enzyme degraded over time in isooctane, whereas in the PEO-containing elastomer the activity was preserved over several consecutive reaction cycles (Figure 2B).

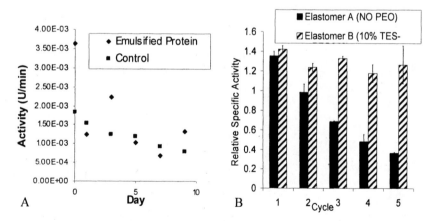

Figure 2. A. Enzyme activity of α-chymotrypsin A. extracted from a w/o emulsion vs control[31]; B. In plain silicone elastomers that optionally contain mono-functional PEO (TES-mPEO350).

Lipase is known to be generally tolerant of hydrophobic media and, in nature, generally resides at cell membranes.[27,28] The enzymatic activity of lipase in hydrophobic media indicated that lipase has a much higher enzymatic activity

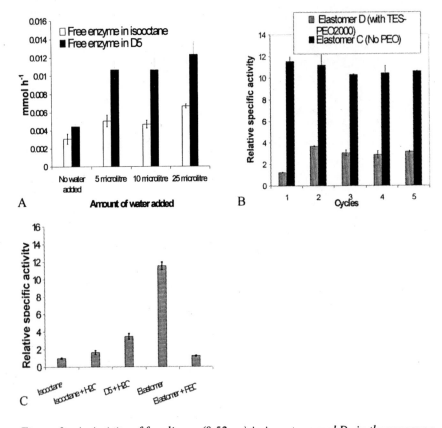

Figure 3. A. Activity of free lipase (0.52mg) in isooctane and D_5 in the presence of different amounts of water. B. Effect of including functional PEO in the elastomers (8%TES-PEO2000) on the activity of the entrapped lipase, measured by catalyzed esterification reaction. C. Comparison of enzyme activity as a function of media.

in D_5, compared to isooctane (Figure 3A), as judged by its ability to convert octanol and lauric acid into octyl dodecanoate[29,30] at different water concentrations.

Silicone elastomers with and without functional PEO were prepared from silanol terminated silicones as noted above with incorporation of lipase as a powder or, when functional PEO was added to the elastomer, as a slurry. After curing, elastomer disks (6 mm diameter, 2 mm thickness) were cut and examined as catalysts for ester formation from octanol and lauric acid over repeated cycles. Activity is expressed relative to the free enzyme activity (production of ester as determined by GC) in isooctane solution (taken as a specific activity of 1.0).

Results showed that lipase activities followed the order: silicone elastomer, no PEO > elastomer with PEO > lipase powder. Thus, the presence of hydrophilic materials, with the exception of minimal amounts of water, were counterproductive with respect to lipase activity (Figure 3B).

Discussion

The behavior of three constitutionally different proteins, with vastly different characteristics, has been examined on contact with silicones. It is apparent that all three proteins can be made to be reasonably compatible with silicones provided that the functionality grafted to the silicone, and the concentration of water, are carefully controlled. Three control elements were used to manipulate protein stability in a silicone environment: local viscosity, water concentration and the presence of PEO.

Lipase, known for its ability to withstand hydrophobic media, was found to be active in silicone oil, and more active in the less mobile elastomer. It is possible that an interface is established between silicone and protein when the elastomer cures, which restricts the ability of the enzyme to unfold. In such a case, the active site of the enzyme[27,28] can anchor at the hydrophobic interface, and be held in an active conformation by the locally high viscosity. The incorporation of hydrophilic media such as PEO in the elastomer reduces the local hydrophobicity of the interface and, consequently, the activity of the enzyme. For this enzyme, the local water concentration was less significant than for other proteins.

HSA unfolds readily at a silicone/water interface.[31] However, with the addition of non-ionic surfactants, the protein is protected from unfolding even though it lies at the PEO surfactant interface, as shown by fluorescence experiments. PEO may be increasing the local viscosity, acting as an osmolyte,[32] or simply insulating the HSA from the silicone. Such protective roles are exploited therapeutically for protein drugs.[33]

A different role must be ascribed to TES-PDMS, which led to a reduction in the rate of HSA unfolding within the emulsion droplets, and to the ability to create and stabilize a silicone/water interface: an emulsion of water and D_4 could not be formed from either HSA or TES-PDMS on their own.[31] Although the HSA is a constituent of the interface, it unfolds at a rate similar to that of the control. Of more interest is the correlation between emulsion stability and protein HSA folding in these emulsions, which suggests the HSA/TES-PDMS interaction may result from a binding event which becomes disfavored as the protein starts to unfold. Such a possibility provides an alternative method to release the contents of the water droplet through controlled HSA denaturation.

α-Chymotrypsin denatures at silicone/water interfaces.[17] When a surfactant such as DC3225 was present, the enzyme sat at the emulsion interface and, as with HSA, was protected from exposure to the silicone as evidenced by its enzymatic activity, which was comparable to control in spite of the mechanical energy used to make and break the emulsion.[17] On one hand, there is a driving force for proteins to adsorb at interfaces.[34] On the other hand, PEO-interfaces are typically protein repellent.[11-13] The balance of these two can account for the observed enzymatic behavior in w/o emulsions.

A similar situation arises with chymotrypsin in silicone elastomers doped with PEO. In this case, the enzyme must be PEO coated to avoid the denaturation that normally occurs on contact with silicone. The rate of denaturation in elastomers is retarded by higher viscosity, and by the very low availability of water; compare this with rapid denaturation in emulsions, when the protein is swollen with water and internally mobile. Water can be used to regulate enzyme activity in these systems; small amounts are needed for catalytic activity, but increased degrees of freedom from too much water lead to unfolding and inactivation.

By controlling the mobility of the interface, and the presence or absence of hydrophilic surface active materials, it was possible to stabilize three proteins of widely varying structure, function, and properties in hydrophobic silicone oil or elastomer domains. It is, therefore, likely that silicone-based environments can similarly be formulated for other proteins, which opens the possibility to use silicone-, rather than organic-based, materials, as biosensors, immobilized enzymes, and drug delivery vehicles.

Conclusions

Silicones are exceptionally hydrophobic materials that interact with enzymes, mainly through hydrophobic interactions, frequently causing denaturation leading to loss of enzymatic activity. Enzymes can be rendered stable and, in some cases, more active in silicone oils or elastomers by using the presence of non-ionic surfactants and water: specific formulations need to be developed for different enzymes.

Acknowledgements

We thank the Natural Sciences and Engineering Research Council of Canada for financial support of this work.

References

1. Owen, M.J. *Surface Chemistry and Applications*, In *Siloxane Polymers*, S. J. Clarson and J. A. Semlyen, Eds., Prentice Hall: Englewood Cliffs, NJ, **1993**, Chap. 7, p. 309.

2. Gekko, K.; Ohmae, E.; Kameyama, K.; Takagi, T. *Biochim. Biophys. Acta* **1998**, *1387*, 195.

3. Taboada, P.; Fernández, Y.; Mosquera, V. *Biomacromolecules* **2004**, *5*, 2201.

4. Yuchun, X.; Das, P.K.; Klibanov, A.M. *Biotech. Lett.* **2001**, *23*, 1451.

5. Rees, D.G.; Gerashchenko, I.I.; Kudryashova, E.V.; Mozhaev, V.V.; Halling, P.J. *Biocat. Biotrans.* **2002**, *20*, 161.

6. Halling, P.J.; Ulijn, R.V.; Flitsch, S. *Curr. Opinion Biotechnol.* **2005**, *16*, 385.

7. Davis, B.G.; Boyer, V. *Nat. Prod. Rep.* **2001**, *18*, 618.

8. Wang, P.; Dai, S.; Waeszada, S.D.; Tsao, A.Y.; Davison, B.H. *Biotechnol. Bioeng.* **2001**, *74*, 249.

9. Dzyuba, S.V.; Klibanov, A.M. *Biotechnol. Lett.* **2003**, *25*, 1961.

10. Zelisko, P.M.; Coo-Ranger, J.J.; Brook, M.A. *Polym. Prep. (Amer. Chem. Soc., Div. Polym. Chem.)* **2004**, *45*(1), 604-605.

11. Malmsten, M.; Van Alstine, J.M. *J. Colloid Int. Sci.* **1996**, *117*, 502.

12. Chen, H.; Brook, M.A.; Sheardown, H. *Biomaterials* **2004**, *25*, 2273.

13. Park, J.H.; Bae, Y.H. *Biomaterials* **2002**, *23*, 1797.

14. Liu, M.; Ragheb, A.; Zelisko, P.; Brook, M.A. *Preparation and Application of Silicone Emulsions Using Biopolymers*, In *Colloidal Biomolecules, Biomaterials, and Biomedical Applications* (Surfactant Science, Vol. 116), Elaïssari, A.; Ed., Mercel Dekker Inc., **2004**, Chap. 11, 309-329.

15. Zelisko, P.; Bartzoka, V.; Brook, M.A. *Exploiting Favorable Silicone-Protein Interactions: Stabilization Against Denaturation At Oil-Water Interfaces*, in *Synthesis and Properties of Silicones and Silicone-Modified Materials*, Clarson, S.J.; Fitzgerald, J.J.; Owen, M.J.; Smith, S.D.; Van Dyke, M.E.; Eds, ACS Symposium Series 838, **2003**, Ch. 19, pp. 212-221.

16. Ragheb, A.; Brook, M.A.; Hrynyk, M. *Chem. Commun.* **2003**, 2314-2315.

17. Zelisko, P.M.; Brook, M. A. *Langmuir* **2002**, *18*, 8982.

18. Ragheb, A. M.; Brook, M. A. *Biomaterials* **2005**, *26*, 6973-6983.

19. Hermanson, G.T. *Bioconjugate Chemistry*, **1996**, San Diego: Academic Press.

20. Klinkesorn, U.; Sophanodora, P.; Chinachoti, P.; McClements, D.J.; Decker, E.A. *J. Agric. Food Chem.* **2005**, *53*, 8365.

21. Gancz, K.; Alexander, M.; Corredig, M. *Food Hydrocolloids* **2006**, *20*, 293.

22. Flora, K.; Brennan, J.D.; Baker, G.A.; Doody, M.A.; Bright, F.V. *Biophys. J.* **1998**, *75*, 1084.

23. Flora, K.K.; Brennan, J.D. *Chem. Mater.* **2001**, *13*, 4170.

24. Hill, R. M.; Lin; Z. US Patent 6,616,934, Clear silicone microemulsions (to Dow Corning), 2003.

25. Shin, J-S.; Luque, S.; Klibanov, A.M. *Biotechnol. Bioeng.* **2000**, *69*, 577.

26. Brook, M.A. *Silicon in Organic, Organometallic and Polymer Chemistry*, Wiley, 2000, Chap. 9.

27. Reetz, M.T.; Zonta, A.; Simpelkamp, J.; Rufinska, A.; Tesche, B. *J. Sol-Gel Sci. Technol.* **1996**, *7*, 35.

28. Brzozowski, A.M.; Derewenda, U.; Derewenda, Z.S.; Dodson, G.C.; Lawson, D.M.; Turkenburg, J.P.; Bjorkling, F.; Hugejensen, B.; Patkar, S.A.; Thim, L. *Nature* **1991**, *351*, 491.

29. Zaks, A.; Klibanov, A.M. *J. Biol. Chem.* **1988**, *263*, 8017.

30. Russell, A.J.; Klibanov, A.M. *Biochem. Soc. Trans.* **1989**, *17*, 1145.

31. Zelisko, P.M.; Flora, K.K.; Brennan, J.D.; Brook, M.A. submitted.

32. Eggers, D.K.; Valentine, J.S. *J. Mol. Biol.* **2001**, *314*, 911-922.

33. Chapman, A.P. *Adv. Drug Delivery Rev.* **2002**, *54*: 531-545.

34. *Biomaterials Science*, B.D. Ratner, A.S. Hoffman, F.J. Schoen, J.E. Lemons, Eds., 2nd Ed., Elsevier: San Diego, 2004.

Silsesquioxanes

Chapter 17

Brewster Angle Microscopy Studies of Aggregate Formation in Blends of Amphiphilic Trisilanolisobutyl-POSS and Nitrile Substituted Poly(dimethylsiloxane) at the Air–Water Interface

Hyong-Jun Kim[1,2], Jennifer Hoyt Lalli[1,3], Judy S. Riffle[1], Brent D. Viers[1], and Alan R. Esker[1,*]

[1]Department of Chemistry and the Macromolecules and Interfaces Institute, Virginia Polytechnic Institute and State University, Blacksburg, VA 24061
[2]Current address: Department of Materials Science and Engineering, University of Michigan, Ann Arbor, MI 48109
[3]Current address: Nanosonic, 1485 South Main Street, Blacksburg, VA 24060

Polar silicone adhesives with a strong affinity for both metal and metal oxide surfaces are known to be good candidates for dispersing heat conducting fillers within the adhesive layer between the circuit and the heat sink of microelectronic circuit housings. Given the well-known surface activity of poly(dimethylsiloxane) and other silicones, polar silicones are a logical choice for modifying nanofiller interactions within the confines of quasi-two-dimensional monolayers and thin films at the air/water (A/W) interface. In this study, mixtures of a polyhedral oligomeric silsesquioxane, trisilanolisobutyl-POSS ($T_7iBu_7(OH)_3$), and a polar silicone, poly[(3-cyanopropyl)-methylsiloxane] (PCPMS), spread as mono-layers at the A/W interface have been used to examine the surface phase behavior and aggregation as a function of $T_7iBu_7(OH)_3$ composition, surface concentration, and compression rate. Polymer-$T_7iBu_7(OH)_3$ interactions are understood through the analysis of surface pressure-area per molecule isotherms and by direct observation with Brewster angle microscopy (BAM). Monolayer films for all PCPMS/T$_7$iBu$_7$(OH)$_3$ blends form homogeneous ideal mixtures prior to film collapse. After the film starts collapsing, BAM studies show that aggregates present in the blends exhibit characteristics of both T$_7$iBu$_7$(OH)$_3$ blends with non-polar PDMS and T$_7$iBu$_7$(OH)$_3$ blends with a polar phosphine oxide substituted PDMS derivative (PDMS-PO). Network structures like those found in non-polar PDMS/T$_7$iBu$_7$(OH)$_3$ blends can form under special conditions and coexist with small isolated aggregates that are more like aggregates found in more polar PDMS-PO/T$_7$iBu$_7$(OH)$_3$ blends. This behavior is attributed to hydrogen bonding and the nearly identical affinity of PCPMS and T$_7$iBu$_7$(OH)$_3$ for water.

Introduction

Polyhedral oligomeric silsesquioxane (POSS) molecules have attracted scientific attention (1-4) because of their hybrid molecular structure (5-8). Physical properties such as small dielectric constants (9-13), increased glass transition temperatures (14-21), improved thermal stability (22, 23), and enhanced resistance to oxidation (5-8) can be obtained by incorporating POSS into polymeric systems. In addition, POSS based systems afford the possibility to create copolymers (24-26) and nanostructured materials (27-33) with unique processing (34, 35) and mechanical (36-42) properties. As a hybrid material, surface and interfacial interactions are vital to POSS applications. Nonetheless, interfacial studies initially lagged behind bulk studies (43-49). Recently, the interfacial properties of two POSS derivatives trisilanolisobutyl-POSS (50, 51) and trisilanolcyclohexyl-POSS (52,53) have been investigated at the air/water (A/W) interface. These studies showed that trisilanol-POSS derivatives are surface active, and self-assemble into uniform monolayers upon compression. Moreover, trisilanolphenyl-POSS has even been reported to form Langmuir-Blodgett multilayer films by Y-type deposition (54).

The surface activity of trisilanol-POSS derivatives is not entirely a surprise. Studies of less well-defined ethyl and phenyl silsesquioxane resins like (45) have been reported to form Langmuir films at the A/W interface. Moreover, poly(dimethylsiloxane) (PDMS) (55-60) and other silicones (61-69) have been widely studied as monolayers at the A/W interface. One key difference between POSS molecules and other silicones appears to be the need for open-cage POSS architectures as octaisobutyl-POSS does not form uniform monolayer films (*50,70*). The interesting bulk and surface properties of silicones, including small thermal expansion coefficients, low glass transition temperatures and low surface energies have led to a wide variety of applications (*71-75*). One particular application that helped inspire this study is the incorporation of polar nitrile moieties into a silicone to increase adhesive strength to metal and inorganic oxide surfaces (*76, 77*). This observation is consistent with the ability of polar silicones to hydrogen bond with inorganic oxide surfaces. This ability is expected to be particularly strong for inorganic oxide surfaces that are partially reduced to hydroxyl surface coatings, such as surface silanol groups on silica.

In this study interactions between silanol groups on trisilanolisobutyl-POSS (identified as $T_7iBu_7(OH)_3$ or POSS for the case of subscripted variable labels in this manuscript) and a polar silicone, poly(3-cyanopropyl, methyl) siloxane (PCPMS) in blend films will be characterized by surface pressure-area per molecule (Π-A) isotherm studies and Brewster angle microscopy (BAM) at the A/W interface. The structures for these molecules are provided in Figure 1. This system represents a model system in that $T_7iBu_7(OH)_3$ can be regarded as a nanofiller within the nearly two-dimensional (2D) PCPMS matrix. For both $T_7iBu_7(OH)_3$ and PCPMS, water represents an attractive surface. Hence this study will be compared to recent studies of three different blend systems: System 1 - $T_7iBu_7(OH)_3$ blends with a non-polar PDMS (*78*); System 2 - $T_7iBu_7(OH)_3$ blends with a more polar phosphine oxide substituted PDMS derivative, (PDMS-PO) (*79*); and System 3 - blends of non-amphiphilic octaisobutyl-POSS with non-polar PDMS (*70*). Assuming the collapse pressures, $\Pi_{collapse}$, of the individual components at the A/W interface can serve as a measure of the relative affinity of the component for the substrate (water), these three blends can be classified as: System 1 - The filler (POSS) and the matrix (polymer) are both amphiphilic but the filler has stronger interactions with the substrate (water); System 2 - Both the filler and the matrix are amphiphilic but the matrix has stronger interactions with the substrate; and System 3 - The filler is non-amphiphilic while the matrix is amphiphilic, hence the matrix has stronger interactions with the substrate; respectively. For the present blend system, $T_7iBu_7(OH)_3$/PCPMS, both the filler and the matrix are amphiphilic with nearly identical $\Pi_{collapse}$ values, i.e. nearly identical affinities for the substrate (water). Hence, intermediate properties between System 1 and System 2 could be possible.

Experimental

Materials. Trisilanolisobutyl-POSS obtained from Hybrid Plastics, Inc. and the synthesized (*77*) polar polysiloxane, poly[(3-cyanopropyl)methylsiloxane]

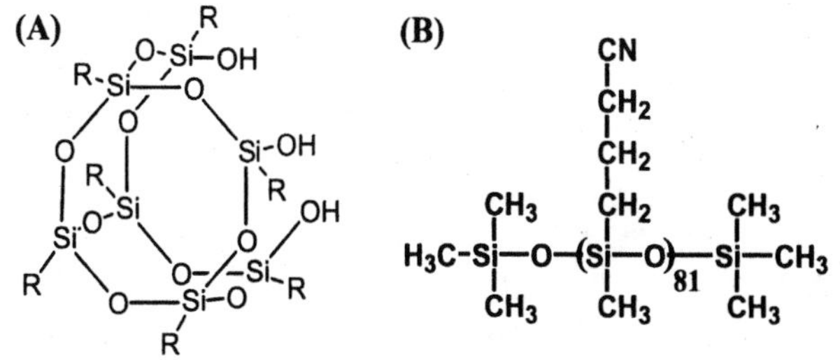

Figure 1. Molecular structures of (A) trisilanolisobutyl-POSS where R = isobutyl, and (B) poly[(3-cyanopropyl)methylsiloxane] (PCPMS).

(PCPMS) with 99 mol% of the cyanopropyl monomer, are shown in Figure 1. ^{1}H NMR yielded a number average molecular weight of M_n = 10.5 kg•mol^{-1} and differential scanning calorimetry yielded a glass transition temperature of T_g = – 65 °C for PCPMS. *sec*-Butylsilyl- and trimethylsilyl-terminated PDMS (M_n = 7.5 kg•mol^{-1}, polydispersity of M_w/M_n =1.09, Polymer Source, Inc.) was used as a "non-polar PDMS" for comparison to PCPMS. HPLC grade chloroform (J. T. Baker) was used without further purification to prepare spreading solutions with concentrations around 0.5 mg•ml^{-1} and all samples were allowed to dissolve for at least 24 hours.

Isotherm and Brewster Angle Microscopy (BAM) Studies. Monolayers were spread onto the surface of a Langmuir trough (500 cm^2, 601BAM, Nima Technology) filled with ultra-pure water (18.2 MΩ, MilliQ Gradient A-10, Millipore) maintained at 22.5 °C in a Plexiglas box with a relative humidity of 70~75%. The surface of the water was further cleaned by suctioning off potential impurities from the interface. Pre-determined volumes of spreading solution were spread onto the water surface using a Hamilton gas-tight syringe to obtain the desired surface concentration. The spreading solvent was allowed to evaporate for at least 20 min before film compression commenced. In order to investigate the thermodynamic properties of the trisilanolisobutyl-POSS/PCPMS mixtures, four methods were used to vary the surface area or surface concentration: 1) "compression" of the barriers at a constant rate, 2) successive "addition" of spreading solution, 3) "hysteresis" loops, and 4) "stepwise compression" (the portions in quotation marks indicate the short form used throughout the manuscript). For compression at a constant rate, each sample was compressed at a rate of 15 cm^2•min^{-1} unless the rate is otherwise noted. For hysteresis loops, each sample was compressed at a rate of 15 cm^2•min^{-1} (unless otherwise noted) to a designated average area per monomer value, <A>, and were immediately expanded at the same rate to a final trough area of 480 cm^2. This method was used to test the reversibility of the different structural transitions in the film. For successive addition, the surface area was held constant and the surface concentration was varied by making successive additions of spreading solution. After each addition, Π was allowed to relax to a constant minimum value ($\Delta\Pi < 0.1$ mN•m^{-1} over a 10 minute period) after evaporation of spreading solvent. For most surface concentrations, the relaxation time was on the order of 20-30 min. For stepwise compression, each sample was initially compressed at a constant rate of 15 cm^2•min^{-1}. After compressing the barriers to a fixed surface area, the barriers were stopped, and the surface pressure, Π, was allowed to relax to a more "equilibrium" value for a specific time period (typically 5 min). After allowing Π to relax, the compression and relaxation process was repeated until the end of the experiment. The surface pressure was measured using a paper Wilhelmy plate to ±0.1 mN•m^{-1} and the surface micrographs were captured using BAM (BAM-601, NanoFilm Technologie GmbH) during compression and expansion of the films. All BAM micrographs are 4.8×6.4 mm^2 and have at least 20 µm lateral resolution. The BAM instrument software uses automatic gain control to obtain an average optical brightness rather than an absolute intensity value. This approach provides much greater sensitivity for detecting domain structures, however; care must be taken when comparing images as they will have independent intensity scales. This feature means that a bright structure in one image may be dull in another image if a more strongly reflecting (brighter) feature is now in the field of view of the instrument.

Results and Discussion

Compression Isotherms of Polysiloxane Films. The surface pressure – area per monomer (Π-A) isotherm of PCPMS is compared to non-polar PDMS in Figure 2. The shape of the Π-A isotherm is related to the balance of hydrophilic and hydrophobic character of the substances. For PDMS, the isotherm is consistent with literature data (*55-60, 80-82*). At $\Pi = 0$ mN•m^{-1}, PDMS is known to form biphasic films of coexisting liquid-like and gas domains (*82*). At the lift-off area, $A_{lift-off,PDMS} = 19$ Å^2•monomer^{-1} (where $\Pi > 0$ mN•m^{-1} occurs), PDMS forms a liquid-like monolayer. The extrapolated limiting area obtained from the steepest part of the isotherm for PDMS, $A_{o,PDMS} = 18$ Å^2•monomer^{-1}, is in close agreement with $A_{lift-off}$. Starting at $A \approx 15$ Å^2•monomer^{-1}, the PDMS isotherm exhibits a plateau ($\Pi \approx 8.4$ mN•m^{-1}) that has been attributed to helix formation (*55*) and more recently to film collapse ($\Pi_{collapse,PDMS} \approx 8.4$ mN•m^{-1}) into multilayer structures (*80-82*). The multilayer formation model is assumed throughout the remainder of this paper. Within the context of this model, the second small transition around $A = 8$ Å^2•monomer^{-1} for PDMS is due to completion of a bilayer, with thicker multilayer films forming at smaller A values. In contrast to the PDMS isotherm, the PCPMS isotherm is simpler. One key difference is that $A_{lift-off,PCPMS} = 36$ Å^2•monomer^{-1} and $A_{o,PCPMS} \approx 28$ Å^2•monomer^{-1} are shifted to higher A values by about 17 and

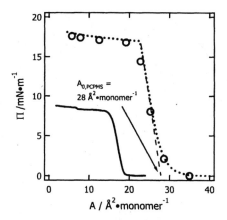

Figure 2. Compression Π-A isotherms of PDMS (——) and PCPMS (·····), and addition data for PCPMS (○) at the A/W interface and 22.5 °C. The dashed line extrapolation to Π = 0 mN•m^{-1} (indicated by the arrow) defines $A_{0,PCPMS}$.

10 Å^2 •monomer^{-1} relative to non-polar PDMS, respectively. These shifts are attributed to the larger size of the nitrile containing repeat units. A second key difference is that the start of film collapse into multilayer structures at A = 23 Å^2•monomer^{-1} occurs at higher surface pressure, $\Pi_{collapse,PCPMS} \approx 17$ mN•m^{-1}, than for non-polar PDMS. This feature is consistent with PCPMS having a stronger affinity for the A/W interface than non-polar PDMS and is also consistent with silicones containing polar end groups (*83*). For the basis of this study, it is important to note that PCPMS does not have as strong an affinity for the A/W interface as a polar phosphine oxide substituted PDMS (PDMS-PO), which exhibits $\Pi_{collapse,PDMS-PO} \approx 38$ mN•m^{-1} (*79*). The final key difference is that PCPMS does not show the transition at A = 8 Å^2•monomer^{-1} seen for PDMS. This observation may be caused by an inability of PCPMS to spread on itself, or the formation of less organized multilayers. One final comparison can be made in terms of addition vs. compression experiments. As seen in Figure 2, there is excellent agreement between the compression and addition isotherms for PCPMS. This feature is similar to non-polar PDMS.

Compression Isotherms of PCPMS/Trisilanolisobutyl-POSS Mixed Films. Figure 3 shows Π-<A> isotherms of PCPMS/trisilanolisobutyl-POSS blends. <A> is the average area per monomer in the mixed film. Starting at <A> = $A_{lift-off,POSS}$ = 230 Å^2•molecule^{-1}, the $T_7iBu_7(OH)_3$ isotherm (far-right curve) shows a gentle rise in surface pressure to a value of about Π ≈ 3 mN•m^{-1} that is consistent with a liquid expanded (LE) type film (*51*). Continuing compression of the $T_7iBu_7(OH)_3$ film from ≈ 180 > <A> > ≈ 140 Å^2•molecule^{-1}, is consistent with a more condensed liquid-like film. Further compression of the $T_7iBu_7(OH)_3$

film to <A> < ≈ 140 Å^2•molecule^{-1} results in film collapse, signified by the cusp in the isotherm at $\Pi_{collapse,POSS}$ ≈ 17.5 mN•m^{-1} for the dynamic compression experiment. In the collapsed regime (plateau at $\Pi_{collapse,POSS}$ ≈ 16.5 mN•m^{-1}), T$_7$iBu$_7$(OH)$_3$ molecules are squeezed out of the air/water interface to form multilayer POSS domains (50, 51). The nature of the aggregates (50, 51) is believed to be influenced by the tendency of POSS to form dimers in the solid state (84). BAM images show that the aggregates formed after the collapse point continue to grow in area fraction with small domains coalescing into larger aggregates (Figure 4H).

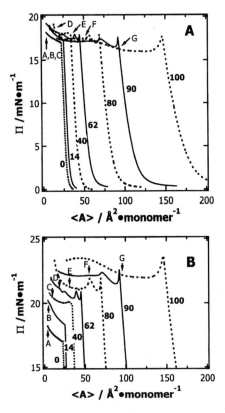

Figure 3. (A) Π-<A> isotherms of PCPMS/T$_7$iBu$_7$(OH)$_3$ mixed thin films at the A/W interface and 22.5 °C. The numbers inside the figure with arrows denote the wt% T$_7$iBu$_7$(OH)$_3$. The letters on the figure signify where the BAM images in Figure 4 were taken. (B) The isotherms in (A) have been replotted whereby each isotherm is offset from the previous isotherm by an additional +1 mN•m^{-1} with respect to increasing T$_7$iBu$_7$(OH)$_3$ content (+6 mN•m^{-1} total for 100 wt% T$_7$iBu$_7$(OH)$_3$ and no off-set for pure PCPMS, i.e. 0 wt% T$_7$iBu$_7$(OH)$_3$).

Moving on to the mixed isotherms, the most striking feature is that qualitatively up to the collapse point, all isotherms look the same until the weight percentage of POSS falls below 40%. Focusing only on the monolayer regime, $\Pi < \Pi_{collapse,PCPMS}$, the normal approach is to carry out a thermodynamic analysis of the excess Gibbs free energy of mixing per unit area, ΔG_{excess}. This process is started by defining the area change upon mixing, ΔA_{mix}, at constant Π, the two-dimensional (2D) analog to the three-dimensional (3D) change in volume upon mixing, ΔV_{mix}:

$$\Delta A_{mix} = \langle A(\Pi) \rangle - \langle A_{ix,ideal}(\Pi) \rangle$$

where $\langle A(\Pi) \rangle$ is the experimental average area per monomer for a mixture and:

$$\langle A_{mix,ideal}(\Pi) \rangle = X_{POSS} A_{POSS}(\Pi) + X_{PCPMS} A_{PCPMS}(\Pi)$$

with the mole fraction of species i represented by X_i and $A_i(\Pi)$ signifying the A value at a given Π value for the pure component i. By analogy to 3D, ΔG_{excess} is then defined as:

$$\Delta G_{excess}(\Pi)/k_B T = \int_0^{\Pi} \Delta A_{mix}(\Pi) d\Pi$$

Applying this analysis to the isotherms in Figure 3A yields $-0.2 < \Delta G_{excess}/kT < 0$ for all Π. These values are smaller in magnitude than those reported in the literature for compatible mixtures of fatty acid compounds, $\Delta G_{excess} \sim 0.5\ k_B T$ (*85-89*). Hence, the homogeneous mixed monolayers of $T_7iBu_7(OH)_3$ and PCPMS observed by BAM are regarded to be ideal mixtures.

Looking at mixtures below 40 wt% $T_7iBu_7(OH)_3$, discernable qualitative features in the vicinity of the collapse point are seen even in Figure 3A. For this reason, the vicinity of the collapse point is expanded in Figure 3B. In order to more clearly see the qualitative features of the collapsed regime in Figure 3B, each film with increasing $T_7iBu_7(OH)_3$ content is offset by $\Pi = 1\ mN \cdot m^{-1}$ from the previous composition. Hence, PCPMS Π values are not shifted and $T_7iBu_7(OH)_3$ Π values are shifted to values 6 $mN \cdot m^{-1}$ higher than their true values. Upon expansion of the isotherms in the vicinity of the collapse point, one clearly sees that below 40 wt% $T_7iBu_7(OH)_3$, the qualitative collapse behavior is more like that of the pure PCPMS film than $T_7iBu_7(OH)_3$. Furthermore, two collapse transitions (cusps) are observed for films with more than 40 wt% $T_7iBu_7(OH)_3$. The observation of two collapse transitions, in close proximity must reflect the fact that one of the two components preferentially collapses first. Given the fact that there is no heterogeneity in BAM images taken between the two collapse transitions, and that pure PCPMS films have a slightly lower $\Pi_{collapse}$ value, it is reasonable to assume that PCPMS starts to squeeze out of the monolayer slightly ahead of the $T_7iBu_7(OH)_3$. After the second cusp, heterogeneity is clearly observed by BAM (Figure 4).

In Figure 4, BAM micrographs of the blends are shown at a specific area per $T_7iBu_7(OH)_3$ molecule of A_{POSS} = 90 $\text{Å}^2 \cdot \text{molecule}^{-1}$, i.e., after pure $T_7iBu_7(OH)_3$ films collapse. The arrows on Figure 3 labeled with letters correspond to the images in Figure 4. Each photo is taken at A_{POSS} = 90 $\text{Å}^2 \cdot \text{molecule}^{-1}$ except for the 14 wt% $T_7iBu_7(OH)_3$ blends because of instrumental limitations associated with mounting the BAM camera to observe highly compressed films. The decision to compare all images at a constant A_{POSS} value is done to ensure that the surface density of $T_7iBu_7(OH)_3$ is constant for different compositions. However, Π is not constant for these comparisons, and the aggregation seen in Figure 4B through 4F occurs at higher Π values. Looking at Figure 4G, the pure $T_7iBu_7(OH)_3$ film is nearly homogeneous. On the basis of Figure 4, one might errantly assume that $T_7iBu_7(OH)_3$ aggregation in PCPMS indicates that PCPMS promotes aggregation. In fact, the situation is exactly the opposite. The qualitative trend in Figure 4 is identical to the trend seen for a more polar PDMS derivative containing phosphine oxide substituents, PDMS-PO by Kim, et al. (79). In that respect the conclusion is that PCPMS delays the onset of aggregation (aggregation occurs at higher Π for the same surface density) relative to pure $T_7iBu_7(OH)_3$, and PCPMS inhibits the formation of large aggregates like the ones seen in Figure 4H upon further compression. It is also clear that the aggregates in Figure 4 are not dispersed as well as the case when the more polar PDMS-PO substituent is used (79). Looking at Figure 4C, it almost appears that some aggregates are starting to coalesce into larger domains. In this respect, Figure 4C is more similar to the behavior seen in $T_7iBu_7(OH)_3$ blends with non-polar PDMS (78).

Reversibility Studies of PCPMS/$T_7iBu_7(OH)_3$ Mixed Films. Figure 5 shows a representative hysteresis loop for an 80 wt% $T_7iBu_7(OH)_3$ blend with PCPMS. In addition to the hysteresis loop (solid line for compression and dashed line for expansion), Figure 5 also contains an addition isotherm (open circles) and a step-wise compression isotherm (dotted lines). The compression isotherm in Figure 5 represents a dynamic situation. The system enters a metastable state during compression for $\Pi > \approx 14.5$ mN·m^{-1}, the plateau pressure of the addition isotherm. As a result, the step-wise compression isotherm oscillates between the dynamic (compression) and equilibrium (addition) limits in the plateau region. For Π values below the metastable state ($\Pi < \approx 14.5$ mN·m^{-1}), all three isotherms, compression, addition and stepwise addition are in reasonable agreement. This behavior is consistent with the behavior seen for pure $T_7iBu_7(OH)_3$ films (50, 51), $T_7iBu_7(OH)_3$ blends with non-polar PDMS (78), and $T_7iBu_7(OH)_3$ blends with PDMS-PO (79). The other feature of Figure 5, is the large hysteresis loop. In the plateau region, $T_7iBu_7(OH)_3$ aggregates form during compression. Upon expansion of the film, respreading of $T_7iBu_7(OH)_3$ aggregates is slower than barrier expansion leading

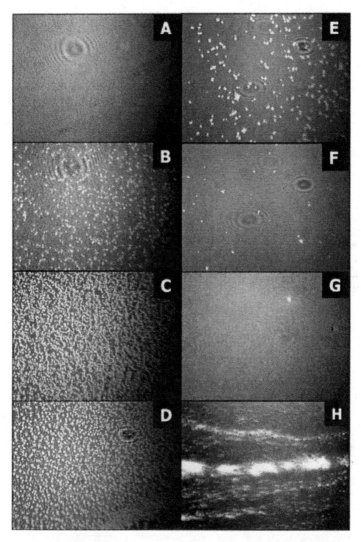

Figure 4. BAM images at A_{POSS} = 90 $Å^2 \cdot monomer^{-1}$ (except for 14 wt% $T_7iBu_7(OH)_3$). Arrows and letters on Figure 3 correspond to the images here. Actual conditions for different wt% $T_7iBu_7(OH)_3$ are (identifying letter, <A> /$Å^2 \cdot monomer^{-1}$, A_{POSS} /$Å^2 \cdot molecule^{-1}$, Π /mN·m^{-1}): pure PCPMS (A, 5, ∞, 18.1), 14 wt% (B, 5.9, 241, 19.0), 40 wt% (C, 8.5, 88, 18.6), 62 wt% (D, 18, 86, 18.2), 80 wt% (E, 36, 91, 17.2), 90 wt% (F, 54, 90, 17.1), and pure $T_7iBu_7(OH)_3$ (G, 90, 90, 16.3), and pure $T_7iBu_7(OH)_3$ at greater compression (H, 40, 40, 17.4) (50, 51). Solidlike domains appear bright in the $4.8×6.4$ mm^2 images.

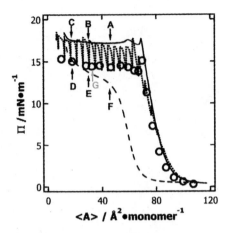

Figure 5. A representative Π-<A> hysteresis loop at 22.5 °C for an 80 wt% T₇iBu₇(OH)₃ blend with PCPMS at a compression and expansion rate of 15 cm²•min⁻¹. A solid line is used to show film compression, while a dashed line depicts film expansion. The dotted line is from step-wise compression with trough area increments of 20 cm² and compression interval of 300 s. The open circles (O) are obtained from addition experiments. Letters on the isotherm correspond to the <A> values where the BAM images in Figure 6 were taken.

to a more rapid drop in Π. Subsequent hysteresis loops are shifted to smaller <A> values because some aggregates never completely respread.

The letters and arrows on Figure 5 signify points where the BAM images in Figure 6 were taken. As seen in the Figure 6A, 6B, and 6C, compression to <A> values as small as 17 Å²•monomer⁻¹ (A_{POSS} = 43 Å²•molecule⁻¹) does leads to significant changes in the size and shape of the POSS aggregates. This behavior is observed for all blend compositions except for the 14 wt% T₇iBu₇(OH)₃ blend. In general the aggregates seem to grow and show a sign of sub-network structure formation. Here sub-network structures would be the combination of small aggregates that precedes the chaining together of individual aggregates into network-like structures like those seen in Figure 6H, and is the terminology that was used to describe aggregate formation in high wt% T₇iBu₇(OH)₃ blends with non-polar PDMS (78). Upon expansion of the film, the weakly interacting aggregates instantly revert to small discrete objects (Figure 6D) and the isolated domains start to re-arrange (Figure 6E) into more elongated structures (Figure 6E), before they finally form sub-network multilayer domains as seen in Figure 6F. This behavior is very different from pure T₇iBu₇(OH)₃ monolayers (50, 51), and blends of T₇iBu₇(OH)₃ with non-polar PDMS (78) or polar PDMS-PO (79). In T₇iBu₇(OH)₃/non-polar PDMS blends with sufficient POSS (> 60 wt%), "network-like" POSS aggregates similar to Figure 6H are observed instantly upon the expansion of films compressed deep into the collapsed regime (78). For the 80 wt% $T_7iBu_7(OH)_3$ blend with PCPMS, network formation almost occurs around $\Pi \approx 12$ mN•m^{-1} during expansion (Figure 6E and 6F), which is roughly the same Π value where network formation occurs for $T_7iBu_7(OH)_3$ blends with non-polar PDMS in Figure 6H (78). In contrast, network formation fails to occur in $T_7iBu_7(OH)_3$/PDMS-PO blends unless the $T_7iBu_7(OH)_3$ content is significantly greater than 90 wt% (79). Considering these observations, one may conclude that the closely matched affinity for water between PCPMS and $T_7iBu_7(OH)_3$ and potential hydrogen bonding between PCPMS and $T_7iBu_7(OH)_3$ hinders aggregation into stable network structures, but not as well as for PDMS-PO/$T_7iBu_7(OH)_3$ blends (79). Hydrogen bonding through the silanol groups on $T_7iBu_7(OH)_3$ with cyanopropyl groups on PCPMS is also consistent with recent findings of strong gas phase interactions between silanol groups in trisilanolphenyl-POSS and organophosphonates (90, 91).

Reversibility Studies of Highly Compressed PCPMS/$T_7iBu_7(OH)_3$ Mixed Films. Figure 7 shows hysteresis loops of PCPMS/$T_7iBu_7(OH)_3$ blends compressed to $A_{PCPMS} < \approx 16$ Å^2•monomer^{-1} ($\Pi > \approx 17.3$ mN•m^{-1}), i.e. very deep into the multilayer regime below the area occupied by a single repeating unit of PCPMS. In this region, the expansion of pure PCPMS thin films at the A/W interface reveals stable (nearly constant) Π values during the early stages of expansion. For the blends (14 and 62 wt% $T_7iBu_7(OH)_3$) expansion in this region also yields relatively stable Π values over a limited A range (A and B on Figure 7) as multilayer PCPMS respreads, before Π drops precipitously. In contrast, 90 wt% $T_7iBu_7(OH)_3$ never achieves as large a Π value during compression and Π falls sharply immediately upon expansion. The hysteresis behavior of the 90 wt% $T_7iBu_7(OH)_3$ blend with PCPMS is similar to $T_7iBu_7(OH)_3$ blends with non-polar PDMS (78) and PDMS-PO (79).

Figure 8 shows BAM images accompanying the hysteresis loops in Figure 7. Figure 8A for the 14 wt% $T_7iBu_7(OH)_3$ blend is taken at the highest Π value. Even though the amount of $T_7iBu_7(OH)_3$ is small, solid-like domains are observed. These domains do not aggregate, and their surface morphology resembles the domains seen in $T_7iBu_7(OH)_3$ blends with polar PDMS-PO, even though $A_{POSS} \approx 240$ Å^2•molecule^{-1} is consistent with submonolayer coverage for a pure $T_7iBu_7(OH)_3$ film (50, 51). As the $T_7iBu_7(OH)_3$ composition increases, the number of multilayer domains increases enough to form structures that resemble interconnected networks as shown for 60 wt% $T_7iBu_7(OH)_3$ in Figure 8B. These structures start to undergo shape changes with further expansion as seen in Figure 8C for 60 wt% $T_7iBu_7(OH)_3$ at larger <A> values. In contrast, the 90 wt% $T_7iBu_7(OH)_3$ blend which was not compressed to Π values as large as the 14 and 60 wt% blends, only showed loose structures (Figure 8D) during the expansion process. In this respect, highly compressed $T_7iBu_7(OH)_3$/PCPMS blends have character that is more consistent with $T_7iBu_7(OH)_3$/non-polar PDMS blends (78). It is worth noting that non-polar PDMS collapses into

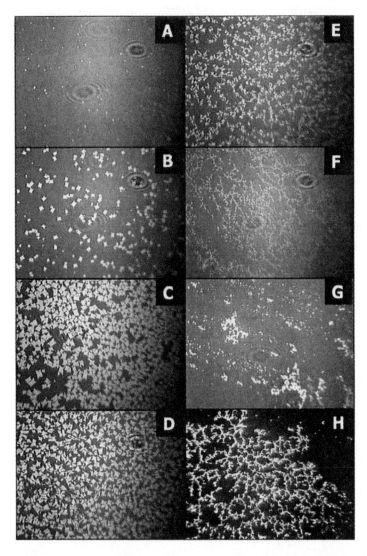

Figure 6. BAM images obtained from an 80 wt% $T_7iBu_7(OH)_3$ blend with PCPMS at 22.5 °C for the hysteresis loop in Figure 5. Micrograph conditions are expressed as (identifying letter, $<A>/Å^2$•monomer^{-1}, $A_{POSS}/Å^2$•molecule^{-1}, Π/mN•m^{-1}) during compression: (A, 46, 116, 17.2), (B, 30, 76, 17.3), (C, 17, 43, 17.6), upon expansion: (D, 20, 51, 14.8), (E, 30, 76, 13.7), (F, 46, 116, 12.2), and by addition (G, 32, 81, 14.4). Aggregation in the PCPMS/$T_7iBu_7(OH)_3$ blend is very different from a 90 wt% $T_7iBu_7(OH)_3$ blend with non-polar PDMS (H, 26, 50, 11.2) obtained during expansion (78).

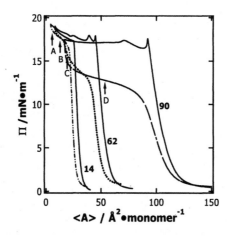

Figure 7. Π-<A> compression (solid lines)-expansion (broken lines) hysteresis loops for 14, 60, and 90 wt% $T_7iBu_7(OH)_3$ blends with PCPMS at 22.5 °C. The numbers on the compression step correspond to the wt% $T_7iBu_7(OH)_3$ in the film and the letters indicate where the BAM images in Figure 8 were taken. All films were compressed and expanded at 15 $cm^2 \cdot min^{-1}$.

multilayers first in $T_7iBu_7(OH)_3$/non-polar PDMS blends, and that compression to A_{PCPMS} < 16 $Å^2 \cdot monomer^{-1}$, is consistent with multilayer formation for the PCPMS component. Hence, silicone multilayer formation may be an important mechanistic feature for the formation of aggregate networks in expanding $T_7iBu_7(OH)_3$/silicone blends at the A/W interface. These studies also show that further insight into collapsed structures and respreading mechanisms may be obtainable for compression to ever smaller <A> values.

Compression Rate Effects on Aggregate Morphology in PCPMS/ $T_7iBu_7(OH)_3$ Mixed Films. Figure 9 shows hysteresis loops for an 80 wt% trisilanolisobutyl-POSS blend with PCPMS for three compression rates. Several significant features are observed. First, the slope of the isotherm in the monolayer state, $(\partial\Pi/\partial A)T$, increases slightly with increasing compression rate. Second, the size of the cusp that signifies the onset of the collapsed regime becomes smaller with increasing compression rate. Third, the separation with respect to <A> between the cusp that signifies the collapse of the first component and the cusp that signifies collapse of the second component (recall the discussion of Figure 3B) becomes smaller with increasing compression rate. As a consequence, increasing compression rates have the apparent effect of broadening the cusp at the on-set of collapse. Finally, the value of the plateau pressure in the collapsed regime (<A> < 70 $Å^2 \cdot molecule^{-1}$) increases slightly with increasing compression rate. These features are equivalent to saying that the compression of the rigid monolayer is faster than the rate of structural relaxation to a multilayer state causing both components to collapse at essentially

282

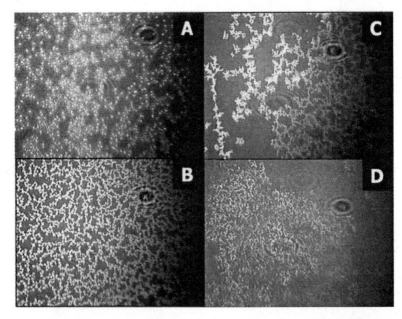

*Figure 8. BAM images obtained during expansion for 14, 60, and 90 wt%
$T_7iBu_7(OH)_3$ blends with PCPMS. The actual conditions for each micrograph
(identifying letter, $<A>$ /$Å^2$•monomer^{-1}, A_{POSS} /$Å^2$•molecule^{-1}, Π/mN•m^{-1}) are as
follows for different wt% $T_7iBu_7(OH)_3$: 14 wt% (A, 5.9, 241, 18.5), 60 wt%
(B, 13, 76, 17.7), 60 wt% (C, 18, 86, 16.2), and 90 wt% (D, 54, 90, 12.7).
These points are also indicated by arrows with matching letters on Figure 7.*

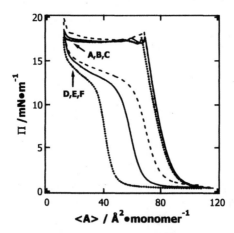

*Figure 9. Π-$<A>$ compression-expansion isotherms for 80 wt% $T_7iBu_7(OH)_3$
blends with PCPMS at 22.5 °C. The compression rates are 8 cm^2•min^{-1} (dotted
line), 15 cm^2•min^{-1} (solid line), and 50 cm^2•min^{-1} (dashed line). The letters on
the isotherms indicate $<A>$ values for the BAM images in Figure 10.*

the same Π value. In essence, compression of the film at a faster rate than the film can collapse produces higher transient surface concentrations and higher surface pressures than one would observe at slower compression rates, or in the stepwise compression and addition experiments.

Another interesting aspect of increased compression rates manifests itself in the morphology of the collapsing film. Figure 10 shows BAM images at A_{POSS} ≈ 47 $\text{Å}^2\cdot\text{molecule}^{-1}$ for different compression rates (8, 15, and 50 $\text{cm}^2\cdot\text{min}^{-1}$) during compression and expansion. At $<A> \approx 18$ $\text{Å}^2\cdot\text{monomer}^{-1}$ (i.e., $A_{POSS} \approx 47$ $\text{Å}^2\cdot\text{molecule}^{-1}$), the appearance of bright domains was attributed to the formation of a multilayer structure where the $T_7iBu_7(OH)_3$ molecules are densely packed and may exist as dimers in the upper phase (50-53). At the lowest compression rate in Figure 9 (A, 8 $\text{cm}^2\cdot\text{min}^{-1}$), a slightly smaller number of irregularly shaped aggregates with larger aggregate sizes form than is seen for higher compression rates. As the compression rate increases to 15 or 50 $\text{cm}^2\cdot\text{min}^{-1}$, thorn-like features on the corners of the aggregates Figure 10 (highlighted in the insets) tend to disappear, with the aggregates in Figure 10C having a nearly square morphology. The effect of increasing compression rate (i.e. higher Π) and the greater polarity of PCPMS relative to non-polar PDMS seem to favor the formation of more uniform aggregates. As seen in Figure 10, the multilayer domains do not have any preferential orientation or elongation. In this respect, the absence of elongated or oriented domains is consistent with $T_7iBu_7(OH)_3$ (50, 51) and are quite different from the unique rod-like aggregates that form in trisilanolcyclohexyl-POSS films (52, 53). Another interesting feature of the aggregates is that the compression/expansion rate seems to influence structural relaxation of the aggregates. As seen in Figure 10A vs. 10D for the compression/expansion cycle at the slowest rate, there is a dramatic difference in both the amount and the shapes of the aggregates at the same surface coverage for the corresponding hysteresis loop in Figure 9. For Figure 10B vs. 10E corresponding to compression at the intermediate rate, the area fraction for aggregates is comparable during compression and expansion; however, changes in domain morphology are clearly visible. As shown in Figure 10C vs. 10F for the compression/expansion cycle at the highest rate, both the area fraction and morphology of the compressed state are retained during film expansion. Hence, the higher applied lateral pressures generated at the highest compression rates appear to produce more regular (ordered) aggregates with stronger intermolecular interactions and longer lifetimes. On the basis of the morphology and aggregate lifetimes, future atomic force microscopy and electron diffraction studies will be performed to determine if the domains represent aggregates or crystals.

Conclusions

Surface morphologies for $T_7iBu_7(OH)_3$/PCPMS blends in Langmuir films exhibit surface morphologies upon collapse of the POSS component whose

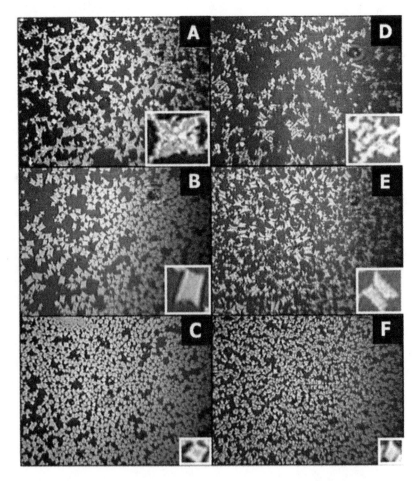

Figure 10. BAM images obtained during compression and expansion for an 80 wt% $T_7iBu_7(OH)_3$ blend with PCPMS at 22.5 °C. The actual conditions for each micrograph (identifying letter, $<A> /Å^2 \cdot monomer^{-1}$, $A_{POSS} /Å^2 \cdot molecule^{-1}$, $\Pi /mN \cdot m^{-1}$) are as follows for different compression rates: 8 $cm^2 \cdot min^{-1}$ (A, 19, 48, 17.3), 15 $cm^2 \cdot min^{-1}$ (B, 19, 48, 17.5), 50 $cm^2 \cdot min^{-1}$ (C, 18, 46, 18.1), and expansion rates: 8 $cm^2 \cdot min^{-1}$ (C, 18, 46, 14.4), 15 $cm^2 \cdot min^{-1}$ (D, 18, 46, 15.1), 50 $cm^2 \cdot min^{-1}$ (E, 19, 48, 15.3). All of the insets have been magnified 5X. The size ratio of the solid bright domains within the insets is ~ 5:3:1 (A:B:C).

characteristics lie between non-polar PDMS/$T_7iBu_7(OH)_3$ blends (*78*) and PDMS-PO/$T_7iBu_7(OH)_3$ blends (*79*). In particular, aggregates in the $T_7iBu_7(OH)_3$/PCPMS blends are less regular than $T_7iBu_7(OH)_3$/PDMS-PO blends (*79*) but do not show irreversible aggregation into large domains (mm scale) or extensive network structures, as is seen in pure $T_7iBu_7(OH)_3$ films (*50, 51*) or $T_7iBu_7(OH)_3$/non-polar PDMS blends (*78*). This behavior is interpreted in terms of favorable interactions (probable hydrogen bonding) between polar PDMS derivatives and $T_7iBu_7(OH)_3$, where the polymer's collapse pressure, $\Pi_{collapse,polymer}$, serves as a measure of the effective hydrogen bonding strength for a given polymer referenced to water and can also be compared to $\Pi_{collapse,POSS}$. In this context, hydrogen bonding strength referenced to water should increase as: $\Pi_{collapse,non\text{-}polar\ PDMS} < \Pi_{collapse,PCPMS} \approx \Pi_{collapse,POSS} < \Pi_{collapse,PDMS\text{-}PO}$. The aggregation behavior is also different from octaisobutyl-POSS/non-polar PDMS blends where dispersion of the POSS component does not involve hydrogen bonding between the components (*70*). Both the density and morphology of the aggregates are a function of how deep the monolayer is compressed into the multilayer regime and the compression rate. For deep compression, to A_{PCPMS} values where PCPMS collapses into relatively stable multilayers, aggregates are even seen at 14 wt% $T_7iBu_7(OH)_3$. This behavior is attributed to the fact that Π values are greater than $\Pi_{collapse}$ for pure $T_7iBu_7(OH)_3$. At moderate $T_7iBu_7(OH)_3$ levels (60 and 90 wt% $T_7iBu_7(OH)_3$), network structures are observed that are similar to the behavior in non-polar PDMS/$T_7iBu_7(OH)_3$ blends (*78*). With increasing compression rates (c.a. 50 cm^2/min), very regular $T_7iBu_7(OH)_3$ aggregates form that retain their shape during expansion and do not undergo aggregation into more highly ordered structures. These studies provide tremendous insight into enhancing POSS/silicone compatibility through the use of trisilanol-POSS derivatives and appropriate polar functional co-monomers for producing better POSS/silicone nanocomposites.

Acknowledgments

The authors appreciate the financial support of the Thomas F. Jeffress and Kate Jeffress Memorial Trust (J-553), Virginia Tech Aspires Program, a 3M Untenured Faculty Award, the National Science Foundation (CHE-0239633), the Post-doctoral Fellowship Program of the Korea Sceince and Engineering Foundation, and DARPA-AFOSR (Contract #F49620-02-1-0408). The authors also thank Dr. Shawn Phillips (Air Force Research Laboratories at Edward's Air Force Base) for material support.

286

References

1. Lichtenhan, J. D. *Comments Inorg. Chem.* **1995**, *17*, 115.
2. Li, G. Z.; Wang, L. C.; Ni, H. L.; Pittman, C. U. *J. Inorg. and Organomet. Polym.* **2002**, *11*, 123.
3. Jeon, H. G.; Mather, P. T.; Haddad, T. S. *Polym. Int.* **2000**, *49*, 453.
4. Feher, F. J.; Schwab, J. J.; Phillips, S. H.; Eklund, A.; Martinez, E. *Organometallics* **1995**, *14*, 4452.
5. Gonzalez, R. I.; Phillips, S. H.; Hoflund, G. B. *J. Spacecraft Rockets* **2000**, *37*, 463.
6. Hoflund, G. B.; Gonzalez, R. I.; Phillips, S. H. *J. Adhes. Sci. Technol.* **2001**, *15*, 1199.
7. Phillips, S. H.; Haddad, T. S.; Tomczak, S. J. *Curr. Opin. Solid State Mater. Sci.* **2004**, *8*, 21.
8. Brunsvold, A. L.; Minton, T. K.; Gouzman, I.; Grossman, E.; Gonzalez, R. *High Perform. Polym.* **2004**, *16*, 303.
9. Hacker, N. P. *MRS Bulletin* **1997**, *22*, 33.
10. Leu, C. M.; Chang, Y. T.; Wei, K. H. *Macromolecules* **2003**, *36*, 9122.
11. Leu, C. M.; Chang, Y. T.; Wei, K. H. *Chem. Mater.* **2003**, *15*, 3721.
12. Lee, Y. J.; Huang, J. M.; Kuo, S. W.; Lu, J. S.; Chang, F. C. *Polymer* **2005**, *46*, 173.
13. Chen, Y. W.; Chen, L.; Nie, H. R.; Kang, E. T. *J. Appl. Polym. Sci.* **2006**, *99*, 2226.
14. Haddad, T. S.; Lichtenhan, J. D. *Macromolecules* **1996**, *29*, 7302.
15. Bharadwaj, R. K.; Berry, R. J.; Farmer, B. L. *Polymer* **2000**, *41*, 7209.
16. Xu, H. Y.; Kuo, S. W.; Lee, J. S.; Chang, F. C. *Macromolecules* **2002**, *35*, 8788.
17. Huang, J. C.; He, C. B.; Xiao, Y.; Mya, K. Y.; Dai, J.; Siow, Y. P. *Polymer* **2003**, *44*, 4491.
18. Pellice, S. A.; Fasce, D. P.; Williams, R. J. J. *J. Polym. Sci., Part B: Polym. Phys.* **2003**, *41*, 1451.
19. Fu, B. X.; Gelfer, M. Y.; Hsiao, B. S.; Phillips, S.; Viers, B.; Blanski, R.; Ruth, P. *Polymer* **2003**, *44*, 1499.
20. Pyun, J.; Matyjaszewski, K.; Wu, J.; Kim, G. M.; Chun, S. B.; Mather, P. T. *Polymer* **2003**, *44*, 2739.
21. Xu, H. Y.; Yang, B. H.; Wang, J. F.; Guang, S. Y.; Li, C. *Macromolecules* **2005**, *38*, 10455.
22. Mantz, R. A.; Jones, P. F.; Chaffee, K. P.; Lichtenhan, J. D.; Gilman, J. W.; Ismail, I. M. K.; Burmeister, M. J. *Chem. Mater.* **1996**, *8*, 1250.
23. Kim, G. M.; Qin, H.; Fang, X.; Sun, F. C.; Mather, P. T. *J. Polym. Sci. Part B: Polym. Phys.* **2003**, *41*, 3299.
24. Lichtenhan, J. D.; Otonari, Y. A.; Carr, M. J. *Macromolecules* **1995**, *28*, 8435.
25. Zhang, C.; Bobonneau, F.; Bonhomme, C.; Laine, R.   M.; Soles, C. L.; Hristov, H. A.; Yee, A. F. *J. Am. Chem. Soc.* **1998**, *120*, 8380.

26. Zheng, L.; Farris, R. J.; Coughlin, E. B. *Macromolecules* **2001**, *34*, 8034.
27. Tsuchida, A.; Bolln, C.; Sernetz, F. G.; Frey, H.; Muelhaupt, R. *Macromolecules* **1997**, *30*, 2818.
28. Laine, R. M.; Zhang, C.; Sellinger, A.; Viculis, L. *Appl. Organomet. Chem.* **1998**, *12*, 715.
29. Choi, J.; Tamaki, R.; Laine, R. M. *Polym. Mater. Sci. Eng.* **2001**, *84*, 735.
30. Wright, M. E.; Schorzman, D. A.; Feher, F. J.; Jin, R. Z. *Chem. Mater.* **2003**, *15*, 264.
31. Choi, J.; Tmaki, R.; Kim, S. G.; Laine, R. M. *Chem. Mater.* **2003**, *15*, 3365.
32. Choi, J.; Kim, S. G.; Laine, R. M. *Macromolecules* **2004**, *37*, 99.
33. Zheng, L.; Hong, S.; Cardoen, G.; Burgaz, E.; Gido, S. P.; Coughlin, E. B. *Macromolecules* **2004**, *37*, 8606.
34. Yoon, K. H.; Polk, M. B.; Park, J. H.; Min, B. G.; Schiraldi, D. A. *Polym. Int.* **2005**, *54*, 47.
35. Zeng, J.; Bennett, C.; Jarrett, W. L.; Iyer, S.; Kumar, S.; Mathias, L. J.; Schiraldi, D. A. *Composite Interfaces* **2005**, *11*, 673.
36. Romo-Uribe, A.; Mather, P. T.; Haddad, T. S.; Lichtenhan, J. D. *J. Polym. Sci., Part B: Polym. Phys.* **1998**, *36*, 1857.
37. Mather, P. T.; Jeon H. G.; Romo-Uribe, A.; Haddad, T. S.; Lichtenhan, J. D. *Macromolecules* **1999**, *32*, 1194.
38. Li, G. Z.; Cho, H.; Wang, L. C.; Tohlani, H.; Pittman, C. U. *J. Polym. Sci. Part A: Polym. Chem.* **2004**, *43*, 355.
39. Lee, A.; Xiao, J.; Feher, F. J. *Macromolecules* **2005**, *38*, 438.
40. Kopesky, E. T.; Haddad, T. S.; McKinley, G. H.; Cohen, R. E. *Polymer* **2005**, *46*, 4743.
41. Liang, K. W.; Toghiani, H.; Li, G. Z.; Pittman, C. U. *J. Polym. Sci. Part A: Polym. Chem.* **2005**, *43*, 3887.
42. Kopesky, E. T.; McKinley, G. H.; Cohen, R. E. *Polymer* **2006**, *47*, 299.
43. Kobayashi, H. *Makromol. Chem.* **1993**, *194*, 2569.
44. Knischka, R.; Dietsche, F.; Hanselmann, R.; Frey, H.; Muelhaupt, R.; Lutz, P. J. *Langmuir* **1999**, *15*, 4752.
45. Ogarev, V. A. *Colloid J.* **2001**, *63*, 445.
46. Mya, K. Y.; Li, X.; Chen, L.; Ni, X. P.; Li, J.; He, C. B. *J. Phys. Chem. B* **2005**, *109*, 9455.
47. Turri, S.; Levi, M. *Macromolecules* **2005**, *38*, 5569.
48. Oaten, M.; Choudhury, N. R. *Macromolecules* **2005**, *38*, 6392.
49. Turri, S.; Levi, M. *Macromol. Rapid Commun.* **2005**, *26*, 1233.
50. Deng, J. J.; Polidan, J. T.; Hottle, J. R.; Farmer-Creely, C. E.; Viers, B. D.; Esker, A. R. *J. Am. Chem. Soc.* **2002**, *124*, 15194.
51. Deng, J. J.; Hottle, J. R.; Polidan, J. T.; Kim, H. J.; Farmer-Creely, C. E.; Viers, B. D.; Esker, A. R. *Langmuir* **2004**, *20*, 109.
52. Deng, J. J.; Farmer-Creely, C. E.; Viers, B. D.; Esker, A. R. *Langmuir* **2004**, *20*, 2527.
53. Deng, J. J.; Viers, B. D.; Esker, A. R.; Anseth, J. W.; Fuller, G. G. *Langmuir* **2005**, *21*, 2375.

288

54. Esker, A. R.; Vastine, B. A.; Deng, J.; Ferguson, M. K.; Morris, J. R.; Satija, S. K.; Viers, B. D. *Polymer Prepr. (Am. Chem. Soc., Div. Polym. Chem.)* **2004**, *45*, 644-645.

55. Fox, H. W.; Taylor, P. W.; Zisman, W. A. *J. Ind. Eng. Chem.* **1947**, 1401.

56. Noll, W.; Steinbach, H.; Sucker, C. *Ber. Bunsenges.* **67**, 407.

57. Noll, W.; Steinbach, H.; Sucker, C. *Kolloid –Z.* **1965**, *204*, 94.

58. Noll, W.; Steinbach, H.; Sucker, C. *J. Polym. Sci.* **1971**, *C34*. 123.

59. Granick, S. Macromolecules **1985**, *18*, 1597.

60. Granick, S.; Clarson, S. J.; Formoy, T. R.; Semlyen, J. A. *Polymer* **1985**, *26*, 925.

61. Noll, W. *Pure and Applied Chemistry*, **1966**, *13*, 101.

62. Noll, W. *Kolloid –Z. u. Z. Polymere*, **1966**, *211*, 98.

63. Noll, W.; Steinbach, H.; Sucker, C. *Kolloid –Z. u. Z. Polymere*, **1970**, *236*, 1.

64. Noll, W.; Steinbach, H.; Sucker, C. *Kolloid –Z. u. Z. Polymere* **1971**, *243*, 110.

65. Bernett, M. K.; Zisman, W. A. *Macromolecules* **1971**, *4*, 47.

66. Noll, W.; Büchner, W.; Steinbach, H.; Sucker, C. *Kolloid –Z. u. Z. Polymere* **1972**, *250*, 9.

67. Noll, W.; Büchner, W.; Lücking, H. J.; Sucker, C. *Kolloid –Z. u. Z. Polymere* **1972**, *250*, 836.

68. Granick, S.; Kuzmenka, D. J.; Clarson, S. J.; Semlyen, J. A. *Langmuir* **1989**, 5, 144.

69. Granick, S.; Kusmenka, D. J.; Clarson, S. J.; Semlyen, J. A. *Macromolecules* **1989**, *22*, 1878.

70. Hottle, J. R.; Deng, J. J.; Kim, H. J.; Farmer-Creely, C. E.; Viers, B. D.; Esker, A. R. *Langmuir* **2005**, *21*, 2250.

71. *Chemistry and Technology of Silicones*; W. Noll, Ed.; Academic Press: New York, NY, 1968.

72. *Silicon Based Polymer Science: A Comprehensive Resource*; J. M. Zeigler and F. W. G. Fearon, Eds.; Advances in Polymer Chemistry v. 224; American Chemical Society: Washington, DC, 1990.

73. *Siloxane Polymers*; S. J. Clarson and J. A. Semlyn, Eds.; PTR Prentice Hall: Englewood Cliffs, NJ, 1993.

74. *Silicones and Silicone-modified Materials*; S. J. Clarson, Ed.; ACS Symposium Series v. 729; American Chemical Society: Washington, DC, 2000.

75. *Synthesis and Properties of Silicones and Silicone-modified Materials;* S. J. Clarson, Ed.; ACS Symposium Series v. 838; American Chemical Society: Washington, DC, 2003.

76. Hoyt, J. K.; Phillips, P.; Li, C. H. H.; Riffle, J. S. *Adhesives Age*, **2000**, *43*, 28.

77. Hoyt, J. K. *Ph.D. thesis*, Virginia Polytechnic Institute and State University, Blacksburg, VA 2002.

78. Hottle, J. R.; Kim, H. J.; Deng, J. J.; Farmer-Creely, C. E.; Viers, B. D.; Esker, A. R. *Macromolecules* **2004**, *37*, 4900.

79. Kim, H. J.; Deng, J.; Lalli, J. H.; Riffle, J. S.; Viers, B. D.; Esker, A. R. *Langmuir* **2005**, *21*, 1908.
80. Mann, E. K.; Langevin, D. *Langmuir* **1991**, *7*, 1112.
81. Lee, L. T.; Mann, E. K.; Langevin, D.; Farnoux, B. *Langmuir* **1991**, *7*, 3076.
82. Mann, E. K.; Henon, S.; Langevin, D.; Meunier, J. *J. Phys. II*, **1992**, *2*, 1683.
83. Jalbert, C.; Koberstein, J. T.; Hariharan, A.; Kumar, S. K. *Macromolecules* **1997**, *30*, 4481.
84. Feher, F. J.; Newman, D. A.; Walzer, J. F. *J. Am. Chem. Soc.* **1989**, *111*, 1741.
85. Gaines, G. L. *Insoluble Monolayers at Liquid-Gas Interfaces*; Wiley: New York, NY, 1966.
86. Sadrzadeh, N.; Yu, H.; Zografi, G. *Langmuir* **1998**, *14*, 151.
87. Dynarowicz-Latka, P.; Kita, K. *Adv. Colloid Interface Sci.* **1999**, *79*, 1.
88. Seoane, R.; Miñones, J.; Conde, O.; Miñones, J.; Casas, M.; Iribarnegaray, E. *J. Phys. Chem.,B* **2000**, *104*, 7735.
89. Seoane, R.; Dynarowicz-Tstka, P.; Miñones, J.; Rey-Gomez-Serranillos, I. *Colloid Polym. Sci.* **2001**, *279*, 562.
90. Ferguson-McPherson, M. K.; Low, E. R.; Esker, A. R.; Morris, J. R. *Langmuir* **2005**, *21*, 11226.
91. Ferguson-McPherson, M. K.; Low, E. R.; Esker, A. R.; Morris, J. R. *J. Phys. Chem. B* **2005**, *109*, 18914.

Chapter 18

Hydrophobic Silsesquioxane Nanoparticles and Nanocomposite Surfaces

An Overview of the Synthesis and Properties of Fluorinated Polyhedral Oligomeric SilSequioxanes (POSS) and Fluorinated POSS Nanocomposites

Joseph M. Mabry[1,*], Ashwani Vij[1,*], Brent D. Viers[1], Wade W. Grabow[1], Darrell Marchant[1], Scott T. Iacono[1], Patrick N. Ruth[2], and Isha Vij[2]

[1]Air Force Research Laboratory and [2]ERC Incorporated, Materials Applications Branch, Edwards Air Force Base, CA 93524
*Corresponding authors: joseph.mabry@edwards.af.mil and ashwani.vij@edwards.af.mil

Fluorinated Polyhedral Oligomeric Silsesquioxanes are hydrophobic nanoparticles. One compound, FD_8T_8, is ultrahydrophobic, possessing a water contact angle of 154°. This is believed to be the most hydrophobic and lowest surface tension crystalline substance known. Analysis of the x-ray crystal structure indicates a large number of Si...F contacts may lead to ultrahydrophobicity. Hydrophobicity of poly(chlorotrifluoroethylene) nanocomposites also increases with the incorporation of FluoroPOSS and will be discussed.

Introduction

Polyhedral oligomeric silsesquioxanes (POSS) continue to be explored for use in many new applications.[1] Applications include space-survivable coatings,[2-4] and ablative and fire-resistant materials.[5-7] POSS compounds have a rigid, inorganic core and have been produced with a wide range of organic functionality. Due to their physical size, POSS incorporation in polymers generally serves to reduce chain mobility, which often results in the improvement of both thermal and mechanical properties.

The addition of fillers to polymeric matrices is of major technological importance. However, the effects of this process are still not fully understood. Filler addition can impart enhanced scratch resistance, increase thermal or mechanical properties, and improve processing parameters. There has been much effort to optimize the factors in the addition of filler. One factor is filler chemistry. Silicate and carbon black based fillers are quite common. They are often inexpensive and their incorporation into many polymer systems is fairly straightforward. When miscibility is a problem, surface modification of the fillers to further enhance their compatibility is widespread. The silylation of surface silanol groups on silica fillers is a good example. Processing is another factor that has been optimized. The use of high shear to break up large agglomerates or aggregates of nanoscopic particles is common. These approaches yield nanoscopic species with large surface areas, which should favor physisorption and/or chemisorption between the polymer chain and the filler.

A number of reports have detailed that POSS materials can act as reinforcing fillers (or reinforcing comonomers) in a number of nanocomposite systems.[8-10] The results reported herein are somewhat different in that the monodisperse POSS building blocks seem to be rather non-interacting. Specifically, the organic functionality surrounding the silsesquioxane core is composed of fluoroalkyl moieties. Fluoroalkyl compounds are known to be basically inert. This is largely because they are non-polarizable and have low surface free energies. Fluoroalkyl chains are often rigid, due to steric and electronic repulsion. These POSS materials are monodisperse and crystalline. The melting point of the POSS is lower than the processing conditions of the fluoropolymers, so one can safely assume that hard filler effects should not be an issue. In this regard, one may expect that these materials could exhibit small molecule, solvent like, characteristics. The POSS could be well dispersed and act as molecular ball bearings. This paper will discuss parameters and surface properties of simple blends of FluoroPOSS materials in poly(chloro-trifluoroethylene) matrices.

Experimental

Materials

POSS compounds, (1H,1H,2H,2H-heptadecafluorodecyl)$_8$Si$_8$O$_{12}$ (**FD$_8$T$_8$**) and (1H,1H,2H,2H-tridecafluorooctyl)$_8$Si$_8$O$_{12}$ (**FO$_8$T$_8$**) were prepared using previously reported methods.[11] Poly(chlorotrifluoroethylene) (PCTFE) was obtained from Daikin.

Composite Preparation

Fluoropolymer composites have been prepared by melt blending five, ten, or fifteen weight percent of **FD$_8$T$_8$** or **FO$_8$T$_8$** into PCTFE. All samples were blended in a DACA Micro Compounder for 3 minutes at 100 rotations per minute. The DACA Micro Compounder is a conical co-rotating twin-screw extruder with a bypass allowing the material to circulate for specified times. The capacity of the mixer is 4.5 cm^3. The mean shear rate is approximately 100 s^{-1} and is reported based on a treatment given in literature.[12] The blends were compounded at 280 °C. Samples for contact angle measurements were prepared into thin films. The films were prepared by compression molding two grams of the polymer-blend extrudate utilizing a Tetrahedron compression molder. The polymer extrudate was placed between two sheets of thick aluminum foil at 10 °C greater than the compounding temperatures for 10 minutes using one ton of force. All films were less than 0.3 mm thick and about 80 mm in diameter and appeared homogenous and similar to the respective unfilled fluoropolymer.

Contact Angle

Contact Angle analyses were performed on a First Ten Angtroms 110 series system using a syringe metering pump. Deionized water was used as the interrogating liquid. Small drops of water were accurately metered onto a flat surface, and the full screen image of the drop was captured with the frame grabbing software coupled to a CCD camera operating at the optimized zoom and contrast. The contact angle was determined via the software suite or via graphical fitting of the contact tangents in the captured image. Both approaches gave the same nominal value within ± 2 degrees. Only the value of the quasistatic advancing angle is reported.

Results and Discussion

FluoroPOSS Synthesis

FluoroPOSS were produced by the base-catalyzed hydrolysis of trialkoxy silanes. In small-scale syntheses, these compounds tend to condense into T_8 cages (Figure 1), rather than cage mixtures, as has been previously observed in the base-catalyzed synthesis. This is significant because the usual method to produce T_8 cages is the acid-catalyzed hydrolysis of trichlorosilanes, which results in a much longer reaction time and the production of an undesirable acidic byproduct.

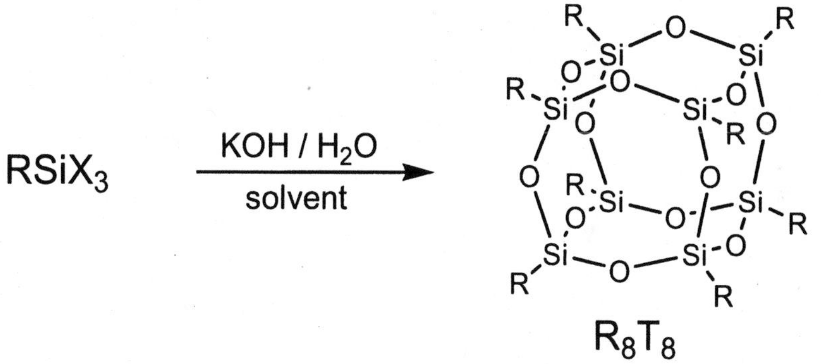

$$RSiX_3 \xrightarrow[\text{solvent}]{\text{KOH / H}_2\text{O}}$$

$$R_8T_8$$

Figure 1. Synthesis of Fluoroalkyl$_8$T$_8$.

A variety of FluoroPOSS compounds have been produced, including **FD$_8$T$_8$** and **FO$_8$T$_8$**. Synthesis is currently underway on a number of others. The yields for these reactions are often nearly quantitative. The byproduct is a resinous material that is formed when the condensation is less controlled. The resin is typically removed by extraction.

An interesting occurrence has been observed during the scale-up of the FluoroPOSS synthetic procedure. During large-scale syntheses FluoroPOSS compounds, cage mixtures are often formed. A cage mixture is a combination of cages with eight, ten, and twelve silicon atoms. However, during a purification step involving extraction of the basic catalyst, the cage mixture is converted exclusively to eight-member cages (Figure 2). Because this step involves the dissolving of the compound into a fluorinated solvent, it is believed that the presence of the catalyst allows conversion of the cage mixture to the most thermodynamically stable product. Apparently, in the fluorinated solvent used, the eight-member cage is the favored product. This conversion can be followed by ^{29}Si NMR spectroscopy. Calculations are underway to confirm this hypothesis.

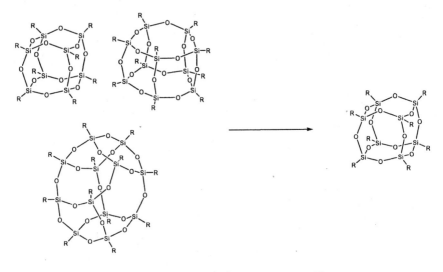

Figure 2. Redistribution of cage mixture to T_8 cages.

FluoroPOSS Properties

The properties of the FluoroPOSS compounds are quite interesting. They tend to volatilize at approximately 300 °C, rather than decompose, as is observed with many POSS compounds. The FluoroPOSS are the highest molecular weight POSS yet produced. The **FD$_8$T$_8$** has a molecular weight of 3993.54 g/mol. The density of these materials is also very high, with crystals of the **FD$_8$T$_8$** having a density of 2.067 g/mL.

Various surface properties of the FluoroPOSS compounds have been examined. Water contact angles are a measure of the surface free energy of a surface. As the surface energy decreases, the contact angle increases to a maximum of 180°. The trend observed in the FluoroPOSS compounds is that the surface energy decreases as the length of the fluoroalkyl chain increases. While this may not be surprising, the observed contact angles are unexpectedly high. The **FD$_8$T$_8$** has a water contact angle of 154° (Figure 3), which is approximately 40° higher than the water contact angle of polytetrafluoroethylene (PTFE). The correlation between the chain length and the contact angle is not linear. The contact angle appears to be increasing at an increasing rate.

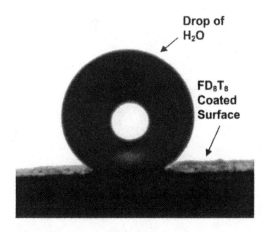

*Figure 3. Water drop on surface of **FD₈T₈** with a contact angle of 154°.*

Solid State Structures of FluoroPOSS

Due to the highly crystalline nature of FluoroPOSS compounds, single crystals were grown from fluorinated solvents. Although these crystals exhibit different morphologies, both materials investigated under this study belong to the triclinic crystal system. Single crystal x-ray diffraction analysis of FD_8T_8 and FO_8T_8 at room temperature did not allow for any reasonable structure solution as there is a large amount of disorder due to the movement of fluoroalkyl chains, which is consistent with observation of large diffused scattering. Our attempts to cool crystals to low temperatures resulted in crystal shattering, probably due to a rapid phase transition. This challenge was overcome by very slow cooling rates. However, at 103 K, the quasi powder-like pattern of diffused scattering disappears, resulting in sharper spots indicating significant ordering due to reduced entropy. The ultrahydrophobic FD_8T_8 exhibits a structure with two molecules within a large asymmetric unit and the fluoroalkyl chains propagating in a zig-zag manner as seen in Figure 4.

The fluoroalkyl chains adopt both gauche and eclipsed conformations as a result of close intra- and intermolecular Si...F contacts (Figure 5). In addition to these contacts, intermolecular H...F contacts are also observed, which result from lower disorder at 103K. The Si...F contacts are in the range of 3.0-3.5 Å, which is below the sum of van der Waals radii of silicon and fluorine. These contacts cause the non-fluorinated methylene groups to lie flat, along the axis of the rigid fluoroalkyl chains, and increase the packing efficiency of the crystals with an almost parallel arrangement of alkyl chains. This results in the formation of a surprising unique structure due to self-assembly.

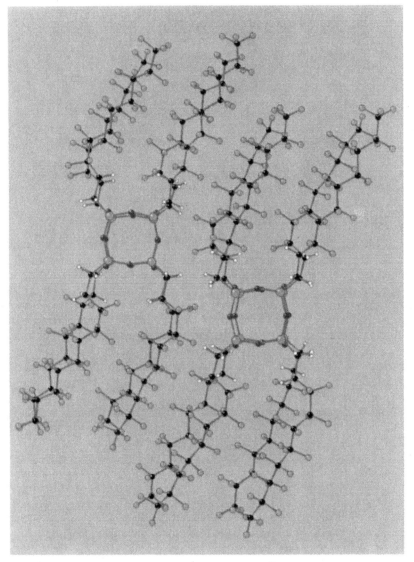

Figure 4. X-ray crystal structure of FD_8T_8.

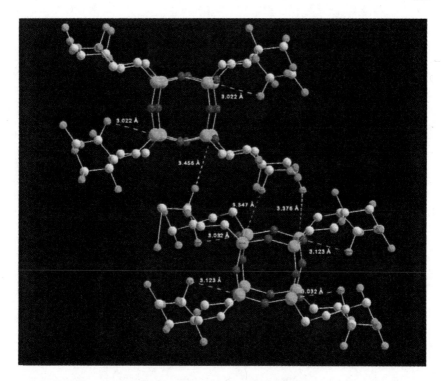

Figure 5. Cage region of FD₈T₈ structures showing Si-F and H-F contacts.

POSS Fluoropolymers

The FluoroPOSS compounds mentioned above were blended into poly(chlorotrifluoroethylene). For the purposes of this paper, **FD₈T₈** and **FO₈T₈** blends in PCTFE will be used to describe processing, thermo-mechanical, and surface properties.

Processability

The processability of the samples were compared using torque and load, a measure of the pressure generated in the compounder. The pressure is generated in the mixer due to its conical design. With a constant volume of material compounded and a fixed screw speed, the pressure generated is proportional to the viscosity of the material. The lower the pressure, the lower the viscosity, and the easier the material is to process. The second measure of processability is the torque output by the motor. This gives an indication of the mechanical energy put into the system and is proportional to the current used by the motor. A similar measure was utilized to characterize the processability of polyethylene and hyperbranched polymer blends.[13] The lower the torque output the more processable the polymer for any given screw speed.

These two measures of the processability of the polymer blends were recorded at 30 second intervals during processing. It was found, within a 95% confidence interval, that the load and torque values were constant for the duration of processing excluding the first 30 sec. Therefore, an average value for both torque and load is assigned to each processing run. In order to investigate the effect of the addition of FluoroPOSS, relative torque and relative load values were computed utilizing the average values in comparison to the average values found for the unfilled resins. Figure 6 shows the relative torque and load values with respect to the weight percentage of POSS added to PCTFE blends. The solid symbols represent the relative torque values whereas the relative load results are illustrated by the open symbols. The square symbols symbolize the results of the FD_8T_8 blends and the circular symbols show the FO_8T_8 blend results. One will notice that the processability of 10 weight percent of either POSS in PCTFE is improved by greater than 30 percent. In addition, the FD_8T_8 exhibits a greater effect on the torque and load for the PCTFE blends.

Surface Properties

While fluoropolymers are known for their hydrophobicity and low coefficients of friction, incorporation of FluoroPOSS may help to improve these

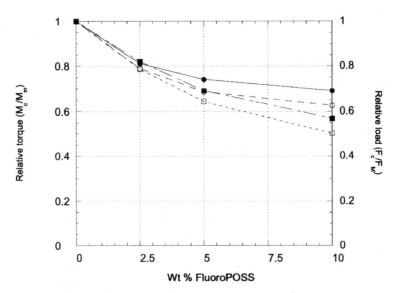

Figure 6. Effect of FluoroPOSS on torque and load for PCTFE blends.

properties even further. Water contact angles are a measure of surface hydrophobicity and provide insight into the free energy of the surface. Contact angles have been obtained on PCTFE nanocomposites containing FD_8T_8 and FO_8T_8. Technologies that may benefit from the blending of FluoroPOSS into fluoropolymers include abrasion resistance, lubricity, anti-icing, and non-wetting applications. Figure 7 shows a drop of water on the surface of a PCTFE film. The contact angle was measured at 88°. Figure 6 also shows a drop of water on the surface of a PCTFE blend containing 10% FD_8T_8. The contact angle for this film was measured at 128°. There is a 40° increase in contact angle with just 10% added FD_8T_8.

Figure 7. Water contact angles of 88° and 128° on PCTFE films.

It has also been observed, as one might expect, that the contact angle increases with increasing weight percent POSS. Contact angles have been obtained on other fluoropolymers as well. All show a similar trend. It should be noted that a surface with a contact angle of 90° or higher is considered a "non-wetting" surface, while a surface with a contact angle below 90° is considered "wetting." Unfilled PCTFE has a contact angle of 88°. Addition of FluoroPOSS produces a "non wetting" surface.

Other Properties

In order to determine the effect of the FluoroPOSS on the mechanical properties on PCTFE, dynamic mechanical analysis (DMTA) was performed. The variation in moduli and glass transition temperatures seen with the addition of FluoroPOSS is small enough to be statistically insignificant.

The level of dispersion of POSS compounds into polymer systems is largely dependent on surface chemistry. Atomic Force Microscopy (AFM) and Scanning Electron Microscopy (SEM), along with the element mapping capability of SEM, were used to determine nanoparticle dispersion. These techniques indicate good to excellent particle dispersion in the polymer matrix.

Conclusions

Two fluorinated Polyhedral Oligomeric Silsesquioxanes (FluoroPOSS) have been produced. The large-scale synthesis results in the production of cage mixtures, which can then be converted to T_8 cages by a redistribution reaction. The FluoroPOSS compounds are hydrophobic, with the **FD_8T_8** possessing a water contact angle of 154°, making it ultrahydrophobic. Analysis of single crystal x-ray data indicates that molecular scale surface roughness may lead to ultrahydrophobicity.

These compounds have been blended into poly(chlorotrifluoroethylene) (PCTFE). These POSS fluoropolymers may be useful as low friction surfaces or hydrophobic coatings. Contact angle measurements of the POSS fluoropolymers show an improvement of water contact angles over the unfilled materials. The **FD_8T_8/PCTFE** composite shows a contact angle improvement of 40° over the unfilled material. The low surface energy POSS compounds also appear to act as a processing aid during fluoropolymer processing, significantly reducing both the torque and load measurements in the extruder. Thermal and mechanical properties of the blended fluoropolymers do not differ significantly from those of the unfilled polymers.

Acknowledgements

The authors would like to thank the Air Force Office of Scientific Research and the Air Force Research Laboratory, Propulsion Directorate for funding.

References

1. POSS is a registered trademark of Hybrid Plastics Inc., Fountain Valley, CA 92708
2. Gonzalez, R. I.; Phillips, S. H.; Hoflund, G. B. J. *Spacecraft and Rockets* **2000**, *37*, 463.
3. Hoflund, G. B.; Gonzalez, R. I.; Phillips, S. H. *J. Adhesion Sci. Technol.* **2001**, *15*, 1199.
4. Gilman, J. W.; Schlitzer, D. S.; Lichtenhan, J. D. *J. Appl. Polym. Sci.* **1996**, *60*, 591.
5. Lu, S-. Y.; Hamerton, I. *Prog. Polym. Sci.* **2002**, *27*, 1661.
6. Lee, G. Z.; Wang, L.; Toghiani, H.; Daulton, T. L.; Pittman, C. U. Jr. *Polymer* **2002**, *43*, 4167.
7. Zhang, W.; Fu, B. X.; Schrag, E.; Hsiao, B.; Mather, P. T.; Yang, N-. L.; Xu, D.; Ade, H.; Rafailovich, M.; Sokolov, J. *Macromolecules* **2002**, *35*, 8029.
8. Fu, B. X.; Gelfer, M. Y.; Hsiao, B. S.; Phillips, S.; Viers, B.; Blanski, R.; Ruth, P. *Polymer* **2003**, *44*, 1499.
9. Deng, J.; Polidan, J. T.; Hottle, J. R.; Farmer-Creely, C. E.; Viers, B. D.; Esker, A. *J. Am. Chem. Soc.* **2002**, *124*, 15194.
10. Fu, B. X.; Hsiao, B. S.; White, H.; Rafailovich, M.; Mather, P. T.; Jeon, H. G.; Phillips, S.; Lichtenhan, J.; Schwab, J. *Polymer Int.* **2000**, *49*, 437.
11. Mabry, J. M.; Vij, A.; Viers, B. D. *Angew. Chem. Int. Ed. Engl.* Manuscript in preparation.
12. Dedecker, K.; Groeninckx, G. *Polymer* **1998**, *39*, 4993.
13. Hong, Y.; Cooper-White, J. J.; Mackay, M. E.; Hawker, C. J.; Malmstrom, E.; Rehnberg, N. *Rheol. J.* **1999**, *43*, 781.

Chapter 19

Formation and Functionalization of Aryl-Substituted Silsesquioxanes

Yusuke Kawakami, Dong Woo Lee, Chitsakon Pakjamsai, Makoto Seino, Akiko Takano, Akio Miyasato, and Ichiro Imae

School of Materials Science, Japan Advanced Institute of Science and Technology [JAIST], Asahidai 1–1, Nomi, Ishikawa 923–1292, Japan

Aryl-substituted polyhedral oligosilsesquioxanes are an important class of compounds for the reinforcement of composite materials and the creation of a new matrix of high tolerance against extreme conditions. They are of even more interest as materials for electronic applications. Reaction conditions to obtain such materials with completely condensed cage structures, incompletely condensed double decker structure are studied. Some functional transformations of such compounds are also illustrated.

Polyhedral oligomeric silsesquioxanes (POSS) have nanometer-size cubic structures with wide surface area and controlled porosity, and can be functionalized with a wide variety of organic groups for the design of nanocomposite materials and hybrid polymer systems. Hydrolysis of trichloro- or trialkoxysilane gives polysilsesquioxane with formal structure $[RSiO_{1.5}]_n$, but the actual product is often a mixture of insoluble gel, crystalline compounds, or oil. The product distribution depends on the reaction conditions, such as solvent, concentration, catalyst, temperature, and organic R group. Octa-functional cubic polysilsesquioxanes (octa functional pentacyclo[$9.5.1.1^{3,9}.1^{5,15}.1^{7,13}$]octasiloxane, T8), such as hydroxy-functionalized T8-H, or dimethylsilyloxy-functionalized T8-OMe$_2$SiH, are not simply a constituent of nano-hybrid systems, but can function as a starting material used in advanced display technology as a support for amorphous functional dye dispersant systems.

In this report, the tendency of cage formation in the hydrolysis of (substituted phenyl)trialkoxysilane or trichlorosilane under various conditions is described, followed by some substitution reactions on aromatic ring systems. Vinyl-functionalized POSS and carbazole-functionalized POSS were synthesized from T8-OMe$_2$SiH by reacting with hexadiene or 9-vinylcarbazole, respectively. POSS-containing polymer soluble in ordinary solvents, and material exhibiting principally monomeric emission was obtained.

Results and Discussion

There is a competition during the condensation process of hydrolyzed products from trialkoxysilanes or trichlorosilanes between "intermolecular" to give randomly branched polymers and ladder polymers, and "intramolecular" to give loop structure, which eventually forms cyclic polyhedral cage products. The products usually contain unreacted silanol groups depending on their structures. Recently, Wallace reported the use of matrix-assisted laser desorption/ionization time of flight mass spectrometry (MALDI-TOF) to identify the products with loop structure. After 4 hours refluxing in the presence of benzyltrimethylammonium hydroxide, precipitate was formed for (4-dimethylaminophenyl)- and phenyltrimethoxysilane, and 4-trimethoxysilyl-biphenyl. Benzene soluble oligosilsesquioxanes from phenyltrimethoxysilane showed 2 principal peaks in the range of δ = -69~-72, and -78~-81 ppm assignable to T^2 and T^3 structure, and MALDI-TOF MS showed several peaks.

The insoluble precipitate shows sharper peaks, and shows only T^3 signal in ^{29}Si NMR at -78.30 ppm, which strongly suggested a perfectly or nearly perfectly closed cage structure of the products. MALDI-TOF MS gave only one sharp peak at 1032 as $[M + H]^+$ signal. This compound was assigned as octaphenyl-pentacyclo[$9.5.1.1^{3,9}.1^{5,15}.1^{7,13}$]octasiloxane, Ph-T8). The yield was

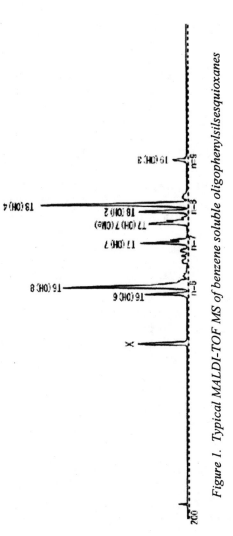

Figure 1. Typical MALDI-TOF MS of benzene soluble oligophenylsilsesquioxanes

about 80%. Biphenyltrimethoxysilane and 4-dimethylaminophenyltrimethoxysilane also gave a high benzene insoluble fraction. 4-Methylphenyl derivative did not give any insoluble product.

If hydrolysis of three methoxy groups is similarly fast, condensation of silanol functions should give a highly branched structure. If hydrolysis of the one is slower than the other two and than condensation between two silanol groups, linear or low branch structure will be basically formed as the initial products. SEC results show that the condensation of initially formed oligosilsesquioxane proceeds to a stage of formation of oligosilsesquioxane with maximum 8 to 10 silicon atoms. The aromatic moiety might have functioned to differentiate the reactivity of three methoxy groups in the hydrolysis, and to give low branch structure that is soluble in benzene.

Oligosilsesquioxane structure is shown by representing with dot (silicon atom) and line (siloxane) symbols. The dot connecting to only 1 line indicates T^1 structure, and dot connecting to 2 and 3 lines T^2 and T^3 structures. The oligosilsesquioxanes consisting of n silicon atoms is abbreviated as T^n. The degree of condensation (f_n) shown in the left of the figure was defined as the extent of intramolecular condensation of silanol groups as typically shown, based on presumed 8-membered ring shown in Figure 2 in the cases of 8 silicon atom containing oligosilsesquioxanes.

In the benzene soluble fraction, oligosilsesquioxane containing 6 to 9 silicon atoms, with loops but incompletely cyclized to cage structure, were observed as the main products. Fractions containing 7 and 8 silicon atoms are the major components of the product, irrespective of the substituent. With 4-dimethylamino substituent, T8 and T9 were formed the most with small amounts of T6 fraction, and the amount of T9 was the highest among the compounds studied. Solubility of the compounds and steric effects might have a big influence on the product distribution.

When methyl substituent is introduced in the *ortho* position, methyl-substituted T8 was obtained in good yield under the same reaction condition used for non-substituted derivative (82%). Contrary, introduction of methyl group at *para* or *meta* position of phenyl group lowered the yields of T8. These isomeric Me-PhT8 showed different elution behavior on the size exclusion chromatograph (SEC), and on different melting points and solubility.

meta-Phenyl T8 showed the highest solubility in ordinary organic solvents and the highest melting point (424°C), *para* derivative intermediate (407°C), and *ortho* derivative showed the lowest solubility and melting point (385°C). Introduction of longer alkyl group at ortho position improved the solubility of the compounds. Thermal decomposition temperatures were almost identical for three isomers (1-3).

When methylene chloride was used as the solvent, along with small amounts of *para*MePhT8, deca(*para*-methylphenyl)-hexaacyclo-[13.5.1.1^{3,13}.1^{5,11}.1^{7,19}.1^{9,17}]decasiloxane, *para*MePh-T10) was obtained as the

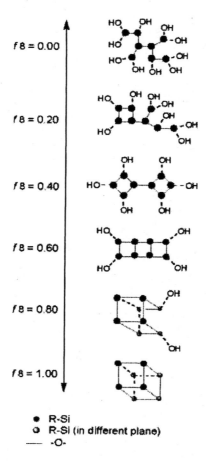

$f8 = 0.00$

$f8 = 0.20$

$f8 = 0.40$

$f8 = 0.60$

$f8 = 0.80$

$f8 = 1.00$

● R-Si
○ R-Si (in different plane)
---- -O-

Figure 2. Loop formation of oligosilsesquioxane by condensation

major product in the presence of tetrabutylammonium fluoride or benzyl-trimethylammonium hydroxide.

The low yield (7%) of *para*MePhT8 from *para*-methylphenyltrichlorosilane could be improved to 22% by using *para*-methylphenyltrimethoxysilane as the starting material in the presence of catalytic amounts of hydrochloric acid. Nitration of these isomers proceeded very nicely with copper (II) nitrate trihydrate in acetic anhydride.

para-Trimethylsilyl-substituted PhT8 was obtained in reasonable yield by benzyltrimethylammonium hydroxide.

When the hydrolysis was carried out under strongly basic condition in isopropanol with sodium hydroxide, so-called double decker type octasiloxane was obtained.

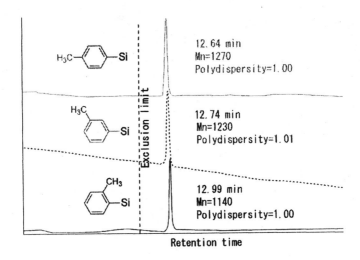

12. 64 min
Mn=1270
Polydispersity=1. 00

12. 74 min
Mn=1230
Polydispersity=1. 01

12. 99 min
Mn=1140
Polydispersity=1. 00

Retention time

Figure 3. Size exclusion chromatograms of isomerically methyl-substituted phenyl T8.

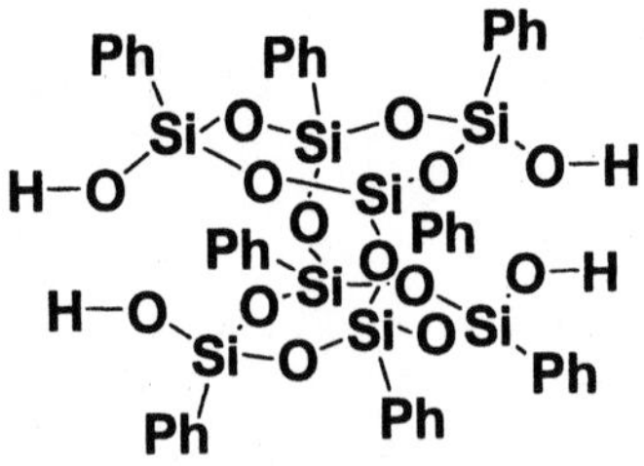

Figure 4. Double decker type T8

Single crystal structures of the silanol (DDOH) and trimethylsilylated derivative (DDOTMS) were studied by X-ray diffraction (4).

The T8 monomers with different content of vinyl groups were synthesized by two-step hydrosilylation. Silane-functionalized cubic silsesquioxane, octakis(dimethylsiloxy)octasilsesquioxane (Q8M8H) was firstly reacted with 6, 5, and 4 equiv. of allylbenzene in the presence of Pt(1,3-divinyl1,1,3,3-teramethyldisiloxane: dvs) complex, followed by the reaction with excess hexadiene. SEC exhibited mainly two fractions of oligomeric products, which were isolated by preparative SEC. In ^{1}H NMR, the signals were observed at 4.92-4.97 and 5.74-5.83 ppm for vinyl protons, at 0.56-0.60, 1.30-1.42 and 2.02 ppm for aliphatic protons of hexenyl groups, and at 0.61-0.70, 1.61-1.69, and

(a)

(b)

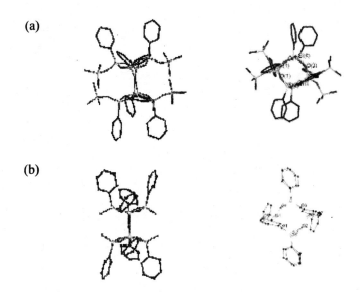

Figure 5. ORTEP Plot of DDOTMS (a) and DDOH (b) determined by a single-crystal X-ray diffraction

2.61 ppm for aliphatic protons of phenylpropyl groups. The methyl proton of SiMe showed complicated signals at 0.08-0.13 ppm, indicating a mixture of compounds with two different substituents. The content of hexenyl groups was estimated by integral ratio of methylene protons at 2.02 and 2.61 ppm. The DEPT spectrum did not show CH-carbon signals arisen from α-addition, indicating that hydrosilylation reaction yielded selectively β-addition products. The signals in ^{29}Si NMR at 12.78 and 12.67 ppm are ascribed to $C_6H_7SiMe_2(O-)$ and $PhC_3H_6SiMe_2(O-)$, and that at -108.84 ppm to a silicon atom of cubic structure $Si(O-)_4$.

Figure 6 shows the MALDI-TOF MS in the 1500-2100 m/z range using DHBA as matrix. Separated products gave spectra corresponding to their $[M + H]^+$ ions.

In all cases after the reaction with hexadiene, several peaks were also observed and could be ascribed to the general formula $Q8M8(C_6H_{12})_m(C_3H_6Ph)_n$. These results indicate that intramolecular cyclization of SiH group and vinyl group was not involved in the reaction. In the spectrum of the product (1:4 ratio), seven species (m = 0 to 6) were observed, and those with 1 to 2 (C_6H_{12}) groups were present in relatively high concentration. In the spectra of products (1:5, 1:6 ratio), all species (m = 0 to 8) were observed. Intensity of these peaks almost agreed with integral value of ^{1}H NMR.

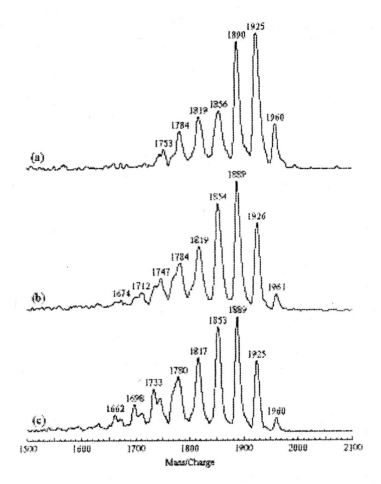

Figure 6. MALDI-MS of the products obtained from Q8M8H with (a) 6, (b) 5, and (c) 4 equiv. of allylbenzene.

The hydrosilylation polymerizations of these compounds with 1,4-bis(dimethylsilyl)benzene (BSB) were carried out in presence of Pt(dvs). Polymer soluble in common organic solvents was obtained in good yield, although the molecular weight was not high (M_w/M_n = 12900/3300) and polydipersity was very broad, indicating that the product is a mixture of oligomeric products with different number of T8. The signals of methyl and phenylene proton of BSB unit were observed at 7.47 and 0.23 ppm. The ^{29}Si NMR showed rather clear signals. In addition of three signals arisen from starting material, the signal of silicon of BSB unit was observed at –3.26 ppm. These results indicated that the polymerization proceeds without side reactions to produce polymer as shown in Scheme 1.

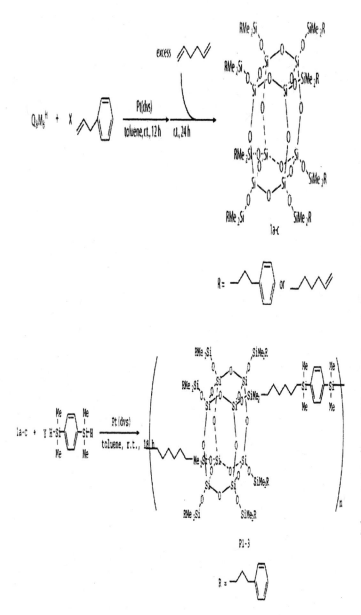

Scheme 1. Synthesis of T8-containing polymers soluble in ordinary organic solvents.

Thermal stability of the polymers was studied by thermo gravimetric analysis (TGA). Thermal stability of the polymer increased by about 10 °C from the monomer. The polymers exhibited on-set temperatures at 446 °C and residues of 24.4 %. The difference of thermal stability by the molecular weight was hardly observed (5).

Carbazole was introduced into the dimethylsiloxy group of Q8M8H by hydrosilylation reaction with 9-vinylcarbazole to obtain octakis[2-(carbazol-9-yl)ethyldimethylsiloxy]silsesquioxane (POSS-Cz). The product was soluble in common organic solvents, such as toluene, THF, chloroform, dichloromethane.

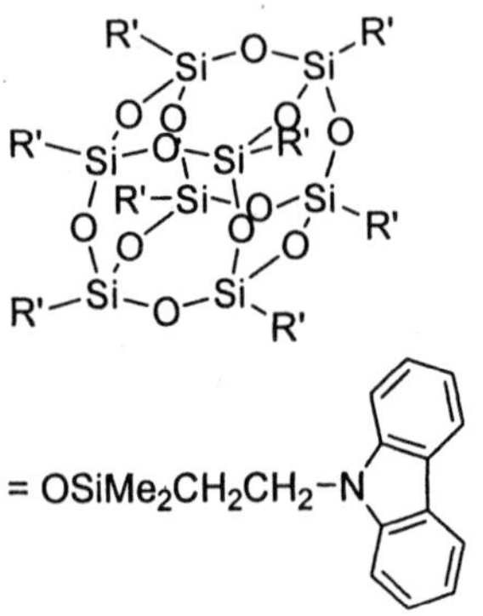

R' = OSiMe$_2$CH$_2$CH$_2$–N

Figure 7. POSS-Cz.

Differential scanning calorimetry (DSC) and thermogravimetry (TGA) were investigated. POSS-Cz was found stable until around 400 °C even in air. When POSS-Cz was heated, it showed only sharp endothermic peak at 166 °C due to melting. On the other hand, when the sample obtained by cooling the melt sample to 0 °C was heated, it showed only glass-transition temperature at 37 °C. Different from 9-ethylcarbazole (EtCz, T_m = 68 °C), POSS-Cz does not crystallize at all once it was melted. The glass-transition temperature of POSS-Cz is much lower than that of poly(9-vinylcarbazole) (PVCz, T_g = 190 °C), because of the flexible spacer between POSS and carbazole. In polarized optical microscopy (POM), the image of glassy sample in cross-Nicol state was dark, suggesting to be isotropic. The results of XRD and POM suggest the glassy POSS-Cz being amorphous.

Figure 8 shows the electronic absorption and photoluminescence spectra of dilute solutions (1 x 10^{-6} mol dm^{-3} in THF) of POSS-Cz, EtCz and PVCz. POSS-Cz showed two peaks at 332 and 347 nm due to the π-π* absorption (B-band) of carbazole, which is quite similar to that of EtCz (λ_{max} = 332 and 346 nm). Photoluminescence spectrum of POSS-Cz showed two peaks at 353 and 370 nm, which was the mirror image of the electronic absorption spectrum. This result is also similar to that of EtCz. The quantum yield of POSS-Cz in air was

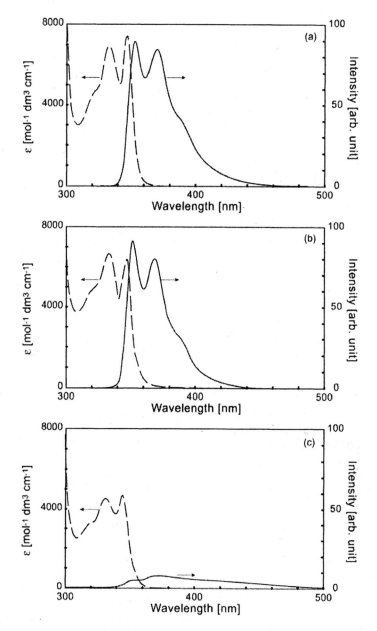

Figure 8. Electronic absorption (dash) and photoluminescence (solid) spectra of (a) POSS-Cz, (b) EtCz, and (c) PVCz.

312

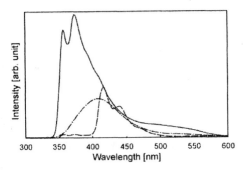

Figure 9. Photoluminescence spectra of POSS-Cz (solid), EtCz (dash), and PVCz (dash-dot) in solid state on Quartz substrate.

estimated by referring that of p-terphenyl ($\phi_{p\text{-terphenyl}}$ = 0.87), to be $\phi_{POSS\text{-}Cz}$ = 0.27, which is almost same as that of EtCz (ϕ_{EtCz} = 0.30). Although the electronic absorption spectrum of PVCz is similar to those of POSS-Cz and EtCz (λ_{max} = 331 and 344 nm), the photoluminescence spectrum of PVCz is quite different and broad, with quantum yield (ϕ_{PVCz} = 0.10) much lower than those of POSS-Cz and EtCz. In the case of POSS-Cz, each carbazole seems to be isolated by the rigid core, so that the formation of excimer is prevented.

This result suggests that the carbazole in POSS-Cz is almost isolated even in the solid state, different from EtCz and PVCz. Interestingly, the photoluminescence spectra of POSS-Cz in the solid film showed mainly similar monomeric emission peak, while those of EtCz and PVCz showed broad peaks due to the formation of excimer by aggregation (Figure 9).

It is reported that POSS having cyanobiphenyl can show liquid crystallinity, when a long alkyl spacer is introduced between the POSS core and the aromatic group. This means that the cyanobiphenyls far from the rigid POSS core can aggregate to align in the same direction. POSS-Cz has an ethylene-siloxy spacer between POSS core and the carbazole, but the spacer length seems too short to allow aggregation, and therefore shows monomeric behavior in photoluminescence (6).

References

1. Pakjamsai, C.; Kawakami, Y. *Polymer J.* **2004**, 36, 455.
2. Pakjamsai, C.; Kobayashi, N.; Koyano, M.; Sasaki, S.; Kawakami, Y. *J. Polym. Sci. A.* **2004**,42, 4587.
3. Pakjamsai, C.; Kawakami, Y. *Design. Monom. and Polym.* **2005**, 8, 423.
4. Lee, D. W.; Kawakami, Y. inpreparation
5. Seino, M.; Kawakami, Y. *Polymer J.* **2004**, 36, 422.
6. Imae, I.; Kawakami, Y. *J. Mater. Chem.* **2005**, 15, 4581.

Chapter 20

Transparent Polymer–Polyhedral Oligomeric Silsesquioxane Composites

Subramanian Iyer[1], Amjad Abu-Ali[1,2], Andrew Detwiler[1], and David A. Schiraldi[*]

[1]Department of Macromolecular Science and Engineering, Case Western Reserve University, 2100 Adelbert Road, Cleveland, OH 44120
[2]Current address: PolyOne Corporation, Avon Lake, OH 44012
[*]Corresponding author: david.schiraldi@case.edu

Polyhedral oligomeric silsesquioxanes (POSS[®]) are an important class of nanoscale inorganic-organic hybrid materials, which have been incorporated into organic polymers via melt blending as well as copolymerization. In this paper, the different criteria necessary for formation of transparent POSS-polymer composites are discussed, as are general concepts governing the thermal/mechanical properties of these composites. A series of nanocomposites were produced from polyhedral oligomeric silsesquioxane materials and four transparent, amorphous, engineering plastics. While the majority of the nanocomposites within this study retained high optical clarity, only cellulose propionate composites exhibited both high clarity and enhanced thermo-mechanical properties. The specific system, which exhibited these desirable properties, is the only one tested which can be expected to possess strong particle-matrix interactions. Such interactions are likely a key to the preparation of stable melt-blended POSS nanocomposites of engineering plastics, which possess enhanced properties.

Introduction

Polyhedral oligomeric silsesquioxane (POSS®) molecules possess a cage-like structure (1–3 nm in size) and a hybrid chemical composition ($RSiO_{1.5}$) which is intermediate between silica (SiO_2) and silicones (R_2SiO).[1] The well-defined structure of these nano materials contains stable inorganic Si–O cores surrounded by substituents which can be modified to present a wide range of polarities and reactivities. POSS molecules can be incorporated into polymer systems through blending,[2] grafting or copolymerization[3,4,5,6,7,8,9] aiming at nanostructured polymeric materials whose properties bridge the property space between organic plastics and ceramics. Recent studies on POSS-containing hybrid copolymers and thermosets have been reported indicating reinforced mechanical[6-8] and thermal properties.[7-9] POSS is expected to act as a nano-scale filler with which to modify polymer matrixes and potentially produce nanocomposites with new or improved properties. The organic shell of substituents can be used to create compatibility by matching polarity with host polymers, simultaneously improving mechanical properties due to reinforcement by the rigid silicate core. When effectively incorporated into polymer systems, these nanoscale fillers have been shown to improve the thermomechanical properties of materials that range from polyethylene to epoxy networks, as well as imparting resistance to singlet oxygen and decreased flammability. In this manner, control of the microstructure of the POSS nanocomposites can be achieved and is key to the performance and properties of such materials.[3,9] It is generally believed that nano-scale POSS domains with ordered and self-assembled features in a polymer matrix are highly desirable and lead to the observed improvement in material properties.[10,11,12,13]

Even though POSS has the ability to be dispersed in polymers, most polymer-POSS systems are opaque due to solubility limits and inadequate dispersion domain sizes. POSS grades exhibit drastically different effects on polymers, ranging from plasticization, no effects, to increases in glass transition temperatures. The production of clear transparent composites is desirable for various applications, such as aviation windshields capable of withstanding the high temperatures associated with supersonic speeds.

Polyhedral oligomeric silsesquioxanes have the general structure shown in Figure 1. The R groups can range from hydrogen, to bulky isooctyl or phenyl groups and can contain functional groups such as epoxides or isocyanates. Due to their size and potential to be compatible with polymers, POSS fillers have the ability to form true molecular composites; compatibility of these materials with polymers will be governed with the nature of their organic peripheries, while their inorganic cores serve to reinforce the polymers.

Though there has been a significant amount of work carried out with POSS copolymers showing enhanced thermomechanical properties of polyurethanes,[14] epoxy thermosets,[15] and dicylcopentadienes,[16] very little has been reported concerning the use of POSS as a reinforcing filler. In this paper we report melt blending of POSS grades with four amorphous polymers, combinations that were carefully chosen in order to facilitate compatibility and interactions between polymers and fillers. The materials investigated in this paper include amorphous copolyester, a cyclic olefin copolymer, polycarbonate and cellulose propionate. Differences between composites containing fully-condensed POSS cages, and those possessing a hybrid inorganic–organic 3D partial cage-like structure bearing three silanol (Si–OH) groups are also examined. It was recently reported that the trisilanol POSS grades, especially isooctyl trisilanol POSS, exhibit enhanced compatibility with PET, as judged by single phase melts and optically transparent extrudate.[17]

Experimental

All materials were used as received with any further modification. Polyethylene terephthalate/cyclohexanedimethanol copolymer, PETG, and cellulose propionate (Tenite®, CP) were supplied by the Eastman Chemical Company. Polycarbonate (PC, Makrolon® 2405) was obtained from Bayer AG. The cyclic polyolefin copolymer (COC, Topas® 8007) was obtained from Ticona GmbH. All materials were dried at 100°C 24 hours prior to extrusion. The extrusion temperatures and the different POSS grades used with different polymers are listed in Table 1. Materials were processed on a DACA co-rotating microextruder (Model 20000) with a residence time of 5 minutes. Post extrusion, the extrudates were compression molded at the respective processing temperatures at approximately 12000 lbs pressure. Samples were heated and held under pressure for about 1 minute before quenching the films between water cooled platens. Films of 0.3 mm thickness were made for characterization. The different POSS grades used in this study were of two forms; either a fully condensed cage structure or an incomplete cage structure with silanol groups. The representative structures for the two forms of POSS are shown in Figure 1.

The blended samples were examined for qualitative differences in transparency, color and toughness. Dynamic mechanical analysis was performed on a Tritec-2000 dynamic mechanical analyzer from Triton Technologies, UK. Scanning electron microscopy was carried out on a Phillips X-30 ESEM, after coating fractured surfaces of films with palladium. An Instron model 5565 universal testing machine was used to obtain the tensile properties of the samples. Testing was conducted at three temperatures, 23, 85 and 120°C; a strain rate of 12.70 mm/min was used for these measurements (5 specimens of each sample were tested). Results are reported for the high temperature runs only.

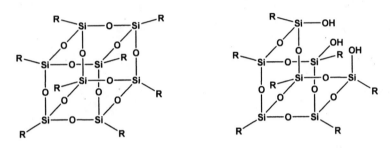

Figure 1: Representative structures of complete and incomplete cage structures of POSS

Table 1: Processing temperatures and POSS grades used with different polymers

Polymer	POSS grades	Processing Temperature °C
PETG	Trisilanol Isooctyl POSS Isooctyl POSS	240
COC	Isobutyl POSS Trisilanol Isobutyl POSS Trisilanol Isooctyl POSS	220
PC	Trisilanol Phenyl POSS Isooctyl POSS	290
CP	Trisilanol Isooctyl POSS Trisglycidyl isobutyl POSS	200

Results

Appearance

The extrusion observations and sample appearances of compression molded polymer/POSS films are summarized in Table 2. In the case of PC/POSS blends, clear composites were obtained only with phenyl trisilanol POSS, while trisilanol isooctyl POSS and octaphenyl POSS both resulted in opaque extrudates. The compression molded films of PC/phenyl trisilanol POSS exhibited a high level of transparency, while the other PC blend films were translucent. It is reasonable to conclude then that only phenyl trisilanol POSS was well dispersed within a polycarbonate matrix. Each of the POSS grades melt blended with COC produced transparent clear films and extrudates, indicating high levels of compatibility/miscibility between these materials. Since there is no likelihood of chemical interactions between the hydrocarbon COC polymer and the POSS grades, we therefore expect the nanoparticles to be molecularly or nano-scale dispersed within the COC copolymer system. When PETG was used as the polymer matrix, opaque extrudates and translucent films were obtained. The absence of compelling polarity matches, and the possibility of POSS thermal degradation under polyester processing conditions apparently lead to poor dispersion in these cases. The appearance of transparent, ductile cellulose propionate was largely unaffected by addition of the POSS grades tested, up to as much as 5 wt% filler, indicating significant compatibility for these systems.

Dynamic Mechanical Analysis and Tensile Testing

Figures 2-5 illustrate the effects of POSS grades on the thermomechanical properties of the four transparent engineering plastics examined in this study. These DMA results can be classified into three distinct behaviors. In the first case, there is a clear plasticization effect as observed by reduction in the glass transition temperatures and moduli of the composites as various POSS grades are added; both polycarbonate and the cyclic olefin copolymer exhibit this behavior. In the second case, the addition of POSS to the glycol modified polyethylene terephthalate has no measurable effect on the glass transition temperature and little on the modulus of the composites. Reinforcement is observed as the third type of behavior - this effect is seen with incorporation of POSS in cellulose propionate. The rubbery modulus of the CP composites increases 100 fold with incorporation of POSS at 5 wt%, while there is no significant shift in the glass transition temperature. Consistent with reinforcement/enhanced rubbery modulus of CP upon addition of POSS, its high temperature tensile modulus increases with filler loadings up to 10 wt%, Figure 6.

Table 2. Extrusion observation of various Polymer/POSS composites

Composition	Extrusion Observation	Compression molded film observation
PETG control	Transparent, Clear	Transparent, Tough
5% Isooctyl POSS	Opaque	Translucent, Tough
5% Trisilanol Isooctyl POSS	Opaque	Translucent, Tough
COC control	Transparent, Clear	Transparent, Ductile
5% Isobutyl POSS	Transparent, Clear	Transparent, Ductile
5% Trisilanol Isobutyl POSS	Transparent, Clear	Transparent, Brittle
5% Trisilanol Isooctyl POSS	Transparent, Clear	Transparent, Ductile
PC control	Transparent, clear	Transparent, Ductile
5% Trisilanol Phenyl POSS	Transparent, clear	Transparent, Brittle
5% Trisilanol Isooctyl POSS	Opaque	Translucent, Ductile
5% Octaphenyl POSS	Opaque	Translucent, Bubbles, Brittle
CP Control	Transparent, Clear	Transparent, Ductile
5% Trisilanol Isooctyl POSS	Transparent, Clear	Transparent, Ductile
5% Trisglycidyl Isobutyl POSS	Transparent, Clear	Transparent, Ductile

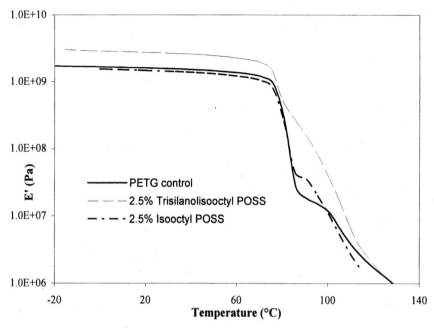

Figure 2: DMA curves for PETG composites

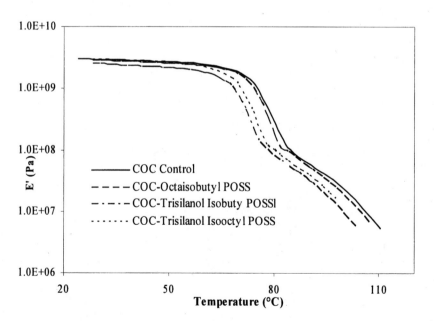

Figure 3: DMA curves for COC composites

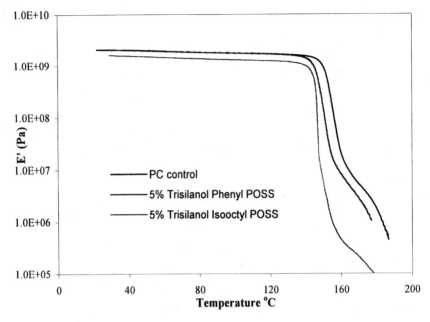

Figure 4: DMA curves for PC composites

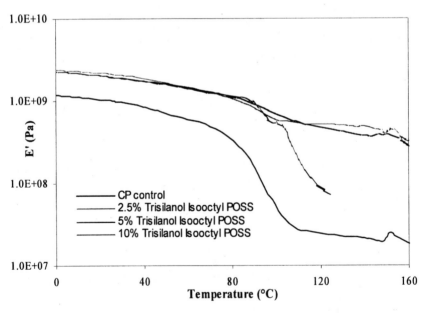

Figure 5: DMA curves for Cellulose Propionate composites

Effect of TPOSS @ 120°C

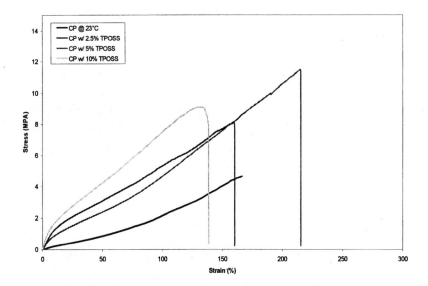

Figure 6: CP/isooctyl trisilanol POSS Tensile testing, 120°C

Scanning Electron Microscopy

The morphology of one of the highly compatible systems, polycarbonate with 5% phenyl trisilanol POSS blends is shown in Figure 7. In case of such PC/POSS blends, few if any filler-related features are evident. Analysis by EDAX confirms that almost equal concentrations of silicon exist between matrix and features (such as they are), consistent with miscibility within the PC/POSS blends.

Discussion and Conclusions

The compatibility or miscibility of POSS particles within matrix polymers clearly depends upon the nature of the R groups which decorate its corners; in some cases, these peripheral organic groups can serve to shield the silicate core from the organic hosts.[17] Incomplete ("T7") POSS cage structures possess three free silanol groups, which are capable of interacting with carbonyl groups as well as hydroxyl groups on the polymer via hydrogen bonding. As a viscous fluid with high viscosity, trisilanol isooctyl POSS and isooctyl POSS grades are different from the crystalline POSS materials examined in this study. It has also been demonstrated that although POSS is relatively stable up to 300°C, given sufficient time and temperature, some degree of nanofiller degradation is

a) PC

b) PC 5% Trisilanol Phenyl POSS

Figure 7: SEM micrograph of a) Polycarbonate and b) 5% phenyl trisilanol POSS

possible.[18] Such degradation can result in an increase in the molecular weight of POSS thus causing aggregation/oligomerization of particles, increasing the possibility of phase separation between these fillers and matrix polymers.

From the DMA results presented herein, a decrease in the glass transition temperatures of polycarbonate and cyclic olefin copolymers was observed. The results from SEM and EDAX further show that in case of polycarbonates, POSS is molecularly dispersed. It has been demonstrated that when POSS is molecularly distributed in polymers there is often a reduction in the glass transition temperature of the composite.[19] We believe that when POSS is molecularly dispersed in the system, individual particles act as molecular lubricants, allowing for chain slippage in the polymer. This slippage results in lowering of the glass transition temperature of the polymer. However, when the compatibility between the polymer and POSS is low, as in the case with PETG, there is no significant effect on the glass transition temperature. Lower compatibility between POSS and the polymer leads to larger particle size of the dispersed phase, which results in opaque/translucent composites. Since there is gross phase separation between the phases, POSS does not significantly affect the polymer properties. The third system investigated was the cellulose propionate/trisilanol isooctyl POSS system. Cellulose propionate possesses a large number of hydroxyl groups capable of interacting with the silanol groups on incompletely condensed POSS cages. The hydrogen bonding interactions between the filler and the polymer can exist beyond the glass transition temperature of the polymer, thereby causing an increase in rubbery modulus and high temperature tensile properties.

Hence, transparent nanocomposites of POSS and amorphous engineering plastics can be obtained in cases where high levels of compatibility exist. Simple solubility of nanofiller in polymer plasticizes the transparent plastic, similar to the action of more traditional organic plasticizers. In such cases when polarities of host and guest are well matched, with additional polymer-filler interactions (such as hydrogen bonding), molecular scale reinforcement of the polymer can also be obtained without loss of optical clarity.

Acknowledgements

Financial support from Hybrid Plastics, and gifts of polymers from Ticona, Bayer and Eastman are gratefully acknowledged.

References

1. http://www.hybridplastics.com
2. Fu, B.X.; Yang, L.; Somani, R.H.; Zong, S.X.; Hsiao, B.S.; Philips, S. *Journal of Polymer Science, Polymer Physics Edition*, **2001**, 39, 2727

325

3. Zhang, W.; Fu, B.X.; Seo, Y.; Schrag, E.; Hsiao, B.; Mather, P.T. *Macromolecules*, **2002**, 35, 8029

4. Fu, B.X.; Hsiao, B.S.; Pagola, S.; Stephens, P.; White, H.; Rafailovich, M. *Polymer*, **2001**, 42, 599

5. Fu, B.X.; Hsiao, B.S.; White, H.; Rafailovich, M.; Mather, P.T.; Jeon, H.G. *Polymer International*, 2000, 49, 437

6. Mather, P.T.; Jeon, H.G.; Romo-Uribe, A.; Haddad, T.S.; Lichtenhan, J.D. *Macromolecules*, **1999**, 32, 1194

7. Romo-Uribe, A.; Mather, P.T.; Haddad, T.S.; Lichtenhan, J.D. *Journal of Polymer Science, Polymer Physics Edition*, **1998**, 36, 1857

8. Kim, G-M.; Qin, H.; Fang, X.; Sun, F.C.; Mather, P.T. *Journal of Polymer Science, Polymer Physics Edition*, **2003**, 41, 3299

9. Lee, A.; Lichtenhan, J.D. *Journal of Applied Polymer Science*, **1999**, 73, 1993

10. Laine, R.C.; Brick, C; Kim, S.G.; Tamaki, R; Chen, H.J.; Choi, H.J. *225*[th] *ACS national meeting, New Orleans, LA. Materials Chemistry*, **2003**, paper 11

11. Schiraldi, D.A.; Zeng, J.; Kumar, S.; Iyer, S; Dong F. *225*[th] *ACS national meeting, New Orleans, LA. Materials Chemistry*, **2003**, paper 9

12. Esker, A.R.; Vastine, B.A.; Deng, J.; Polidan, J.T.; Viers, B.D.; Satija, S.K. *225*[th] *ACS national meeting, New Orleans, LA. Materials Chemistry*, **2003**, paper 2

13. Coughlin, E.B. *227*[th] *ACS national meeting, Philadelphia, PA. Polymer Chemistry*, **2004**, paper 275

14. Fu, B.X.; Hsiao, B.S.; White, H.; Rafailovich, M.; Mather, P.T.; Jeon, H.G.; Phillips, S.; Licthenhan, J. Schwab, J. *Polymer International*, **2000**, 49(5), 437

15. Yong, N.; Sixun, Z. *Macromolecular Chemistry and Physics*, **2005**, 206,20, 2075

16. Constable, G.S.; Lesser, A.J.; Coughlin, B.E. *Macromolecules*, **2004**, 37,1276

17. Zeng, J.; Iyer, S.; Gonzalez, R.; Kumar, S; Schiraldi, D.A. *High Performance Polymer Journal*, **2005**, 17,403

18. Zeng, J.; Bennett, C.; Jarrett, W.L.; Iyer, S.; Kumar, S.; Mathias, L.J. *Composite Interfaces*, **2005**, 11, 673

19. Kopesky, E.T.; Haddad, T.S.; Cohen, R.E.; McKinley, G.H. *Macromolecules*, **2004**, 37(24), 8992

Silica and Related Systems

Chapter 21

The Interaction of 'Silicon' with Proteins: Part 2. The Rold of Bioinspired Peptide and Recombinant Proteins in Silica Polymerization

Siddharth V. Patwardhan[1], Kiyotaka Shiba[2], Heinz C. Schröder[3], Werner E. G. Müller[3], Stephen J. Clarson[4], and Carole C. Perry[1,*]

[1]Division of Chemistry, School of Biomedical and Natural Sciences, Nottingham Trent University, Clifton Lane, Nottingham NG11 8NS, United Kingdom
[2]Department of Protein Engineering, Cancer Institute, Japanese Foundation for Cancer Research, Kami-Ikebukuro, Toshima, Tokyo 170–8455, Japan
[3]Institut fur Physiologische Chemie, Abteilung Angewandte Molekularbiologie, Universitat Duesbergweg 6, 55099 Mainz, Germany
[4]Department of Chemical and Materials Engineering, University of Cincinnati, Cincinnati, OH 45221
*Corresponding author: Carole.Perry@ntu.ac.uk

Various organisms are known to produce hierarchically organised ornate biosilica structures in vivo with great fidelity. In several recent studies it has been proposed that the controlled moulding of biosilicas is regulated by biomolecules such as peptides, proteins and polysaccharides. However, the mechanisms underpinning biomolecule mediated biosilici-fication and the interactions of 'silicon' with biomolecules remain unclear. The aim of this paper is to carry out model studies of silicification with the intention of probing the interactions between biological molecules derived from biosilicifying organisms with silica. We present a detailed study of silica formation in the presence of a recombinant silicatein protein (derived from the sponge *Suberites domuncula*) and an R5 peptide (derived from silaffin proteins isolated from the diatom *Cylindrotheca fusiformis*). We have also undertaken model studies of silica formation using recombinant proteins of known primary and secondary structure and pI to investigate the effect that structure has on structure development from solution. A silicon-catecholate complex was used that liberates silicic acid under circumneutral pH. The kinetics of silicic acid polymerisation was monitored using a colorimetric method. The aggregation of polysilicic acid oligomers and particle growth was studied using dynamic light scattering. The nature of the solid materials formed was investigated by electron microscopy - transmission electron microscopy to study the sizes of the fundamental particles and scanning electron microscopy to study the morphology of aggregates. In order to assess surface area, porosity and pore size distributions, gas adsorption analysis was performed. Thermal analysis was used to quantify the amount of organic material entrapped into silicas, which may originate from the catechol released after decomposition of the silicon complex or from the protein or peptide added to the reaction mixture. It was found that these functional additives affect all aspects of silicic acid polymerisation. In particular, the aforesaid proteins/peptides were able to control the kinetics of silica polymerisation. In addition, most of these additives modulated the surface area and pore size distribution of the silicas prepared in their presence.

Introduction

The high degree of sophistication achieved in synthesising hierarchical biominerals, often under mild processing conditions is intriguing.[2-4] Silica, which is a technologically important mineral, is synthesised *in vitro* under conditions of high temperature and extremes of pH.[5-8] On the other hand, various organisms are known to produce biosilica *in vivo* with great fidelity.[9,10]

Several recent studies have proposed that the controlled moulding of ornate and organised biosilica is regulated by biomolecules such as peptides, proteins and polysaccharides (see ref.[11] for a review). However, the exact mechanism governing biomolecule-mediated biosilicification remains unclear. In order to gain insights into biosilica formation and to devise novel synthetic approaches, several *in vitro* studies investigating biomolecule-silica interactions have been undertaken.[12-19]

Sumper, Kroger and co-workers have identified a set of biomolecules, which they later proposed to facilitate (bio)silica formation based on their silica precipitating abilities *in vitro*.[20] One diatom species – *Cylindrotheca fusiformis* – was used for the investigations. It was found that hydrofluoric acid treatment of silica cell walls released silaffin proteins – silaffin-1A (4 kDa), silaffin-1B (8 kDa) and silaffin-2 (17 kDa). Using the NH_2-terminus amino acid sequence of silaffin-1B and the cDNA library of *C. fusiformis*, the *sil1* gene (795 bp) was identified. The *sil1* gene encoded *sil1p* peptide contains R1-R7 highly homologous peptides as one of its key features. It was further found that the R5 peptide may correspond to silaffin-1A$_1$. Model studies using the synthetic R5 peptide were undertaken *in vitro* to study their silica precipitation activity by various researchers.[21-25] When silaffin-1A (mixture of -1A$_1$ and -1A$_2$) was compared with pR5 (of unknown length) for silica precipitation activity, it was observed that the latter can only precipitate silica at pH > 6, while the former shows a maximum activity around pH 5.[21] This difference in silica precipitation activity was proposed to be due to the presence of post-translational modifications that are present only on silaffin proteins. It was suggested that post-translational modifications may not be "required" for silicification *in vitro*, while the modifications might have important roles to play *in vivo*.[23] In further investigations of silaffin proteins, it was proposed that the C-terminus KRRIL, KRRNL and RRIL groups were/are proteolytically removed and all the serine residues phosphorylated.[20]

In order to validate the structural requirements of silaffins and related peptides, the R5 peptide and its mutants have been investigated for their activity in silica precipitation. The study presented, for the first time, a close scrutiny of structural motifs of peptides derived from *sil1p* that may be essential for *in vitro* silica precipitation.[25] It was proposed that the terminal RRIL motif helps in the formation of self-assembled aggregates of the peptides in solution. Similar self-assembly has been previously reported in the case of silica precipitation using biomimetic block co-polypeptides,[26] polyamines,[27,28] and native silaffin.[29] It is interesting to note that all the silica precipitating peptides considered in the study did not contain any post-translational modifications. However, there is no direct evidence so far that indicates siloxane bond catalysis in silica polymerisation in the presence of R5 peptide. In order to gain further insights into the R5-mediated silica synthesis, we present herein the effects of R5 on silica polymerisation and the final product.

The sponges – *Tethya aurantia* and *Suberites domuncula* – produce silica in the form of needle like spicules and were thus studied to isolate proteins that facilitate biosilicification.[30-32] It was found that each spicule contained a central filament of protein. After various treatments to dissolve the mineral silica from the sponge, silica proteins were isolated and were named as *silicatein*. Silicatein proteins were found to be highly similar to members of the cathepsin L family of proteases. Silicatein proteins were able to precipitate silica *in vitro* when tetraethoxysilane (TEOS) was used as the silica precursor. It was proposed that the serine, histidine and asparagine triad was catalysing the hydrolysis of TEOS, which would otherwise require acid/base catalysis.[30-32] Further evidence was observed that illustrates the silicon-silicatein interaction: silicate was found to regulate the expression of silicatein.[32] In order to understand the interactions of 'silicon' and silicatein, and to study in detail the role of silicatein protein in silicic acid condensation, we have carried out a detailed systematic study of silica polymerisation in the presence of a recombinant silicatein protein (corresponding to silicatein isolated from *S. domuncula*).

Bioextracts isolated from various biosilicifying organisms have been studied for their sequences and structures. However, there is still not enough information available about these bioextracts, for example their exact secondary structures, and hence it is difficult to correlate the silica precipitation ability of these isolates with their structures. It has been speculated that proteins involved in biomineralisation control, through their secondary structural motifs, biomineral/precursor-protein interactions. Furthermore, biomineralising proteins have been found to contain repetitive amino acid sequences.[33] In order to address these issues, we have designed and synthesised recombinant proteins with known primary and secondary sequences. In addition, these proteins have been extensively investigated for their properties (see below). Furthermore, since it has been shown previously that cationic peptides and polymers facilitate silica formation under ambient conditions,[12] the proteins used in the current study were designed to contain high amount of cationic amino acid residues (see Figure 1). The primary amino acid sequence of these proteins contains repetitive peptide motifs. In particular, we have used three proteins of distinct and known structures – α-helical, β-sheet and random coil. It should be noted that related recombinant proteins have been shown to regulate the crystal growth of inorganic salts (sodium chloride, potassium chloride, copper sulphate) and sucrose,[34] thus it will be interesting to observe the effect of the aforesaid proteins in silica formation. Previously, we have shown that one of these proteins (the α-helical protein) facilitates silica formation in an alkoxysilane system; however, details of the interactions between the proteins and silica were not reported.[35] We present here the role of these proteins and their structures on *in vitro* silica synthesis under ambient conditions and circumneutral pH.

In brief, we present a detailed study of silica formation in the presence of a recombinant silicatein protein (derived from the sponge *Suberites domuncula*)

R5 peptide

SSKKSGSYSGSKGSKRRIL

Silicatein

DYPEAVDWRT	KGAVTAVKDQ	GDCGASYAFS	AMGALEGANA
LAKGNAVSLS	EQNIIDCSIP	YGNHGCHGGN	MYDAFLYVIA
NEGVDQDSAY	PFVGKQSSCN	YNSKYKGTSM	SGMVSIKSGS
ESDLQAAVSN	VGPVSVAIDG	ANSAFRFYYS	GVYDSSRCSS
SSLNHAMVVT	GYGSYNGKKY	WLAKNSWGTN	WGNSGYVMMA
RNKYNQCGIA	TDASYPTL		

pYT320 α helical	pYT338 β sheet	pYT287 Random coil
MRGSHHHHHHSSGWVD	MRGSHHHHHHSSGWVDP	MRGSHHHHHHSSGWVD
PENLQAE	----------RGVVWT	----------PKSAVHTR
RKVLQGRMENLQAE	RSRMDAESYGRGVVWT	---------GAKSAVHTR
RKVLQGRMENLQAE	RSRMDAESYGRGVVWT	GLECTRSRMDAKSAVTPV
RKVLQGRMENLQAE	RSRMDAESYGRGVVWT	GLECTRSRMDAKSAVHTR
RKVLQGRMENLQAE	RSRMDAESYGRGVVWT	GLECTRSRMDAKSAVHTR
RKVLQGRMENLQAEP	RSRMDAESG	GLECTRSRMDAKSAVHTR
QSIAGSYGKPASGG		GLECTRSR---KICSPHQ
MW 12.3	MW 11.09	GGVN
pI 9.86	pI 10.7	MW 13.6
alpha-rich in CD	beta-rich in CD	pI 10.6
		Disordered in CD

Figure 1. Information on the primary sequence of the peptides/proteins studied.

and an R5 peptide (derived from silaffin proteins isolated from the diatom *Cylindrotheca fusiformis*). We have also undertaken model studies of silica formation using recombinant proteins of known primary and secondary structure and pI to investigate the effect that biomolecule structure has on silica structure development from solution.

Experimental

Materials

The silicic acid precursor dipotassium silicon tris catecholate $(K_2[Si(C_6H_4O_2)_3]).2H_2O$ (97%) and anhydrous sodium sulphite (97%) were purchased from Sigma Aldrich Chemicals. Ammonium molybdate, hydrochloric acid (37%) and sulphuric acid (98%) were purchased from Fisher Scientific. Oxalic acid (99%) and p-methyl amino phenol sulphate (99%) were purchased from Acros Chemicals and standard stabilised silicate solution (1000 ppm as SiO_2) was purchased from BDH. R5 peptide (SSKKSGSYSGSKGSKRRIL) was obtained from New England Peptides, MA, USA. Recombinant proteins used – silicatein, YT320 (α helical), YT338 (β sheet rich) and YT287 (random coil) were prepared as described below. All chemicals were used without further treatment. Distilled deionised water (DI water) having a conductivity less than <1 µs cm^{-1} was used.

Protein Synthesis

Recombinant silicatein was prepared as described previously.[36] The silicatein-cDNA from *S. domuncula* (accession number CAC03737.1), spanning the putative mature silicatein segment from amino acid$_{113}$ to amino acid$_{330}$ (nucleotide$_{337}$ to nucleotide$_{993}$), was inserted into the oligohistidine expression vector pQE-30 via *Bam*HI (5-end) and *Hin*dIII (3-end). After transformation of *Escherichia coli* strain Origami, the expression of the fusion protein was induced with isopropyl 1-thio-D-galactopyranoside. The protein was extracted and purified on Ni-NTA-agarose resin. The purity of the material was checked by electrophoresis in 10% polyacrylamide gels, containing 0.1% sodium dodecyl sulfate. The purified fusion protein was cleaved from the histidine-tag by digestion with recombinant enterokinase; the recombinant polypeptide had a M_r of 26 kDa.

The YT320, YT338 and YT287 proteins were produced, purified and characterised as previously reported.[37,38] Briefly, proteins were expressed in *E. coli* cells, purified by using TALON (Clontech) resin under denaturing conditions. The denaturant (urea) was removed by dialysis against water.

Silica Synthesis

Silicic acid polymerisation experiments were carried out at room temperature. The initial silicic acid concentration used was kept constant at 30 mM. A precalculated amount *ca.* 2M HCl was added to each experiment in order to dissociate the silicon complex thus releasing orthosilicic acid. The resulting pH was found to be 6.8 ± 0.30, unless otherwise specified. Known amounts of each additive (*ca.* 1-5 wt% of anticipated sedimentable silica) were added in each case immediately prior to acidification and the reaction mixtures shaken gently for proper mixing of reactants. Thereafter the solutions were left to condense silica for seven days.

Characterisation

Kinetics

The chemical reagents and procedures used for molybdosilicate analysis were as reported previously.[5,39,40] "Molybdenum blue" reagent (MBR) and reducing agent were prepared by a well-established protocol.[39] The kinetic studies were carried out by removing 10 μL aliquots of reaction mixture at given time intervals. Each aliquot was added to 1.5 ml MBR solutions and were then diluted with 15 ml DI water. The solutions were then left for 15 minutes to form the molybdosilicate yellow complex. Reducing agent (8 ml) was then added to these complexes which gives a blue colour to the solutions. The absorbance at 810 nm was recorded using a UV spectrophotometer. Calibration using 10 ppm silicic acid indicated that absorbance varies linearly with the concentration of the molybdosilicate complex and hence with the silicic acid concentration in the concentration range used in this study.

Aggregation

Aggregation and particle formation behaviour of the samples with and without additives was monitored using Dynamic Light Scattering (DLS). Samples were prepared in similar fashion as described above and were placed in a 1 cm polymethylmethacrylate cell. Aggregation and particle growth of silica particles was then monitored over a 40 hour period using a Coulter N4 plus photon correlation spectrometer with a He-Ne (632.8nm) laser supply. All measurements were carried out at an angle of 90° at a constant temperature

(298K). Measurements obtained were averages of the data collected over intervals of 5 to 60 minutes.

Gas Adsorption

Surface area measurements were obtained from precipitated silica samples isolated after 168 hours (7 days) reaction time. The samples were collected by centrifugation and were washed twice with DI water and once with ethanol in order to remove untrapped materials and were then lyophilised. Samples were additionally calcined at 823 K to remove organic material and reanalysed. Nitrogen gas adsorption/desorption analysis was performed using a Quantachrome Nova3200e surface area and pore size analyser. Samples were first degassed overnight at 80-100°C under vacuum. Surface areas were then measured *via* the BET method[41] over the range of relative pressures 0.05 – 0.3 at which the monolayer is assumed to assemble. Pore radii were determined by the BJH method[42] using the desorption branch of the isotherm.

Electron Microscopy (EM)

The morphologies and the fundamental particle sizes for all the samples before and after calcination were studied using Scanning Electron Microscopy (SEM) and Transmission Electron Microscopy (TEM) respectively. For EM studies, lyophilised samples were dispersed onto respective sample holders. For SEM analysis the silica samples were gold coated with argon plasma. Images were acquired using a JEOL JSM-840A SEM with an accelerating voltage of 20 kV and JEOL 2010 TEM operating at 200 kV.

Thermal gravimetric analysis (TGA)

The organic material associated with the precipitated silica samples was detected by thermal gravimetric analysis. Analysis was performed in air using a Stanton Redcroft TG 760 furnace, balance controller and UTP temperature controller by heating up to 1100 K at a rate of 10 K min^{-1} with data being sampled every 30 seconds. Entrained organic content was calculated by comparison of weight loss between 400 and 600 K relative to the silica produced from the catecholate complex alone.

Results and Discussion

In order to study different aspects of silicic acid polymerisation and the effects of various biomolecules on silica formation, a model system has been used in this study. In this system, a silicon complex was used that liberates silicic acid under circumneutral pH as shown in Scheme 1.[39] Concentrations of orthosilicic acid and disilicic acid were monitored using a colorimetric method, aggregation and particle growth using dynamic light scattering and the nature of the solid materials formed by electron microscopy (transmission electron microscopy to study the sizes of the fundamental particles and scanning electron microscopy to study the morphology of aggregates), gas adsorption analysis (to assess surface area, porosity and pore size distributions) and thermal analysis to quantify the amount of organic material (from the catechol released from decomposition of the complex or from the protein or peptide added to the reaction mixture). All experiments were conducted at a silicon concentration of 30 mM and protein/peptide was added to *ca.* 1-5% by weight in relation to the anticipated silica precipitate.

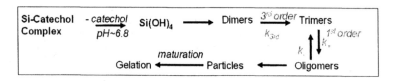

Scheme 1. Reactions showing the dissociation of the silicon-complex liberating silicic acid and its subsequent polymerisation.

Bioinspired strategies employed for silica synthesis, such as that described herein, make use of organic biomolecules and their synthetic analogues. Silica syntheses in the presence of such additives obey certain set of rules, and are discussed elsewhere.[12,43] It has been shown that the additives control the *in vitro* bioinspired/biomimetic silicification by either catalysis, aggregation, structure direction (templating/ scaffolding) or a combination thereof. Each of these aspects in silicification was studied systematically and the observations are presented below.

It has been postulated that silicatein proteins are globular proteins.[44] The amino acid sequences for the highly basic synthetic model proteins are given in Figure 1. YT320 had a solution structure that was largely alpha helical, YT338 had a structure rich in beta sheet and YT287 had a disordered structure.[34,37,38] Synthetic R5 peptide (based on findings from studies of the diatom *C. fusiformis*) was also studied for comparison. It is noted that R5 peptide was found to possess a random coil-like open structure when it was monitored at different pH values using Circular Dichroism.[45]

Measurements of silicic acid concentration yielded information on the following distinct stages in the polymerisation process: (1) formation of trimers, (2) formation of oligomers that serve as nuclei (a first order reversible reaction with the forward reaction being described as growth and the reverse step being described as dissolution) and (3) particle growth (Scheme 1).

All experiments were conducted at the same pH except for experiments involving silicatein where experimental data is available for two distinct starting pH values, 6.1 (significantly lower than for the blank system) and 7.1 (a little higher than for the blank system). The kinetic data obtained is shown below. All the rate constants are plotted relative to the blank system (Figure 2).

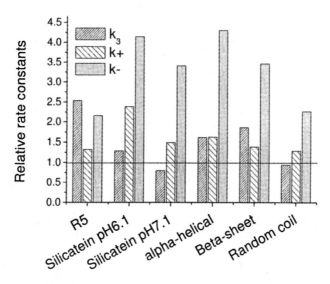

Figure 2. Kinetics of silicic acid condensation.

All recombinant proteins showed an effect on the initial stages of silica formation. A large increase in the rate of the dissolution process was observed for all additives (Figure 2), suggesting that structural reorganisation was promoted in the presence of these additives. The addition of the R5 peptide showed the strongest and statistically the most significant effect on the condensation kinetics at the very early stages of silica formation. Such an effect has been speculated in previous studies.[21,23] However, it should be noted that there was no direct evidence provided in the literature demonstrating siloxane bond formation in the presence of the R5 peptide. Addition of these proteins/peptides had a dramatic effect on the decondensation step (Figure 2).

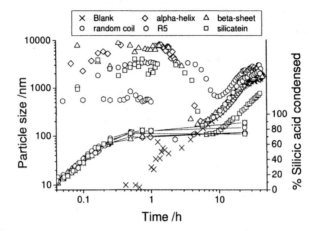

Figure 3. Particle growth in the presence of additives plotted as a log-log graph. Total silicic acid polymerised is also shown (symbols+lines).

The aggregation behaviour of selected samples was studied using DLS. The data for silicatein, the R5 peptide and that obtained from the recombinant proteins is shown in Figure 3 as a log-log plot. From the initial high values of particle size (these values could be reproducibly produced), it was found that the R5 peptide, recombinant proteins and silicatein formed self-assembled structures, as seen from the initial section of the curves. R5 solutions in DI water, for example, were found to form stable self-assembled structures of sizes *ca.* 500 nm as determined by DLS (data not shown), which correspond with the initial structures formed in the presence of silicic acid (Figure 3). These assemblies were destroyed as silica formation proceeded (Figure 3). This observation provides a direct evidence of silica-protein/peptide interactions and might suggest that the formation of silica oligomers and particles, which possess negative charges at neutral pH, disturb the protein/peptide self-assemblies due to electrostatic interactions. After the protein/peptide assemblies were destroyed, the particle formation behaviour followed the growth observed for the blank sample, which may be due to the low amounts of the proteins present (1-5 wt %) in the system.

The mature silica samples (7 days reaction time) were collected, washed, lyophilised and further analysed by thermal analysis, nitrogen gas adsorption and electron microscopy. Samples were additionally calcined at 823 K to remove organic material and reanalysed by nitrogen adsorption analysis. The presence of the additives even at *ca.* 1-5% by weight had a very significant effect on the physical nature and properties of the silicas produced. Data obtained from gas adsorption are shown for the blank system (Figure 4) and for additions of the silicatein protein (Figure 5), the R5 peptide (Figure 6) and the recombinant proteins (Figure 7), before and after calcination.

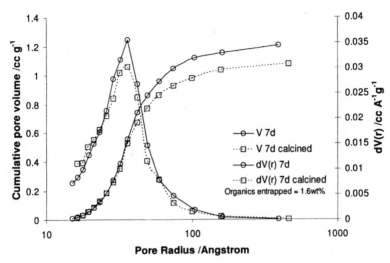

Figure 4. Pore size and pore volume data for blank system.

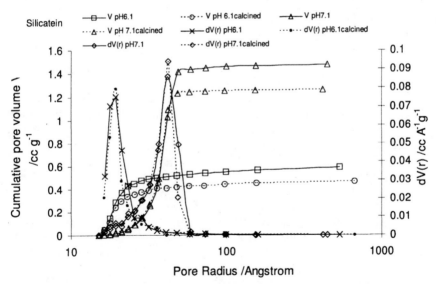

Figure 5. Data obtained from gas adsorption on silica prepared using silicatein protein before and after calcination.

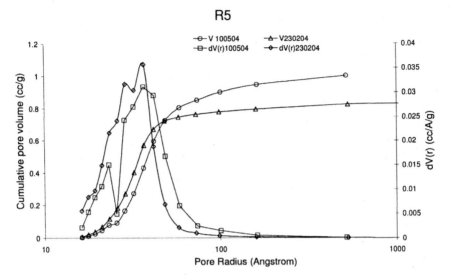

Figure 6. Data obtained from gas adsorption on silica prepared using R5 peptide before and after calcination.

For the blank system, the silica produced changed little with calcination with a small reduction in surface area (probably statistically insignificant) and a small decrease in pore volume but no change in average pore diameter (Figure 4). The level of entrapped organic was *ca.* 1.6% (from the catechol) and which varied little when silicatein was used as the additive (see Figure 9 for TGA data). However, materials prepared in the presence of silicatein at pH 6.1 exhibited a much higher surface area with much smaller pores and a much smaller pore volume. Calcination resulted in a large decrease in surface area and a reduction in pore volume but no change in pore diameter (Figure 5). For the reactions initiated at pH 7.1 in the presence of silicatein protein, surface areas were more akin to that expected for the blank system but the silica exhibited little change in pore volume but had bigger pores with a narrow range (Figure 5). Calcination led to little change in surface area (small decrease) and pore radius but to a very large change in pore volume. The addition of the R5 peptide significantly decreased the surface area and pore volume but no effect was observed on the pore radius (Figure 5). The decrease in surface area was the most significant in the presence of the R5 peptide and suggests the formation of a dense silica. It should be noted in this regard that R5 also had the most significant impact on the initial kinetics of the silica formation reaction (Figure 2).

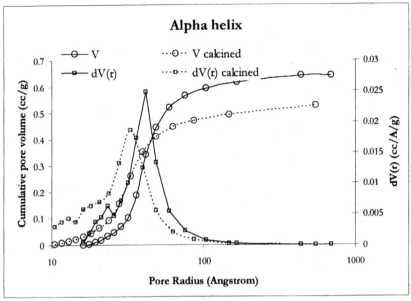

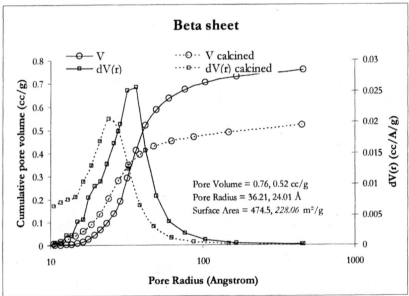

Figure 7. Data obtained from gas adsorption on silica prepared using recombinant proteins before and after calcination. Shaded bars represent samples after heat treatment. Continued on next page.

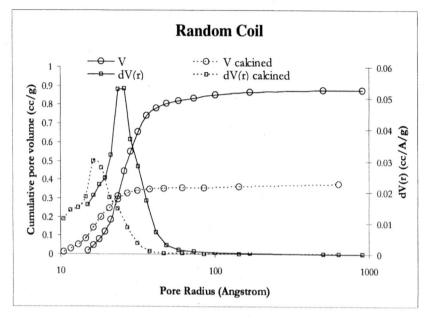

Figure 7. Continued.

Additions of the recombinant proteins with largely different secondary structures generated silicas with a wide range of surface areas and porosities. A decreasing trend in surface area, pore volume and average pore diameter could be loosely correlated with increasing levels of organic material entrapped within the precipitated silicas (most entrapped material, lowest surface area, pore volume and pore diameter); Figure 7-9. The trend was alpha helical < beta sheet < random coil. On calcination, reductions in surface area, pore volume and average pore diameter were all observed. In addition, the reduction in surface area could be correlated with the reaction kinetics in the presence of these proteins. The α helical and β sheet-like proteins increased the rate of the principally trimer forming reaction (Figure 2) and produced silicas with low surface areas. Materials with a wide range of porosities could be prepared using the simple model system in the presence of low (*ca.* 1-5% by weight) amounts of an 'active' additive. The summary of the data obtained from gas adsorption analysis is presented in Figure 8.

The entrapment of the peptides/proteins into final silica structures, if any, was studied using Thermogravimetric Analysis (TGA) and the data presented in Figure 9. In the case of the recombinant protein mediated silicification, variable amounts of proteins were found to be entrapped. In order to investigate the

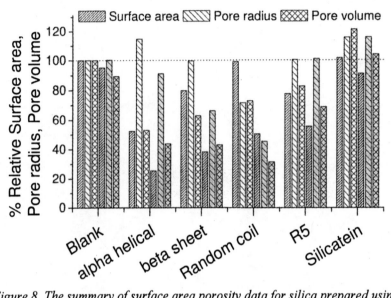

Figure 8. The summary of surface area porosity data for silica prepared using peptides/proteins and the effect of thermal treatment.

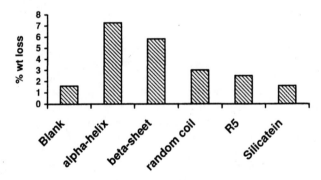

Figure 9. Thermogravimetric Analysis (TGA) of silica prepared using peptides/proteins.

effects of these proteins on the silica nano- and micro-structures, the analysis of the silica structures using Scanning and Transmission electron microscopy was undertaken. The morphologies of the samples prepared in the presence of the additives were not significantly different, probably because of their low amounts (Figure 10). The results also indicated that, when compared with the blank system, there were no significant variation in the fundamental particle sizes as observed by using TEM (data not shown).

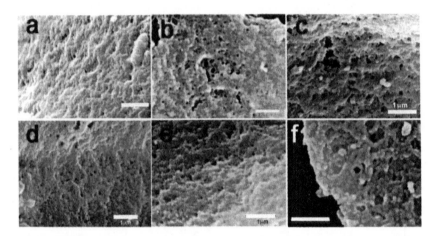

Figure 10. SEM images obtained from silica synthesized without any additive (a), and in the presence of the R5 peptide (b), silicatein protein (c), the alpha-helical protein (d), the beta-sheet protein (e), and the random coil protein (f).

Conclusions

We present a study of silica formation in the presence of a recombinant silicatein protein (derived from the sponge *Suberites domuncula*), an R5 peptide (derived from silaffin proteins isolated from the diatom *Cylindrotheca fusiformis*) and three recombinant proteins of known primary and secondary structure and pI to investigate the effect that structure has on structure development from solution. All of the recombinant proteins showed an effect on the initial stages of silica formation. A large increase in the rate of the dissolution process was observed for all additives, suggesting that structural reorganisation was promoted in the presence of these biomolecules. The addition of the R5 peptide showed the strongest and statistically the most significant effect on the condensation kinetics at the very early stages of silica formation. The particle growth in the presence of additives followed the growth observed in the blank sample, which may be due to the low amounts of the proteins present (1-5 wt %) in the system. The presence of the additives had a very significant effect on the physical nature and properties of the silicas produced.

Acknowledgement

We would like to thank the European Union for funding our research as a part of the SILIBIOTEC project. We also thank Dr. David Belton for his kind help with the gas adsorption analysis.

References

1. Eglin, D.; Shafran, K. L.; Coradin, T.; Livage, J.; Perry, C. C. *in preparation.*
2. Mann, S.; Webb, J.; Williams, R. J. P., Eds. *Biomineralization*; VCH: Weinheim, 1989.
3. Lowenstam, H. A.; Weiner, S. *On Biomineralization*; Oxford University Press: New York, 1989.
4. Simkiss, K.; Wilbur, K. M. *Biomineralization*; Academic Press: San Diego, 1989.
5. Iler, R. K. *The Chemistry of Silica*; John Wiley & Sons: New York, 1979.
6. Brinker, C. J.; Scherer, G. W. *Sol-Gel Science: The Physics and Chemistry of Sol-Gel Processing*; Academic Press: Boston, 1990.
7. Stober, W.; Fink, A.; Bohn, E. *J. Colloid Interface Sci.* **1968**, *26*, 62-69.
8. Kendall, T. *Industrial Minerals* **2000**, *March*, 49-59.
9. Simpson, T. L.; Volcani, B. E. *Silicon and Siliceous Structures in Biological Systems*; Springer-Verlag: New York, 1981.
10. Müller, W. E. G., Ed. *Silicon Biomineralization*; Springer: Berlin, 2003.
11. Perry, C. C. *Rev. Minerology Geochem.* **2003**, *54*, 291-327.
12. Patwardhan, S. V.; Clarson, S. J.; Perry, C. C. *Chem. Commun.* **2005**, *9*, 1113-1121.
13. Roth, K. M.; Zhou, Y.; Yang, W.; Morse, D. E. *J. Am. Chem. Soc.* **2005**, *127*, 325-330.
14. Delak, K. M.; Sahai, N. *Chem. Mater.* **2005**, *17*, 3221-3227.
15. Belton, D.; Patwardhan, S. V.; Perry, C. C. *Chem. Commun.* **2005**, 3475-3477.
16. Jin, R.-H.; Yuan, J.-J. *Chem. Commun.* **2005**, 1399-1401.
17. Jin, R.-H.; Yuan, J.-J. *Adv. Mater.* **2005**, *17*, 885.
18. Knecht, M. R.; Wright, D. W. *Langmuir* **2004**, *20*, 4728-4732.

346

19. Coffman, E. A.; Melechko, A. V.; Allison, D. P.; Simpson, M. L.; Doktycz, M. J. *Langmuir* **2004**, *20*, 8431-8436.
20. Sumper, M.; Kroger, N. *J. Mater. Chem.* **2004**, *14*, 2059-2065.
21. Kroger, N.; Deutzmann, R.; Sumper, M. *Science* **1999**, *286*, 1129-1132.
22. Brott, L. L.; Pikas, D. J.; Naik, R. R.; Kirkpatrick, S. M.; Tomlin, D. W.; Whitlock, P. W.; Clarson, S. J.; Stone, M. O. *Nature* **2001**, *413*, 291-293.
23. Whitlock, P. W. *Silicon-Based Materials in Biological Environments*: Ph.D. Dissertation, University of Cincinnati, 2005.
24. Patwardhan, S. V. *Silicification and Biosilicification: Role of Macro-molecules in Bioinspired Silica Synthesis*; Ph.D. Dissertation, University of Cincinnati, 2003.
25. Knecht, M. R.; Wright, D. W. *Chem. Commun.* **2003**, 3038-3039.
26. Cha, J. N.; Stucky, G. D.; Morse, D. E.; Deming, T. J. *Nature* **2000**, *403*, 289-292.
27. Patwardhan, S. V.; Mukherjee, N.; Steinitz-Kannan, M.; Clarson, S. J. *Chem. Commun.* **2003**, *10*, 1122-1123.
28. Coradin, T.; Durupthy, O.; Livage, J. *Langmuir* **2002**, *18*, 2331-2336.
29. Kroger, N.; Lorenz, S.; Brunner, E.; Sumper, M. *Science* **2002**, *298*, 584-586.
30. Shimizu, K.; Cha, J.; Stucky, G. D.; Morse, D. E. *Proc. Natl. Acad. Sci. USA* **1998**, *95*, 6234-6238.
31. Cha, J. N.; Shimizu, K.; Zhou, Y.; Christiansen, S. C.; Chmelka, B. F.; Stucky, G. D.; Morse, D. E. *Proc. Natl. Acad. Sci. USA* **1999**, *96*, 361-365.
32. Krasko, A.; Lorenz, B.; Batel, R.; Schröder, H. C.; Müller, I. M.; Müller, W. E. G. *Eur. J. Biochem.* **2000**, *267*, 4878-4887.
33. Evans, J. S. *Curr. Opin. Colloid Int. Sci.* **2003**, *8*, 48-54.
34. Shiba, K.; Honma, T.; Minamisawa, T.; Nishiguchi, K.; Noda, T. *EMBO reports* **2003**, *4*, 148-153.
35. Patwardhan, S. V.; Shiba, K.; Raab, C.; Hüsing, N.; Clarson, S. J. In *Polymer Biocatalysis and Biomaterials*; Cheng, H. N.; Gross, R. A., Eds.; Oxford University Press, 2005.
36. Müller, W. E. G.; Krasko, A.; Le Pennec, G.; Steffen, R.; Ammar, M. S. A.; Wiens, M.; Müller, I. M.; Schröder, H. C. *Prog. Mol. Subcell. Biol.* **2003**, *33*, 122-195.
37. Shiba, K.; Takahashi, Y.; Noda, T. *J. Mol. Biol.* **2002**, *320*, 833-840.
38. Shiba, K.; Shirai, T.; Honma, T.; Noda, T. *Protein Engn.* **2003**, *16*, 57.
39. Belton, D.; Paine, G.; Patwardhan, S. V.; Perry, C. C. *J. Mater. Chem.* **2004**, *14*, 2231-2241.
40. Harrison, C. C.; Loton, N. *J. Chem Soc. Faraday Trans.* **1995**, *91*, 4287-4297.
41. Brunauer, S.; Emmett, P. H.; Teller, E. *J. Am. Chem. Soc.* **1938**, *60*, 309-319.

42. Barrett, E. P.; Joyner, L. G.; Halenda, P. P. *J. Am. Chem. Soc.* **1951**, *73*, 373-380.
43. Coradin, T.; Lopez, P. J. *ChemBioChem* **2003**, *4*, 251-259.
44. Shimizu, K.; Morse, D. E. In *Biomineralization*; Baeuerlein, E., Ed.; Wiley-VCH: Chichester, 2000; p 207.
45. Farmer, R.; Patwardhan, S. V.; Kiick, K. L. *unpublished results.*

Chapter 22

Synthesis of Styryl- and Methacryloxypropyl-Modified Silica Particles Using the Hydrolytic Sol–Gel Method and the Miniemulsion Polymerization Approach

Anna Arkhireeva and John N. Hay

Chemistry Division, SBMS, University of Surrey, Guildford, Surrey GU2 7XH, United Kingdom

The hydrolytic sol-gel route has been used to obtain a range of organically modified silica particles of sub-micron diameter. A modified Stöber method has been adapted to prepare styryl- and methacryloxypropyl-modified silica particles with diameters in the range of 200-1000nm by base-catalyzed hydrolysis and condensation reactions of organotrialkoxysilane precursors at ambient temperature. The size of these particles is considerably larger than that of silsesquioxane particles produced in this way from organically modified precursors containing less bulky organic groups, such as methyl or ethyl groups. The particles have been stabilized against aggregation by the use of surfactants. Nearly monodisperse methacryloxypropyl- modified silica particles with diameters in the range of *ca.* 40-900 nm were produced by the miniemulsion polymerization method using a range of surfactants and hexadecane to suppress Ostwald ripening. A range of potential applications for these hybrid nanocomposite materials includes their use as modifiers of the properties of selected polymer systems, such as acrylic and styrenic polymers or coatings.

Introduction

The widely used sol-gel process has provided convenient methods for the production of inorganic oxides and their hybrids, such as organically modified silicas. A significant amount of attention has focused recently on the hydrolytic and non-hydrolytic sol-gel routes to organically modified silica nanoparticles[1-7]. One truly promising commercial application for the hybrid nanoparticles that can be obtained using these sol-gel methods is their use as nanofillers to modify the properties of selected polymer systems, such as acrylic and styrenic polymers or coating systems. The surface of the organically modified silica particles is expected to be more hydrophobic than that of inorganic particles due to the presence of organic groups and the reduction of the number of residual silanol groups in the product[1,7]. This may result in an improved compatibility between the filler and the host polymer.

The main objective of this work is to establish a method to produce silsesquioxane nanoparticles with the empirical formula $[R_2Si_2O_3]_n$ containing polymerizable organic groups, such as methacryloxypropyl or styryl. It is believed that the use of such nanofillers will provide a better compatibility between the nanoparticles and certain polymeric systems. The use of the sol-gel process as a route to such hybrid materials has been limited until very recently. Krishnan and He have reported a stepwise coupling polymerization synthesis of the ladder-like silsesquioxanes containing methacryloxypropyl groups, which resulted in production of viscous materials[8]. For the synthesis of organically modified silica nanoparticles containing bulky organic groups, such as aminopropyl[9] or methacryloxypropyl[10,11], the most common approach involves grafting of organic groups by reacting inorganic silica nanoparticles with organotrialkoxysilane coupling agents. It has been shown that in this way the surface of the nanofillers is made more hydrophobic, which improves their dispersibility in acrylic coating materials[10] and leads to a better compatibility with the host polymeric systems, such as polystyrene[11]. Bourgeat-Lami and Lang[11] have noted that styrene monomer may react with the reactive end groups of the coupling agent, so that polymeric chains are created that are chemically attached to the surface of the nanoparticles. However, this grafting of organic groups to the surface of silica nanoparticles is confined to a limited number of reactive groups on the surface of the particles, the access to which is significantly hindered.

The hydrolytic modified Stöber method[2] has been adapted here to prepare styryl and methacryloxypropylsilsesquioxane particles of sub-micron diameter. This low temperature method, which involves base-catalyzed hydrolysis and condensation reactions of organotrialkoxysilane precursors in a mixture of water, ammonia and ethanol, has proved to be a useful route to silsesquioxane nanoparticles containing compact organic groups, such as methyl, ethyl, phenyl and vinyl[1,2]. The rate of hydrolysis and condensation reactions is expected to be very low, so that the reaction time has been increased from 30 minutes, as in the case of precursors containing less bulky organic groups[1,2], to several days. Another approach to produce silsesquioxane nanoparticles containing bulky organic groups may involve the use of surfactants in order to provide additional stabilization against aggregation of the particles. For example, Ottenbrite et al.[3] have described a Stöber-like synthesis of silsesquioxane nanoparticles with diameters in the range of 0.1-0.5 μm by hydrolysis and cocondensation reactions of vinyltrimethoxysilane and aminopropyltriethoxysilane precursors in a water/surfactant solution. Moreover, the emulsion polymerization approach has proved to be a useful alternative method for the production of silsesquioxane particles[12-14]. Noda et al.[12,13] have reported the emulsion polymerization synthesis of methylsilsesquioxane particles with diameters in the range of 0.2-5.0 μm using a range of surfactants. Baumann and Schmidt[14] have prepared strictly spherical methylsilsesquioxane particles of controlled size in the range of 3-15 nm using benzethonium chloride as a surfactant. Here we report the Stöber synthesis of styryl and methacryloxypropylsilsesquioxane particles in the presence of benzethonium chloride and Igepal CO 210 surfactants.

The use of organotrialkoxysilane precursors containing polymerizable organic groups offers the possibility to produce hybrid particles comprising an inorganic siloxane matrix and a polymeric organic phase. The siloxane matrix imparts good mechanical strength, light-scattering power and thermal and chemical stability, whereas the polymeric component is responsible for introduction of such properties as flexibility and processability, desirable surface characteristics, as well as elasticity and the ability to form films. Several groups have focused on the formation of organically modified silica nanoparticles of a core-shell structure by direct encapsulation of inorganic silica particles in polymer matrices, such as polystyrene[11,15,16] and polymethylmethacrylate[16,17]. However, since the surface of inorganic particles is hydrophilic, it needs to be made more hydrophobic to provide encapsulation by polymerization of hydrophobic monomers, which is usually achieved by chemical grafting of organotrialkoxysilane coupling agents[11]. One of the methods to encapsulate inorganic particles in polymer matrices is the miniemulsion polymerization approach[16,18,19]. This approach involves the use of a surfactant dissolved in water and a cosurfactant dissolved in the monomer phase. After these two phases are mixed together, the mixture is subjected to highly efficient homogenization using a high shear mixer, such as an ultrasonifier or a homogenizer, which breaks the oil phase into sub-micron size droplets. In this way, the surface area of the monomer droplets is larger than that of the micelles, so that the probability for a growing radical to enter the monomer droplets is high enough for droplet nucleation to prevail over micelle nucleation[18-20]. The cosurfactants, such as hexadecane and cetyl alcohol, are used to prevent droplet degradation by monomer diffusion (Ostwald ripening), when the monomer diffuses from small to large droplets owing to a higher chemical potential of the monomer in small droplets. The surfactants are used to minimize destabilization of miniemulsions by droplet coalescence by providing electrostatic or steric stabilization of the droplets. This method has been adapted here to prepare organically modified silica particles directly from organotrialkoxysilane precursors. In this work, hexadecane was used to suppress Ostwald ripening and cetyltrimethylammonium bromide, dodecyl sodium sulfate and Igepal CO 210 were used as surfactants. The miniemulsions were produced using an ultrasonifier and the miniemulsion polymerization syntheses were carried out by stirring these emulsions at 80^0C for 20 hours. The siloxane matrix is expected to be formed by the hydrolysis and condensation reactions of alkoxysilane groups, whereas the polymer component is expected to be produced by polymerization of styryl or methacryloxypropyl groups. The properties of the materials prepared from styryl and methacryloxypropyltrialkoxysilane precursors using the hydrolytic sol-gel method and the miniemulsion polymerization approach are compared and contrasted here.

Experimental

Materials

α-(Triethoxysilyl)styrene (98%, Aldrich), trimethoxysilyl propylmethacrylate (98%, Aldrich), sodium silicate solution ($Na_2O \cdot SiO_2$, 27 wt.% SiO_2, Riedel de Haën), hexadecane (99%, Aldrich) and sodium persulfate (98+%, Aldrich) were used as received. Absolute ethanol (99.86%, Hayman) and ammonium hydroxide (35%, Fisher) were of analytical reagent quality. Benzethonium chloride (97%, Aldrich), cetyltrimethylammonium bromide (Aldrich), dodecyl sulfate, sodium salt (98%, Aldrich) or Igepal CO 210 (Aldrich) were used as surfactants.

Syntheses of organically modified silica particles

Hydrolytic sol-gel method. Silsesquioxanes containing polymerizable organic groups were synthesized using the modified Stöber method as described previously[1,2]. The exemplified procedure for a molar ratio of $H_2O/NH_3/$

organotrialkoxysilane precursor/C_2H_5OH/seed SiO_2 of 1000/500/1/160/0.044 is shown below:

Sodium silicate solution (2.2 wt.% SiO_2) (0.6 ml) was mixed, under vigorous stirring, with absolute ethanol (40.0 ml; 0.68 mol). This dispersion was added to a mixture of ethanol (7.0 ml; 0.12 mol), ammonia (35% NH_3) (134.9 ml; 2.5 mol), water (1.8 ml) and an organotrialkoxysilane precursor (0.005 mol). This mixture was stirred at room temperature using a magnetic stirrer for 4 days or 11 days. The products were then freeze dried for *ca.* 24 hours to yield fine white powders.

To provide additional stabilization against aggregation of the particles, benzethonium chloride or Igepal CO 210 surfactants (25 wt.%) were used. These experiments were carried out in the absence of ethanol with a molar ratio of H_2O/NH_3/organotrialkoxysilane precursor/seed SiO_2 of 1500/350/1/0.044 and the reaction time was 4 days.

Miniemulsion polymerization synthesis. The miniemulsion polymerization approach was used to prepare methacryloxypropyl and styryl-modified silicas. The exemplified procedure for the synthesis of methacryloxypropyl-modified silica is as follows:

An oil phase consisting of trimethoxysilyl propylmethacrylate (3 ml; 12.6 mmol), sodium persulfate (0.08 g; 2.5 wt.%) and hexadecane (0.162 ml; 4 wt.%), which was used as a hydrophobe to suppress Ostwald ripening, was added to an aqueous solution (30 ml) of a surfactant (0.31 g; 10 wt.%). Sodium silicate solution (2.2 wt.% SiO_2) (1.51 ml; precursor/SiO_2 (mol.) = 1/0.044) was then added and the mixture was stirred using a magnetic stirrer for 30 minutes at *ca.* 4^0C. The miniemulsion was produced by sonicating this mixture for 3 minutes at an amplitude of 90% using a Branson 450 Digital Sonifier operating at 20 kHz. The miniemulsion polymerization synthesis was then carried out by stirring this emulsion under nitrogen at 80^0C for 20 hours. The product was isolated by centrifugation, washed several times with water and centrifuged again in order to remove a surfactant, together with the unreacted monomer. The resulting material was dried in air for approximately 3 days and under vacuum for 24 hours to yield a fine white powder.

Characterization techniques

Transmission electron microscopy (TEM) investigations were performed on a Philips 400T transmission electron microscope at a voltage of 120 kV. Carbon-coated copper carrier grids were used to prepare the samples. The micrographs were taken at a number of random locations on the grid. *Ca.* 100-150 particles were measured for each sample.

Energy filtered TEM was used to study the spatial distribution of Si, O and C atoms within the particles. Quantitative images were collected using a Philips CM200 electron microscope operating at 200 kV. The three-window technique was used to obtain the images: one window was used to obtain a signal from a particular element and the other two were used to remove the background.

Scanning electron microscopy (SEM) was performed on a Hitachi S-3200N scanning electron microscope operating at 10 kV. The SEM samples were prepared by applying a layer of dry powder on a stub covered with a carbon film. The samples were coated with a fine gold film in order to avoid charging effects. Micrographs were taken at a number of random locations on the stub. Ca. 50-100 particles were measured for each sample.

C and H elemental analyses were carried out on a Leeman Laboratories Inc. CE 440 elemental analyzer.

Transmission infrared (IR) spectra were recorded on a Perkin-Elmer 2000 FT-IR spectrophotometer. The spectra were obtained using KBr discs produced by pressing a mixture of a finely ground sample and potassium bromide. The following abbreviations were used to describe the absorptions: vs – very strong, s – strong, m – medium, w – weak, b – broad, sp – sharp and sh – shoulder.

Solid-state nuclear magnetic resonance (NMR) spectroscopy was carried out at the University of Durham on a Varian Unity Inova spectrometer using a 7.5 mm magic angle spinning (MAS) probe. The ^{13}C cross-polarized (CP) experiments were performed at room temperature at a resonance frequency of 75.39 MHz. These spectra were obtained using an acquisition time of 20.0 ms, a contact time of 1.0 ms, a pulse delay of 1.0 s and a spin-rate of 5000 Hz (1000-2000 repetitions). The ^{29}Si direct-polarized (DP) spectra were obtained at ambient temperature using a 90^0 pulse angle, a 120.0 s recycle, an acquisition time of 30.0 ms, a spin-rate of 5000 Hz and a frequency of 59.55 MHz (100-1000 repetitions). The Q^n and T^n notations were used to denote tetrafunctional and trifunctional silica species, respectively, where n is the number of siloxane groups bonded to the silicon atom.

X-ray photoelectron spectroscopy (XPS) was carried out on a Thermo VG Scientific Sigma Probe spectrometer. A microfocused monochromated Al Kα source operating at 140 W was used. For the high resolution spectra of C1s, O1s and Si2p, the analysis was performed at a pass energy of 20 eV. For the survey spectra, the analysis was performed at 100 eV. The background subtraction was performed using the Advantage data system. The quantitative surface compositions were calculated from the integrated peak areas. The CC/CH component was introduced at 285.0 eV in order to correct the sample charging effects.

Results and discussion

Stöber synthesis

The hydrolytic Stöber method[2] has been extended to the synthesis of silsesquioxane particles containing polymerizable organic groups, such as styryl and methacryloxypropyl. The chemistry of the hydrolytic sol-gel process can be represented by the following general equation:

$$2 \text{ RSi(OR')}_3 + 3 \text{ H}_2\text{O} \rightarrow [\text{R}_2\text{Si}_2\text{O}_3]_n + 6 \text{ R'OH} \qquad (1)$$

It was noted that the size of the particles obtained in this study (ca. 200-1000 nm) was considerably larger than that of the silsesquioxane nanoparticles synthesized from organically modified precursors containing less bulky organic groups[1,2] (ca. 50-200 nm) (Tables 1 and 2). The particles produced using triethoxysilyl styrene and trimethoxysilyl propylmethacrylate appeared to be more aggregated, irregular and polydisperse than those formed from phenyl, methyl, ethyl and vinyl trialkoxysilanes, especially in the case of methacryloxypropylsilsesquioxanes, for which only few irregular particles were often observed on the micrographs. The size of the particles measured for dispersions using TEM was of the same order as that obtained for dried powders

Table 1. Data on products of Stöber synthesis of methacryloxypropyl and styrylsilsesquioxane particles obtained in the absence of surfactants.

Material	Reaction time / days	Yield (%)	Average diameter / nm		Elemental analysis (wt.%)	
			TEM	SEM	C content	H content
Methacryloxy-propyl-silsesquioxane	Theoretical				46.9	6.1
	4	48	580±1 35	~700-800*	35.3	6.2
	11	59	550±1 80	~800-900*	37.7	6.0
Styryl-silsesquioxane	Theoretical				61.9	4.5
	4	27	430±1 30	500±1 15	44.5	4.3
	11	26	230±1 00	300±1 00	48.9	4.8

*Irregular or coalesced particles

Table 2. Data on products of Stöber synthesis of methacryloxypropyl and styrylsilsesquioxane particles obtained in the presence of surfactants.

Material	Surfactant	Yield (%)	Average diameter / nm		Elemental analysis (wt.%)	
			TEM	SEM	C content	H content
		Theoretical			46.9	6.1
Methacryloxy-propyl-silsesquioxane	Igepal[a]	43.1	7.0	820 ±250*	35.3	6.2
	B[b]	38.5	5.8	770 ±200*	37.7	6.0
		Theoretical			61.9	4.5
Styryl-silsesquioxane	Igepal[a]	47.0	4.1	730 ±100	44.5	4.3
	B[b]	48.1	4.3	410±50 and 1230 ±160	48.9	4.8

[a] *Igepal CO 210,* [b] *Benzethonium chloride.* *Irregular or coalesced particles*

using SEM, suggesting that aggregation of the particles was insignificant during the freeze drying of the samples. Although the average particle diameters measured for the materials produced in the absence of surfactants (Table 1) were similar to those for the materials synthesized in the presence of benzethonium chloride or Igepal (Table 2), the micrographs recorded for these silsesquioxanes confirmed that the use of the surfactants had provided additional stabilization of the particles against aggregation. The particles obtained by the Stöber synthesis using the surfactants were more regularly shaped, less polydisperse and exhibited a lesser degree of aggregation than those produced without the surfactants (Figure 1).

Relatively low yields (Tables 1 and 2) were obtained for these materials, presumably because the rate of hydrolysis and condensation reactions was quite low for the precursors containing such bulky organic groups. The empirical formulae of the products were used to calculate theoretical C and H percentages. The discrepancies between the predicted and experimental values were considerable, especially in the case of styrylsilsesquioxanes (Tables 1 and 2). It was believed that these were due to a more complex condensation mechanism. However, the presence of incompletely condensed species was only marginal, as confirmed by the results of solid-state NMR spectroscopy.

The formation of Si-O-Si bonds and the retention of Si-C linkages have been confirmed by the results of solid-state NMR spectroscopy, together with IR analysis. The presence of methacryloxypropyl and styryl organic groups was confirmed by the results of ^{13}C NMR spectroscopy (Figure 2). The ^{13}C CP-MAS

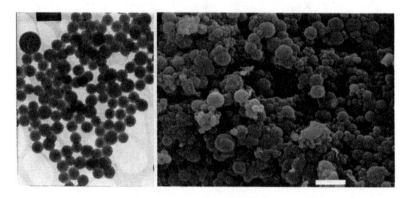

Figure 1. (Left)Transmission electron micrograph of styrylsilsesquioxane particles obtained using the Stöber method in the presence of benzethonium chloride (magnification 13000). (Right) Scanning electron micrograph of styrylsilsesquioxane particles obtained using the Stöber method in the absence of surfactants, reaction time 11 days (bar = 1μm).

spectra obtained for methacryloxypropylsilsesquioxanes produced in the presence and in the absence of surfactants (Figure 2a) showed intense carbon resonances at ca. 167 ppm (C=O), 137 ppm (-C(CH₃)=CH₂), 125 ppm (-C(CH₃)=CH₂), 66 ppm (Si-CH₂-CH₂-CH₂-), 23 ppm (Si-CH₂-CH₂-), 18 ppm (-C(CH₃)=CH₂) and 9 ppm (Si-CH₂-)[10,11]. A peak at around 51 ppm was assigned to the presence of some residual unhydrolyzed methoxy groups. For styrylsilsesquioxanes, the ^{13}C spectra exhibited strong signals at ca. 128 ppm, which were assigned to CH aromatic carbons, and relatively weak resonances at ca. 142 ppm, which were attributed to C1 atoms of the aromatic rings[21,22] (Figure 3b). It is believed that the peaks characteristic of the two vinylic carbons (-C(C₆H₅)=CH₂ and -C(C₆H₅)=CH₂) at around 130-140 ppm have been masked by the signals from the aromatic carbons. No peaks characteristic of residual ethoxy groups were seen in any of these spectra at ca. 18 and 58 ppm, thus confirming that these groups were almost completely hydrolyzed during the reaction.

The presence of methacryloxypropyl and styryl groups was also confirmed by the results of IR analysis performed for the materials synthesized in the presence and in the absence of surfactants (Figure 3). The retention of styryl groups was supported by absorptions at ca. 3055-3059 cm⁻¹ (m, b) characteristic of aromatic C-H stretching vibrations, 1600 cm⁻¹ (m, b) characteristic of Si-C=C linkages, 1444 and 1494 cm⁻¹ (m, sp) characteristic of aromatic C-C stretching vibrations, and 705 cm⁻¹ (s, sp), which was assigned to aromatic C-H bending vibrations[23-25] (Figure 3a). The aromatic ring absorption at ca. 1590-1595 cm⁻¹ was obscured by the double bond absorption band. The IR spectra obtained for methacryloxypropylsilsesquioxanes exhibited well-defined absorptions at 2890-2955 cm⁻¹ (m, sh) and 1455-1458 cm⁻¹ (w), which were assigned to C-H

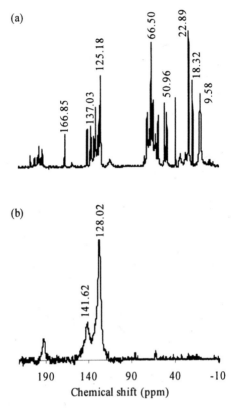

Figure 2. ¹³C CP-MAS NMR spectra of (a) methacryloxypropyl and (b) styrylsilsesquioxanes obtained using the Stöber method in the presence of benzethonium chloride.

linkages, 1720 cm⁻¹ (s, sp) characteristic of C=O groups, 1638 cm⁻¹ (m, sp) characteristic of C=C groups, and 1298-1322 cm⁻¹ (m, b, sh), which was attributed to the presence of C-O bonds[8,11,24,25] (Figure 3b).

The ²⁹Si NMR spectra (Figure 4) obtained for the materials synthesized by the Stöber method showed the presence T (mono-substituted) and Q (unmodified) silica species. For methacryloxypropylsilsesquioxanes, T³ and T² peaks were seen at *ca.* -68 and -60 ppm, respectively (Figure 4a). The ²⁹Si spectra obtained for styrylsilsesquioxanes exhibited well-defined peaks at *ca.* 82 and 73 ppm, which were assigned to T³ and T² silica species, respectively (Figure 4b). The results of deconvolutions of the ²⁹Si data suggested that the amount of fully condensed T³ species was always larger (*ca.* 80-90%) than that of T² species. It can be inferred that, for the materials synthesized using the Stöber method, the condensations were almost complete and the silsesquioxane structures were produced as the final products. The presence of unmodified Q⁴

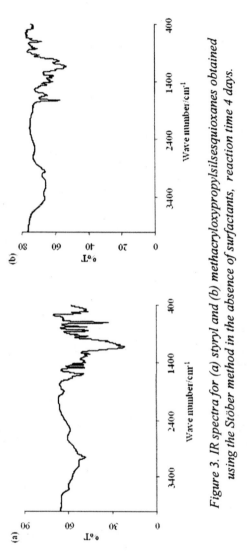

Figure 3. IR spectra for (a) styryl and (b) methacryloxypropylsilsesquioxanes obtained using the Stöber method in the absence of surfactants, reaction time 4 days.

silica species, as indicated by chemical shifts at *ca.* 111-112 ppm, was insignificant and was attributed to the presence of unreacted seed SiO_2. The formation of Si-O-Si linkages was also confirmed by the results of IR spectroscopy (Figure 4). The presence of these bonds was supported by absorptions at 1126-1130 cm^{-1} (vs, b), 1030-1090 cm^{-1} (s, sh) and 440-480 cm^{-1} (m)[26,27].

Miniemulsion polymerization synthesis

The miniemulsion polymerization approach was used to produce styryl- and methacryloxypropyl-modified silicas. Hexadecane was used to suppress Ostwald ripening and dodecyl sodium sulfate, cetyltrimethylammonium bromide and Igepal CO 210 were used as surfactants. In the case of trimethoxysilyl propylmethacrylate precursor, the products were isolated by centrifugation, whereas in the case of triethoxysilyl styrene, it was not possible to isolate the products mechanically or by treating the emulsions with hydrochloric acid and hot steam. For this precursor, the formation of well-defined spherical particles was not confirmed by TEM analysis. Therefore, the results presented below are for methacryloxypropyl-modified silica only.

The size of the particles produced using the miniemulsion polymerization approach was in the range of *ca.* 40-900 nm (Table 3). These particles (Figure 5) were nearly monodisperse, more regularly shaped and exhibited a lesser extent of aggregation than those obtained using the modified Stöber method (Figure 1). Once again, the average particle diameters obtained using TEM and SEM analyses were almost identical, thus suggesting that particle aggregation during centrifugation and subsequent drying was insignificant. It was noted that the size of the particles decreased with increasing the amount of Igepal, as the larger the surfactant concentration used, the better electrostatic or steric stabilization of the droplets was provided. The size of the particles also decreased with increasing the amount of seed SiO_2, as the larger the number of seeds introduced into the reaction, the larger the number of particles of smaller size created. The smallest particles were produced using cetyltrimethylammonium bromide (*ca.* 60 nm) and dodecyl sodium sulfate (*ca.* 45 nm) as surfactants. However, the particles observed for these materials were more aggregated or even coalesced than those produced using Igepal, as supported by both TEM and SEM micrographs.

Theoretical C and H percentages were calculated using the empirical formulae of the products based on the assumptions that either both Stöber and emulsion polymerization syntheses had been accomplished or emulsion polymerization only had taken place. The two sets of data were almost identical and the experimental values were in reasonable agreement with both of them (Table 4).

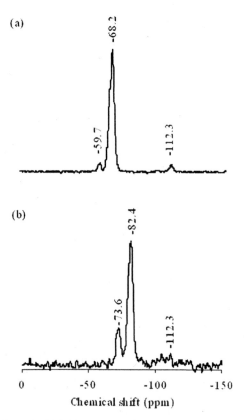

Figure 4. ^{29}Si DP-MAS NMR spectra of (a) methacryloxypropylsilsesquioxane obtained in the presence of benzethonium chloride and (b) styrylsilsesquioxane obtained in the absence of surfactants using the Stöber method, reaction time 4 days.

Solid-state ^{13}C and ^{29}Si NMR spectroscopy, together with IR analysis were used to establish whether Si-O-Si linkages were formed following the hydrolysis and condensations reactions and whether the organic groups polymerized to produce polypropylmethacrylate chains. The ^{13}C NMR spectra (Figure 6a) showed the presence of 'unsaturated' methacryloxypropyl groups, as confirmed by peaks at *ca.* 167 ppm (C=O), 137 ppm (-$\underline{C}$(CH$_3$)=CH$_2$) and 125 ppm (-C(CH$_3$)=$\underline{C}$H$_2$)[10,11], together with somewhat stronger signals at *ca.* 179 ppm, 54 ppm and 46 ppm, which were assigned to the same carbon atoms in polymethacrylate chains containing no C=C double bonds[28,29]. These spectra exhibited intense carbon resonances characteristic of both polymerized and

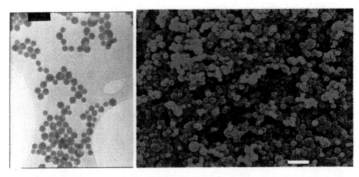

Figure 5. (Left) Transmission electron micrograph of methacryloxypropyl-modified silica particles obtained using the miniemulsion polymerization method, Igepal (10 wt.%), precursor/SiO₂ (mol.) = 1/0.132 (magnification 36000). (Right) Scanning electron micrograph of methacryloxypropyl-modified silica particles obtained using the miniemulsion polymerization method, Igepal(20 wt.%) (bar = 1 μm).

Table 3. Data on products of the miniemulsion polymerization synthesis of methacryloxypropyl-modified silica particles

Reaction details	Yield (%)	Average diameter / nm	
		TEM	SEM
Igepal CO-210 (10 wt.%), no seed SiO₂	81	880±240	800±110
Igepal CO-210 (10 wt.%) precursor/SiO₂ (mol.) = 1/0.044	86	410±120	500±150
Igepal CO-210 (10 wt.%), precursor/SiO₂ (mol.) = 1/0.132	78	120±25	150±23
Igepal CO-210 (20 wt.%)	91	250±50	280±35
Cetyltrimethylammonium bromide (10 wt.%)	95	75±15	60±12
Dodecyl sodium sulfate (10 wt.%)	87	115±60	45±11

Table 4. Elemental analyses of methacryloxypropyl-modified silicas obtained using the miniemulsion polymerization method

Reaction conditions		C content (wt.%)	H content (wt.%)
Stöber + Emulsion polymerization	$(Si_2O_3[(CH_2)_3CO_2CCH_3]_2)n$ CH_2	46.9	6.1
Theoretical Emulsion polymerization	CH_3 $(-CH_2-C-)n$ $OCO(CH_2)_3Si(OCH_3)_3$	48.4	8.1
Igepal CO-210 (10 wt.%)		48.9	7.2
Igepal CO-210 (20 wt.%)		51.9	7.8
Igepal CO-210 (10 wt.%), precursor/SiO$_2$ (mol.) = 1/0.132		49.4	7.2
Igepal CO-210 (10 wt.%), no seed SiO$_2$		44.4	6.9
Cetyltrimethylammonium bromide (10 wt.%)		49.0	7.4
Dodecyl sodium sulfate (10 wt.%)		46.8	7.1

unreacted methacryloxypropyl units at 66 ppm (Si-CH_2-CH_2-$\underline{C}H_2$-), 23 ppm (Si-CH_2-$\underline{C}H_2$-), 18 ppm (-C($\underline{C}H_3$)=CH_2 or -C($\underline{C}H_3$)-CH_2-) and 9 ppm (Si-$\underline{C}H_2$-)[10,11].

The ^{29}Si NMR spectra obtained for methacryloxypropyl-modified silicas produced using the miniemulsion polymerization approach (Figure 6b) were almost identical to those obtained for the Stöber method (Figure 4a). The presence of mono-substituted T^3 and T^2 species was confirmed by chemical shifts at *ca.* -68 and -60 ppm, respectively. The results of deconvolutions of the ^{29}Si spectra indicated that fully condensed T^3 silica species were more abundant (*ca.* 60-90%) than T^2 species (*ca.* 10-40%).

The IR spectra (Figure 7) exhibited intense absorptions at *ca.* 1110-1130 cm^{-1}, which were assigned to the presence of Si-O-Si linkages[26]. The presence of C=O groups was confirmed by intense sharp absorptions at *ca.* 1720-1730 cm^{-1}. These bands were stronger then the analogous bands in methacryloxypropylsilsesquioxanes obtained by the Stöber method (Figure 3). The absorptions at 1638 cm^{-1} characteristic of C=C groups were weaker indicating that although polymerization was not complete, the presence of C=C linkages was only marginal compared to the silsesquioxanes produced using the Stöber method, in which the intensity of C=C and C=O bands was almost the

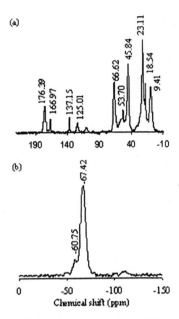

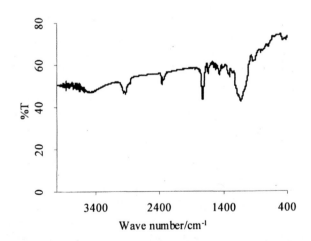

Figure 6. (a) ^{13}C CP-MAS NMR spectrum and (b) ^{29}Si DP-MAS NMR spectrum for methacryloxypropyl-modified silica obtained using the miniemulsion polymerization method in the presence of cetyltrimethylammonium bromide.

Figure 7. IR spectrum of methacryloxypropyl-modified silica obtained using the miniemulsion polymerization method in the presence of Igepal (10 wt.%).

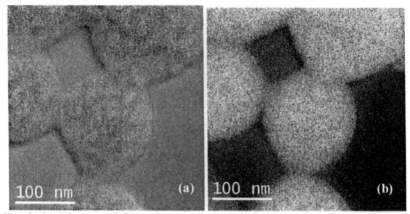

Fig. 8. (a) Silicon and (b) carbon maps obtained using energy filtered TEM for methacryloxypropyl-modified silica particles prepared using the miniemulsion polymerization method in the presence of Igepal (20 wt.%).

same. The absorptions characteristic of C-H linkages at 2932-2959 cm^{-1} (m, sh) and *ca.* 1456-1470 cm^{-1} (w), together with those characteristic of C-O groups at *ca.* 1298 cm^{-1} (m) were clearly seen in these IR spectra[8,11,25] (Figure 7).

These observations suggested that the particles of silsesquioxane structure ($[R_2Si_2O_3]_n$) were obtained in this work. The siloxane matrix contained some incompletely condensed silica species, whereas the organic phase consisted of both polypropylmethacrylate chains and unreacted methacryloxypropyl groups.

In order to establish whether the core-shell structure was attained in the particles produced by the miniemulsion polymerization method, X-ray photoelectron spectroscopy and energy filtered TEM analyses were used. The images obtained using the energy filtered TEM technique (Figure 8) indicated that silicon and carbon atoms were uniformly distributed within the particles, thus suggesting that no particles with the core-shell structure were produced. The elemental composition on the surface of the particles was studied using XPS analysis with a sampling depth of *ca.* 2-10 nm. The results obtained from the C1s, O1s and Si2p spectra were compared with the values calculated using the empirical formula with no hydrogen taken into consideration, which corresponded to the structure of the particles where all the elements were uniformly distributed within the particles (Table 5). It was noted that the near-surface carbon concentration was somewhat higher and the experimental silicon content values were consistently lower than the values calculated from the molecular formula. However, the results obtained for the near-surface elemental composition of inorganic silica particles (Nyasil) suggested that these observations were probably due to the presence of organic contamination on the surface of the particles. In this way, the XPS results also failed to provide any significant evidence in favor of the core-shell structure of the particles synthesized using the miniemulsion polymerization approach.

Table 5. XPS analysis of methacryloxypropyl-modified silicas obtained using the miniemulsion polymerization approach

Material		C content (wt.%)	O content (wt.%)	Si content (wt.%)
SiO$_2$	Calculated	0.0	53.3	46.7
	Nyasil 20	9.9	67.3	22.8
Methacryloxy propyl-modified silica	Calculated (no hydrogen taken into consideration)	49.9	33.3	16.8
	Igepal CO-210 (10 wt.%)	65.2	28.2	6.6
	Igepal CO-210 (20 wt.%)	67.4	26.8	5.8
	Cetyltrimethylammonium bromide (10 wt.%)	69.6	24.8	5.6
	Igepal CO-210 (10 wt.%), precursor/SiO$_2$ (mol.) = 1/0.132	60.7	33.2	6.1

Conclusions

The modified Stöber method has been adapted to produce silsesquioxane particles containing polymerizable organic groups. Methacryloxypropyl and styrylsilsesquioxane particles with diameters in the range of *ca.* 200-1000 nm were obtained in the absence and in the presence of surfactants. These particles were larger, more polydisperse and exhibited more aggregation than particles synthesized from organically modified precursors containing less bulky organic groups. The silsesquioxane particles obtained in the presence of benzethonium chloride or Igepal were more regularly shaped and less aggregated than those produced in the absence of surfactants. Although the discrepancies between the predicted and experimental values for C and H elemental analyses were considerable, the formation of Si-O-Si bonds and the retention of methacryloxypropyl and styryl groups were confirmed by the results of solid-state ^{13}C and ^{29}Si NMR spectroscopy, together with IR analysis. The presence of incompletely condensed silica species was only marginal. A much more detailed study would be needed to develop a procedure to produce more regularly shaped monodisperse silsesquioxane nanoparticles containing polymerizable organic groups and this will be the subject of future work.

The miniemulsion polymerization approach has been used to prepare methacryloxypropyl- and styryl-modified silica particles. Hexadecane was used as a hydrophobe and dodecyl sodium sulfate, cetyltrimethylammonium bromide and Igepal CO 210 were used as surfactants. The formation of particles was not observed in the case of triethoxysilyl styrene precursor. Nearly monodisperse methacryloxypropyl-modified silica particles with diameters in the range ca. 40-900 nm were produced. These particles exhibited a lesser degree of aggregation than those obtained using the Stöber method. The results of IR analysis and NMR spectroscopy suggested that particles of silsesquioxane structure were produced, where the siloxane matrix contained some incompletely condensed silica species and the organic phase consisted of both polypropylmethacrylate chains and unreacted methacryloxypropyl groups. Both X-ray photoelectron spectroscopy and energy filtered TEM failed to provide any significant evidence in favor of the core-shell structure of the particles synthesized using the miniemulsion polymerization approach.

Acknowledgements

This work is part of the MoD/EPSRC Joint Grants Scheme whose support we gratefully acknowledge. The authors thank Dr David Apperly (University of Durham) for provision of solid-state NMR services, Mr Steve Greaves for undertaking XPS analysis and Ms Judith Peters for elemental analysis. We also thank Dr Vlad Stolojan for his help in acquiring energy filtered TEM images. The technical assistance provided by the Microstructural Studies Unit (University of Surrey) is also greatly appreciated.

References

1. Arkhireeva, A.; Hay, J.N. *Polym. & Polym. Composites* **2004**, *2*, 101.
2. Arkhireeva, A.; Hay, J.N. *J. Mater. Chem.* **2003**, *13*, 3122.
3. Ottenbrite, R.M.; Wall, J.S.; Siddiqui, J.A. *J. Am. Ceram. Soc.* **2000**, *83*, 3214.

367

4. Katagiri, K.; Hasegawa, K.; Matsuda, A.; Tatsumisago, M.; Minami, T. *J. Am. Ceram. Soc.* **1998**, *81*, 2501.

5. Choi, J.Y.; Kim, C.H.; Kim, D.K. *J. Am. Ceram. Soc.* **1998**, *81*, 1184.

6. Masters, H.; Hay, J. N.; Lane, J. M.; Manzano, M.; Shaw, S. J. Proceedings of *Organic-Inorganic Hybrids*: Guildford, UK, 2002; paper 5.

7. Arkhireeva, A.; Hay, J.N. *Chem. Mater.* **2004**, submitted for publication.

8. Krishnan, P.S.G.; He, C. *Macromol. Chem. Phys.* **2003**, *204*, 531.

9. Van Blaaderen, A.; Vrij, A. *J. Colloid. Interface Sci.* **1993**, *156*, 1.

10. Bauer, F.; Sauerland, V.; Ernst, H.; Gläsel, H.-J.; Naumov, S.; Mehnert, R. *Macromol. Chem. Phys.* **2003**, *204*, 375.

11. Bourgeat-Lami, E.; Lang, J. *J. Colloid. Interface Sci.* **1998**, *197*, 293.

12. Noda, I.; Kamoto, T.; Yamada, M. *Chem. Mater.* **2000**, *12*, 1708.

13. Noda, I.; Isikawa, M.; Yamawaki, M.; Sasaki, Y. *Inorganica Chimica Acta* **1997**, *263*, 149.

14. Baumann, F.; Schmidt, M. *Macromolecules* **1994**, *27*, 6102.

15. Reculusa, S.; Poncet-Legrand, C.; Ravaine, S.; Mingotaud, C.; Duguet, E.; Bourgeat-Lami, E. *Chem. Mater.* **2002**, *14*, 2354.

16. Tiarks, F.; Landfester, K.; Antonietti, M. *Langmuir* **2001**, *17*, 5775.

17. Luna-Xavier, J.-L.; Bourgeat-Lami, E.; Guyot, A. *Colloid. Polym. Sci.* **2001**, *279*, 947.

18. Asua, J.M. *Prog. Polym. Sci.* **2002**, *27*, 1283.

19. Antonietti, M.; Landfester, K. *Prog. Polym. Sci.* **2002**, *27*, 689.

20. Guyot, A.; Chu, F.; Schneider, M.; Graillat, C.; McKenna, T.F. *Prog. Polym. Sci.* **2002**, *27*, 1573.

21. Newman, R.H.; Patterson, K.H. *Polymer* **1996**, *37*, 1065.

22. Grobelny, J. *Polymer* **1997**, *38*, 751.

23. Skoog, D.A.; Holler, F.J.; Nieman, T.A. *Principles of Instrumental Analysis, 5th Ed.*; Harcourt Brace & Company: Florida, 1998.

24. Craver, C. D. *The Coblentz Society Deskbook of IR Spectra, 2nd edn.*; The Coblentz Society Inc.: 1982.

25. Bellamy, L.J. *The Infra-red Spectra of Complex Molecules, 3rd Ed.*; Chapman and Hall: London, 1979.

26. Iller, R. K. *The Chemistry of Silica*; Wiley: New York, 1979.

27. Anderson, D. R. *Analysis of Silicones, ed. Smith, A. L.*; Wiley-Interscience: New York, 1974.

28. Hiraga, T.; Tanaka, N.; Hayamizu, K.; Mito, A.; Takarada, S.; Yamasaki, Y.; Nakamura, M.; Hoshino, N.; Moriya, T. *Jpn. J. Appl. Phys.* **1993**, *32*, 1722.

29. Chiu, Y.S.; Wu, K.H.; Chang, T.C. *Eur. Polym. J.* **2003**, *39*, 2253.

Indexes

Author Index

Subject Index

A

Aggregation and particle formation, silica formation with recombinant silicatein protein, 334–335

Ahmed-Rolfes-Stepto (ARS) theory, gel point interpretation and prediction, 191–201

Alcohols, separation from aqueous alcohol solution with PDMS membrane, 209, 210t–211f

Amine-catalyzed silicon-hydrogen bond activation, 78–79

Anti-foul and foul-release properties, α,ω-bis(glucidyloxypropyl)-pentasiloxanes, cross-linked with α,ω-diaminoalkanes films, 45–46f

Applications
anti-foul and foul-release properties, films, 45–46f
epoxide and oxetane monomers, 27–28, 29
epoxy resins, 41
membrane separations, 166
outdoor insulators, surface instability, PDMS, 226-227
polyhedral oligomeric silsesquioxanes, 291
silicone polymers, 113
sulfur containing organosilicon compounds, 49-50

ARS theory. *See* Ahmed-Rolfes-Stepto theory

Aryl-substituted silsesquioxanes, formation and functionalization, 301–312

Asymmetric stretching, in-plane, with UVO treatment, PDMS, strain profiles, 235

Atomic force microscopy, end-linked polysiloxane networks, 143–144, 158–159

ATR-FTIR spectra for mechanically assisted polymerization assembly, 232–234f

Attenuated total reflection mode. *See* ATR

B

BAM. *See* Brewster angle microscopy

Basicity, octamethyl-1,4,-dioxatetrasilacyclohexane, 14–16

Bimodal effect in networks, 146–147

Bimodal-type phase separation, atomic force microscopy images, 159

Biosilicification, 328–347

α,ω-Bis(glucidyloxypropyl)-pentasiloxanes, non-fluorinated and fluorinated, 38–46f

Bis(silylpropyl)polysulfanes, formation by phase transfer catalysis, 51f, 52–54f

Bis(triethoxyethyl)dimethylsilicone (TES-PDMS), surfactant in enzyme studies, 257–264

Bis(triethoxysilylpropyl)sulfanes, UV-absorption spectra, 55, 58

Bis(triethoxysilylpropyl)tetrasulfane, HPLC chromatogram, 55, 56f

Bond energy, siloxane, 4–5

Brewster angle microscopy (BAM) and isotherm, 271–284f

Brush formation, polymers, ATR-FTIR spectra, 232–233

Buckled siloxane surfaces, 237–241